Climate Denial in American Politics

Climate Denial in American Politics is a detailed examination of the rise within American politics of climate denialism, the counter movement which challenges the accepted science of climate change.

Organized around the administrations of American presidents from Roosevelt to Biden, this book provides an unprecedented account of climate denial within both the White House and Congress, and the 'climate brawls' that followed. This volume is a rebuke to discredit the climate denier, their propaganda, and their sources. Gerald Kutney examines the evolution of American political thought on climate change and provides a comprehensive survey and analysis of the sordid history of the propaganda which has promoted climate denial and corrupted politicians in America. He uses direct quotes from primary sources, such as government records, to show the extreme and pervasive nature of anti-science opinions made by political climate deniers and limit any misinterpretation that might result from paraphrasing. Weaving the account of climate denialism in American politics with anecdotes from Kutney's own decade-long experience of challenging climate deniers on Twitter using #ClimateBrawl, this book provides a valuable insight into the world of climate obstruction.

Climate Denial in American Politics will be of great interest to students and scholars of climate change, environmental politics and American politics more broadly.

Gerald Kutney is the author of *Carbon Politics and the Failure of the Kyoto Protocol.* He was an adjunct professor at the University of Northern British Columbia and taught the graduate course "Climate Change & Global Warming," and he has presented several guest lectures at Carleton University on climate denialism.

Routledge Advances in Climate Change Research Series

Kick-Starting Government Action against Climate Change
Effective Political Strategies
Ian Budge

The Social Aspects of Environmental and Climate Change
Institutional Dynamics Beyond a Linear Model
E. C. H. Keskitalo

Climate Change and Tourism in Southern Africa
Jarkko Saarinen, Jennifer Fitchett and Gijsbert Hoogendoorn

Climate Cultures in Europe and North America
New Formations of Environmental Knowledge and Action
Edited by Thorsten Heimann, Jamie Sommer, Margarethe Kusenbach and Gabriela Christmann

Urban Planning for Climate Change
Barbara Norman

Climate Action in Southern Africa
Implications for Climate Justice and Just Transition
Edited by Philani Moyo

Climate Change Action and The Responsibility to Protect
A Common Cause
Ben L. Parr

Climate Security
The Role of Knowledge and Scientific Information in the Making of a Nexus
Matti Goldberg

For more information about this series, please visit: www.routledge.com/Routledge-Advances-in-Climate-Change-Research/book-series/RACCR

Climate Denial in American Politics

#ClimateBrawl

Gerald Kutney

Designed cover image: Getty images

First published 2024
by Routledge
4 Park Square, Milton Park, Abingdon, Oxon OX14 4RN

and by Routledge
605 Third Avenue, New York, NY 10158

Routledge is an imprint of the Taylor & Francis Group, an informa business

British Library Cataloguing-in-Publication Data
A catalogue record for this book is available from the British Library

ISBN: 978-1-032-59607-5 (hbk)
ISBN: 978-1-032-59279-4 (pbk)
ISBN: 978-1-003-45541-7 (ebk)

DOI: 10.4324/9781003455417

Typeset in Times New Roman
by codeMantra

To my dear wife Denise who patiently proofread and edited the manuscript with great skill and dedication and was my muse throughout its preparation.

Contents

Tables

Acknowledgements

I wish to express my gratitude to the friends of #ClimateBrawl; during our challenging of the climate deniers on Twitter, they shared many valuable references and experiences on climate denial. A special thanks goes out to the climate change pundits that reviewed the manuscript and provided thoughtful and constructive feedback: the science historian Spencer Weart, the science writer Stephen Leahy, the professor of education Gale Sinatra, environmentalist Bill McKibben, economist Steve Keen, and the former Republican representative Carlos Curbelo who was a founder of the bipartisan Climate Solutions Caucus.

Introduction

The resolution of our climate crisis resides within the power of political leadership. Science had unequivocally demonstrated decades ago that carbon dioxide from the burning of fossil fuels was warming the planet. Yet, politicians did not listen – never a good omen in that every disaster movie starts with the government ignoring a scientist. The climate scientist Michael E. Mann (1965–) stated[1]: "But if you don't know the solution to the politics, you don't know the solution to this problem, because the obstacles are all political." Finding the "solution to the politics" of climate change is what this book is about. The lack of political action has brought us to what may very well be the global crisis of the century. We have polluted the atmosphere with such titanic quantities of carbon dioxide that we are changing the climate, enhancing extreme weather events and other "natural" disasters.[2] And we continue to do so at our peril.

My previous book on the politics of the climate crisis – *Carbon Politics and the Failure of the Kyoto Protocol* – had dealt with the intrigue of international politics, the operations of the IPCC, and the complex negotiations associated with the Kyoto Protocol.[3] *Climate Denial in American Politics: #ClimateBrawl* is a detailed examination of the rise within American politics of climate denialism,[4] a counter movement which challenges the accepted science of climate change. This book investigates how the political system was manipulated in the most powerful democracy in the world, the United States of America. Examining the public record made it clear that the American government was under the control of special interest groups associated with climate denialism.

Since the climate sciences were bad for business, the energy-industrial complex (fossil fuel and related industries) had orchestrated the most massive propaganda operation in history against the science of climate change, and their seditious campaign of climate denialism even infiltrated the upper echelon of the U.S. government. Climate denialism is now entrenched within the Grand Old Party. The climate crisis can only be resolved through political action, especially in Washington, but you must first have the political will to act, and this has been missing, primarily in one party. Consequently, before we can even argue over climate policy solutions, Congress needs to first accept climate change is a problem which demands action.

A more recent counter movement has since evolved to challenge climate denialism, which I have dubbed "climate brawl" that focuses on hard-core climate

DOI: 10.4324/9781003455417-1

deniers among the political elite, contrarian scientists, and major Twitter influencers who are promoting climate denial propaganda. This counter movement aims to neutralize the political influence of climate denialism through an aggressive, yet civil, rebuttal to discredit climate deniers and their sources. The "solution to the politics" of the climate crisis, then, must include a climate brawl.

Outline

Climate Denial in American Politics: #ClimateBrawl begins with a brief history of science denialism – the denial of accepted science – including the attacks against Galileo, Newton, Darwin, and Einstein which illustrate the folly of the cranks challenging the greatest minds of science. The term "denier" in relation to the climate sciences first showed up in the 1990s, but the terminology only came into regular use in the 2010s. As climate denialism gained ground, so did the climate brawl movement. One group that has stood up to climate denial on Twitter, a major staging platform for climate denialism, is the friends of #ClimateBrawl who challenge the anti-science propaganda, and I share our engagements with climate deniers, exposing their strategies and tactics which are similar to those in the political arena.

Chapter 2 addresses the issue of climate change in the White House and Congress up to and including the administration of President Reagan. Government interest in the science of climate change had grown indirectly out of the research on the atomic bomb and the Cold War, and one government agency particularly played a supportive role – the Office of Naval Research. The coming to Washington of leading climate experts to testify in congressional hearings started in the mid-1950s. By the end of the decade, public media were already picking up news stories on the threat of climate change. Presidents of the United States started to pay attention to the science of climate change, especially Lyndon Johnson, Richard Nixon, and Jimmy Carter, and the first congressional hearing specifically on climate change took place during the term of Gerald Ford. After the first decades of scientific testimony, Washington appeared to be taking climate change seriously but that came to an abrupt halt when Ronald Reagan came to office. While even President Reagan was mentioning climate change at the end of his term, this was merely to help George H. W. Bush get elected. A tragic paradigm shift in the political history of climate change occurred over the Reagan and Bush administrations.

We can never say that scientists did not warn us. Enough was known about climate change to justify legislation but then climate denialism arrived. Chapter 3 begins with George H. W. Bush and continues through to the early term of Joe Biden. The division between chapters two and three represents the transition from climate skepticism to climate denialism over the Reagan to Bush administrations which was fostered by the contrarians – a small number of scientists opposed to the science of climate change who had first arrived in Washington at the end of the 1980s, upon the invitation by Republicans in Congress. Climate denialism began to seep into American politics with the presidency of George H. W. Bush and had fully infiltrated the Republican Party by the end of George W. Bush's presidency. President Obama had done the most to combat climate change, including being a force behind the Paris

Climate Agreement, but his efforts were largely negated by President Trump, the most notorious climate denier ever to hold the office. Near the end of the chapter, there is a review of the early term of Joe Biden and the Inflation Reduction Act, the most important (and only) climate-related legislation ever passed by Congress.

Chapter 4 reveals the extent of hard-core climate denialism among the political elite, which included Schlesinger, Bush (GHW), Sununu, Bush (GW), Cheney, Inhofe, Barton, Rohrabacher, Hall, Smith, Trump, and Manchin, ultimately forming a "denial cabal." The one who stood out from the pack was Senator James Inhofe who established many of the talking points and political spin of climate denialism that are still heard today. As climate denialism spread through the Republican ranks, climate brawls developed as Democrats fought back. Presidents would also become involved in the climate brawl, and one in particular would aggressively challenge climate denialism.

The last chapter covers climate brawls with the energy-industrial complex and its "denial machine." By the time of President Reagan, Big Oil was at a crossroads because of the risk of climate change legislation to their business; they could transition away from fossil fuels or fight the science, and they chose the latter. The corrupting influence of propaganda by the denial machine was funded by the energy-industrial complex, and a polished PR campaign scarred American politics by easily winning over conservatives. An all-out reign of terror was underway against the climate sciences. History has documented the government-led persecution of scientists, but only in totalitarian states, such as Stalinist Russia and Nazi Germany. How could something like this have happened in a democratic country like the United States? Propaganda is the voice of tyranny, not democracy, and the "oiligarchs" had established a deep state of denial.

I had originally set out in *Climate Denial in American Politics: #ClimateBrawl* to find the "solution to the politics," and the title of the book gives away the short and simple answer. There is no doubt that progress cannot be made on climate policy development without muting climate denialism in Congress and the White House; otherwise, there can be no political will to act. A climate brawl, though civil, is not a cordial discussion or debate about climate science, but a rebuke to discredit the hard-core climate deniers, their propaganda, and their sources. Challenging climate denial with climate brawls is not the whole "solution to the politics" but a crucial step to getting us there.

Methodology

The book, organized by the administration of an American president, is mainly a research monograph on climate denialism and the counter movement climate brawl, as well as a memoir of my decade long experience challenging climate deniers on Twitter. *Climate Denial in American Politics: #ClimateBrawl* examines the evolution of American political thought on climate change and is a comprehensive survey and analysis of the sordid history of the propaganda by the energy-industrial complex which promoted climate denialism and corrupted politicians in America. The study focuses on federal politics, and state politics has been deliberately omitted; I would not expect the latter to identify any major new features of the problem. Also, I did not explore the rank and file in Washington but concentrated on the political elite which has the greatest influence over the party. The work includes an unprecedented account of climate denial within both the White House and Congress, and the climate brawls that followed with the contrarians, the denial cabal, and the energy-industrial complex.

This political issue can be emotionally charged. Scholarly research, however, requires an impartial approach, and an examination of climate denialism, therefore, cannot exclude consideration of any positive aspects; in this study, which traces the evolution of climate denialism, none were found, which would not surprise the majority of physical scientists who study the climate. Social scientists would likewise generally agree but have identified psychological and sociological factors to account for the rise of the climate denialism movement. Academic studies, along with my decade-long Twitter experience, have been applied in this extensive study of climate denialism, which show that climate denial is waning, but the remaining supporters, especially within the G.O.P., are disproportionately active and still resistant to accepting the science.

Direct quotes from primary sources,[5] such as government records, are liberally used to show the extreme and pervasive nature of anti-science opinions made by political climate deniers and to limit any misinterpretation that might result from paraphrasing. To my knowledge, such an in-depth collection of climate denialism within Washington political circles has never been prepared before.

There is a challenge when examining the "climate change" records from the Federal Government because there are a lot of them: >34,000 records (available from govinfo; up to mid-2022). If you restrict the search to only congressional hearings containing "climate change," there are 2,460 records in the House of Representatives and 2,238 records in the Senate. In this study, more attention has been given to these records. Mining through thousands of pages of congressional hearings on climate change proved highly informative, even shocking at times, considering how anti-science rhetoric was being so casually expressed within these prestigious institutions. As for climate denial and related terms, my searches in the congressional hearings turned up only 119 records. These were preferentially examined and studied in greater detail. The intent of this review is to show how endemic and entrenched climate denialism continues to be in Washington and how to break through the policy malaise on climate change.

Notes

1 Dori 2021.
2 Kutney 2021.
3 Kutney 2014.
4 For previous studies on American policy development on climate change see Howe 2014; Bailey 2015; Mildenberger 2020; Speth 2021; these have not focused on climate denialism.
5 References appear at the end of each chapter, so that each chapter is a complete unit; hyperlinks were confirmed to the end of 2022.

References

Bailey, Christopher J. (2015). *US Climate Change Policy*. London: Routledge.

Dori, Roni (2021). *Climate Change Obstacles Are All Political, Not Technological*. CTECH, December 23; https://www.calcalistech.com/ctech/articles/0,7340,L-3925682,00.html

Howe, Joshua P. (2014). *Behind the Curve: Science and the Politics of Global Warming*. Seattle: University of Washington Press.

Kutney, Gerald W. (2014). *Carbon Politics and the Failure of the Kyoto Protocol*. London: Routledge.

Kutney, Gerald W. (2021). "Natural Disasters Aren't as Natural as They Used to Be." *National Observer*, November 30; https://www.nationalobserver.com/2021/11/30/opinion/natural-disasters-arent-natural-they-used-be

Mildenberger, Matto (2020). *Carbon Captured, How Business and Labor Control Climate Politics*. Cambridge: MIT Press.

Speth, James Gustav (2021). *They Knew, The US Federal Government's Fifty-Year Role in Causing the Climate Crisis*. Cambridge: MIT Press.

1 Climate Denialism

Climate denialism is a modern problem, but it is rooted in science denialism which has been around for ages. Individuals are attracted to science denialism out of a fear that accepting the science threatens something personally important, such as livelihood, social status, lifestyle, religious beliefs and/or political dogma, and this fear can grow into a social movement if fed by propaganda.

We should be clear on the nefarious nature of propaganda which presents itself under a variety of less menacing monikers like misinformation, disinformation, gaslighting, alternative facts, or fake news. Propaganda is an organized campaign of outright lies and half-truths repeated ad nauseam. A chilling alarm rang out from the Bulletin of the Atomic Scientists that propaganda attacks on science were undermining democracies everywhere[1]:

> Science provides the world's searchlight in times of fog and confusion. Furthermore, focused attention is needed to prevent information technology from undermining public trust in political institutions, in the media, and in the existence of objective reality itself. Cyber-enabled information warfare is a threat to the common good. Deception campaigns—and leaders intent on blurring the line between fact and politically motivated fantasy—are a profound threat to effective democracies, reducing their ability to address nuclear weapons, climate change, and other existential dangers.

In the end, decisions with far-reaching consequences are based on falsehoods that serve and protect a particular agenda, which can be harmful to the greater good.

A brief history of science denialism, the baseless rejection of accepted knowledge, exposes the various symptoms of this affliction and is followed by a closer look at modern climate denialism. In the past, a counter movement has inevitably emerged challenging the propaganda of science denialism – these counter movements I have dubbed "science brawl," or "climate brawl" when applied specifically to climate denialism.

DOI: 10.4324/9781003455417-2

Science Denialism

Cranks and Their Junk Science

Science denial is related to junk science, and its origins can be traced back to the earliest Greek literature in what was first called "paradoxes" – a mixture of truth and fiction, history and entertainment, science and pseudoscience – as for example in the writings of the poet Homer and the historian Herodotus (c. 484–c. 425 BC).[2] Thomas Browne (1605–82) was an early scholar of the science brawl, as he challenged false paradoxes, and one could even say that he identified the very first science denier[3]: "The first contriver of Error, and professed opposer of Truth, the Devil."

The British mathematician Augustus De Morgan (1806–71) was a keen investigator into science denialism and an avid collector of false paradoxes (several of which are mentioned below)[4]: "I use the word in the old sense: a paradox is something which is apart from general opinion, either in subject-matter, method, or conclusion." In this context, "general opinion" referred to accepted science. De Morgan's work was popular with the historian and philosopher John Fiske (1842–1901) who introduced the term "cranks" to describe De Morgan's false paradoxers. John Fiske mocked the works of the cranks which had been known as "insane literature," though he preferred the phrase "eccentric literature."[5] Cranks and their science denial continued to be ridiculed; in 1951, the philosopher of science Laurence Lafleur (1907–66) described cranks as[6]: "Most, from whatever educational level, are ignorant of the principles and facts of the field they write about, but a few have acquired an extensive though usually ill-assorted compendium of information." Shortly afterwards, the popular science writer Martin Gardner (1914–2010) also mocked the folly of the cranks[7]:

> In most cases the crank is not well enough informed to write a paper with even a surface resemblance to a significant study. As a consequence, he finds himself excluded from the journals and societies, and almost universally ignored by the competent workers in the field … The eccentric is forced, therefore, to tread a lonely way. He speaks before organizations he himself has founded.

A term related to the practice of the cranks is "pseudo-science;" in 1796, the English historian James Petit Andrews (1737–97) introduced the term,[8] and the noted polymath Oliver Wendell Holmes (1809–94) defined pseudo-science as follows[9]:

> A Pseudo-science consists of a nomenclature, with a self-adjusting arrangement, by which all positive evidence … is admitted, and all negative evidence … is excluded … Its professors and practitioners are usually shrewd people; they are very serious with the public, but wink and laugh a good deal among themselves.

More recently, "crank pseudoscientists" had appeared in the Congressional Record[10]: "Our colleagues seem content to ignore the climate crisis, to hold hearings with discredited, crank pseudoscientists bought and paid for by corporate interests, or to deny the value of scientific thinking altogether."

The related term "junk science" is more modern, having been introduced by Peter Burrowes in his work on socialism in 1903,[11] and in more recent times, the term had been briefly mentioned in a House of Representatives' hearing in 1984,[12] and again in the Congressional Record, six years later.[13]

"Cult of Ignorance"

Junk science is often related to willful ignorance. The artist Apelles (4th century BC) had famously expressed his frustration with an unqualified critic of his work[14]: "Let a shoemaker stick to his last" (*Ne sutor ultra crepidam*); the phrase led to the delightfully insulting term "ultracrepidarian" to describe those who give an opinion beyond their level of knowledge. The "Father of Medicine" Hippocrates (c. 460–c. 370 BC) had warned[15]: "There are in fact two things, science and opinion; the former begets knowledge, the latter ignorance." His contemporary Socrates (470–399 BC) expressed similar views[16]: "There is … only one good, that is knowledge, and only one evil, this is, ignorance," and further added[17]: "have you not observed that opinions divorced from knowledge are ugly things?"

Over two millennia later, the modern writer Isaac Asimov (1920–92) complained that the "cult of ignorance" predominant in America had been[18]: "nurtured by the false notion that democracy means that 'my ignorance is just as good as your knowledge.'" He later concluded[19]: "It is these ignorant people, the most uneducated, the most unimaginative, the most unthinking among us, who would make of themselves the guides and leaders of us all." More recently, the same disturbing trend has been captured by the writer Tom Nichols (1960–) in his article "The Death of Expertise,"[20] where experts are looked upon as an evil influence on society, especially in America, with outcries of "elitism." Going still further, the science historian Robert Proctor (1954–) introduced a new term, "agnotology," which is the study of the use of junk science and other propaganda to foster ignorance within society.[21]

Psychology has recognized this ignorance problem among science deniers and given the affliction an official name, the "Dunning-Kruger effect."[22] In their original analysis, psychologists Justin Kruger and David Dunning concluded[23]: "We propose that those with limited knowledge in a domain suffer a dual burden: Not only do they reach mistaken conclusions and make regrettable errors, but their incompetence robs them of the ability to realize it."

Modern-day science deniers will cast themselves in the role of the persecuted expert, known as the "Galileo Gambit:"[24] "Yes, the authorities thought Galileo was wrong, and they also think that you are wrong—but the fact that he turned out to be right doesn't mean that you are." The foolhardy comparison to Galileo has been a favourite claim with "cranks of all sorts"[25] pining for similar vindication as persecuted geniuses. Conveniently forgotten by the science deniers is that Galileo was himself a victim of science denialism.

In reviewing the attacks against Galileo and other famous scientists discussed next, some humorous and troubling accounts of science denial are recounted. The persecution of Galileo, Newton, Darwin, and Einstein by the cranks is worth looking at more closely as there is so much overlap with present-day climate denial.

Sun Denial

Science denialism arose among conservative elements of the Catholic Church zealously defending the ultimate authority of the Bible. An early disciple of science denialism was the Dominican Giovanni Tolosani (1470/1–1549) who attacked Nicolaus Copernicus (1473–1543).[26] The latter was well acquainted with the cranks, who he fittingly called "babblers:"[27] "There may be babblers, wholly ignorant of mathematics, who dare to condemn my hypothesis, upon the authority of some part of the Bible."

On February 24, 1616, the teaching of the Copernican theory was found by the Inquisition to be "foolish and absurd," and under threat of imprisonment by the science-denial tribunal, Galileo (1564–1642) agreed to stop teaching it.[28] This censorship was one of the first political acts of science denial. Later, in 1632, Galileo used the term "denied" in reference to those who believed in junk science in his famous *Dialogue Concerning the Two Chief World Systems, Ptolemaic and Copernican*[29]: "But once you have denied the principles of the sciences and have cast doubt upon the most evident things, everybody knows that you may prove whatever you will, and maintain any paradox." Galileo had been the first to associate denial with those who refused to accept "the principles of the sciences," and his definition of science denial is as valid now as it was then.

However, an aged Galileo had pushed the restrictions of his court-ordered censorship too far. The *Dialogue* was a thinly veiled promotion of the Copernican theory, which did not go unnoticed by Church officials; on June 22, 1633, a guilty verdict against Galileo was handed down by the science deniers of the Inquisition for breaking the gag order.[30] When the verdict came down, Galileo was said to have defiantly uttered[31]: "Yet it [the Earth] does move." Galileo's brief rebuttal was an early example of a science brawl, as he pushed back against the Inquisition. The science deniers of the Church had flexed their political muscle to win a victory over Galileo and others, but the heliocentric theory soon became accepted science anyways.

Gravity Denial

Augustus De Morgan, in his comprehensive collection of false paradoxes from the Victorian era, identified the science deniers who had challenged Isaac Newton (1643–1727); among the numerous examples he presented were the exploits of Captain Forman who had written a book in 1833 against Isaac Newton and was outraged that the scientific community had not taken his work seriously[32]:

> He then sent a manuscript letter to the Astronomical Society, inviting controversy: he was answered by a recommendation to study dynamics. The above pamphlet was the consequence, in which, calling the Council of the Society "craven dunghill cocks," he set them right about their doctrines. From all I can learn, the life of a worthy man and a creditable officer was completely embittered by his want of power to see that no person is bound in reason to enter into controversy with every one who chooses to invite him to the field.

In other words, Captain Forman believed that the Astronomical Society, instead of advising him to learn some science, should have been obliged to engage with him. He was behaving exactly like modern climate deniers who complain, using angry, aggressive, and vulgar language, about how peer-reviewed journals will not publish their "science."

Two years later, the publisher Sir Richard Phillips (1767–1840) created conspiracy theories about politicians providing stipends for supporters of Isaac Newton to "deceive the world," and he attacked universities for teaching Newton's theories.[33] Of this outspoken critic of Isaac Newton, Augustus De Morgan wrote[34]: "Sir Richard Phillips had four valuable qualities; honesty, zeal, ability, and courage. He applied them all to teaching matters about which he knew nothing; and gained himself an uncomfortable life and a ridiculous memory." A similar fate awaited many science deniers.

Extreme denial outbursts were popular during this period. In 1839, a rant by John Walsh (1786–1847) had been sent to the politician Henry Brougham (1778–1868) for supporting Newtonian science[35]:

> Man! what are you about? You come forward now with your special pleading, and fraught with national prejudice, to defend, like the philosopher Grassi, the persecutor of Galileo, principles and reasoning which, unless you are actually insane, or an ignorant quack in mathematics, you know are mathematically false. What a moral lesson this for the students of the University of London from its head! Man! demonstrate corollary 3, in this note, by the lying dogma of Newton, or turn your thoughts to something you understand.

If one didn't know any better, one might mistake his rant for an anti-science tweet on Twitter (albeit the character count would be too long), as I have been on the receiving end of similar vitriol from modern-day climate deniers. Overall, De Morgan's work was an illuminating study of the science brawl during the Victorian period.

Unlike most other cases of science denialism, there was no obvious ideology or vested interest threatened by Newton's theory of gravity; yet, during the 19th century, Isaac Newton had been a popular target of science deniers which was explained by John Fiske[36]:

> The desire to prove great men wrong is one of the crank's most frequent and powerful incentives. The name of Newton is the greatest in the history of

science: how flattering to one's self it must be, then, to prove him a fool! In eccentric literature the books of Newton are legion.

Famous scientists are still a favourite target of cranks.

Evolution Denial

Another theory that got into trouble with conservative church authorities was evolution.[37] Charles Darwin (1809–82) had written the *Origin of Species* in 1859, and the new theory was immediately caught up in criticism, inside and outside of scientific circles. Later, the noted writers George Bernard Shaw (1856–1950)[38] and Samuel Butler (1835–1902) had both written works of evolution denial.[39] In his *Descent of Man*, Charles Darwin lamented[40]: "Ignorance more frequently begets confidence than does knowledge."

An historic science brawl took place on June 30, 1860, at a meeting of the British Association for the Advancement of Science at Oxford, where Darwin's recently released *Origin of Species* had been hotly debated. A special member of the panel was the church official Bishop Samuel Wilberforce (1805–73), and among the defenders of science was the biologist Thomas Henry Huxley (1825–95); Charles Darwin was too ill to attend. A month before the debate, Bishop Wilberforce had published a criticism of Darwin's work, where he strove to present himself as fair minded and scientifically oriented[41]:

> Our readers will not have failed to notice that we have objected to the views with which we are dealing solely on scientific grounds ... We have no sympathy with those who object to any facts ... because they believe them to contradict what it appears to them is taught in Revelation.

The ideas he advanced in that article formed the basis of his criticisms at the debate. Thomas Huxley countered with[42]:

> Darwin's theory was an explanation of facts; and his book was full of new facts, all bearing on his theory. Without asserting that every part of the theory had been confirmed, he maintained that it was the best explanation of the origin of species which had yet been offered.

A legendary exchange from this debate is often cited (but without any direct evidence that it had taken place)[43]:

BISHOP WILBERFORCE: "Is it on your grandfather's or grandmother's side that you claim descent from the apes?"

THOMAS HUXLEY: "I would rather be descended from an ape than a bishop."

Though the nasty exchange is likely bogus, it may have reflected the tension of the debate.

Later, on November 3, 1864, Thomas Huxley formed the "X Club," a group of nine eminent scientists [including John Tyndall (1820–93), a discoverer of the greenhouse effect] who were committed to science being[44]: "pure and free, untrammelled by religious dogmas." Their major mission was to defend the theory of evolution from the growing propaganda challenging the science, especially from religious corners; in a sense, the X Club was an early group of science brawlers. If I had found this out earlier, I might have used the "#ClimateXClub" instead of #ClimateBrawl on Twitter, for those challenging propaganda against climate change.

Evolution denialists, known as "creationists," who uphold the special creation of humans and the universe by divine decree, were popular among the religious right in America. The best-known creationist was the prominent politician William Jennings Bryan (1860–1925) (Democrat from Nebraska), who espoused his evolution denial in a brochure, *The Prince of Peace*.[45] Mr. Bryan was a prosecutor arguing against Clarence Darrow (1857–1938) in the 1925 Scopes Trial which was a media sensation over the teaching of evolution in Tennessee schools. As part of his defense, Mr. Darrow told the court[46]: "I know there are millions of people in the world who look on [the Bible] as being a divine book … It is not a book of science." Despite his elegant defense, the jury found the teacher, John Scopes, guilty of teaching the theory of evolution for which he received a fine. However, the real trial took place in the court of public opinion where Clarence Darrow had triumphed over evolution denialism.[47] As with the Oxford debate, the trial was a grand example of a science brawl.

Nevertheless, this important public victory for science did not mean that evolution denial in America disappeared. The Congressional Record from 1979, for example, reports the following statement from Senator Orrin Hatch (1934–2022) (Republican from Utah)[48]:

> These teachers [creationists] take the approach of presenting the facts, both for evolution and for creation, and permitting the students to decide for themselves, individually. I believe this is the right way to do it. I say this as one who believes in the theory of creation, and who believes that the twin purposes of academic liberty and excellence are best served by an intelligent presentation of both viewpoints in the classroom.

Even in the 21st century, creationism still flourishes in America.[49] For example, Paul Broun (1946–) (Republican from Georgia), a member of the House Committee on, of all things, Science, Space and Technology, gave a speech where he stated[50]: "God's word is true … All that stuff I was taught about evolution and embryology and Big Bang theory, all that is lies straight from the pit of hell." And during the Trump administration, both Vice President Mike Pence (1959–) and Secretary of Education Betsy DeVos (1958–) supported the idea that schools should "teach the controversy" around evolution and creationism. Some expressed concern over this disturbing development among the political elite[51]:

> The Trump administration, the most notoriously anti-science administration in modern memory, has only added fuel to the creationist fire … The Trump Administration may be heartening creationists, but it is important that the legal system maintain secular values in governmental institutions despite any pressure exerted by the Executive Branch.

In 2013, almost two-thirds of Americans denied evolution,[52] and a poll six years later found that 40% of Americans were creationists.[53] Unlike the sun denial and gravity denial campaigns which have largely dissipated, Darwin's theory of evolution is still fodder for the cranks and their junk science. Evolution denial is one of the longest surviving forms of science denialism.

Genetics Denial

A field of study related to evolution was genetics, which was founded by the scientist and friar Gregor Mendel (1822–84) who had discovered the rules of heredity. While Charles Darwin and Gregor Mendel were contemporaries, a connection between their work was not made at the time.[54] Both theories faced the hostility of the willfully ignorant, but genetics was targeted by, arguably, the most dangerous science denier of all time.

The Soviet biologist Trofim Lysenko (1898–1976) had denied genes existed,[55] and his brand of science denial was called "Lysenkoism." He argued[56]: "We biologists do not take the slightest interest in mathematical calculations … We maintain that biological regularities do not resemble mathematical laws." Science brawls emerged against his contrarian claims. The British botanist Sydney Harland (1891–1982), for example, portrayed him as[57]: "completely ignorant of the elementary principles of genetics and plant physiology … To talk to Lysenko was like trying to explain differential calculus to a man who did not know his 12-times table." And similar comments came from his colleague Julian Huxley (1887–1975), who was none other than the grandson of Thomas Huxley who championed evolution (the propensity for science brawls must have been genetic)[58]: "Sometimes he appears ignorant of the scientific facts and principles involved, sometimes he misunderstands them, sometimes he distorts them, sometimes he counters them with bare assertions of his own beliefs." Huxley's description of Trofim Lysenko fits with that of most science deniers, and he would have been dismissed as just another foolish crank if not for the backing of Joseph Stalin (1878–1953).

Lysenko's notoriety came not from his "science" but his darkly Rasputin-like influence on the Soviet state.[59] By 1935, he had used political ideology to his advantage by framing his biology as "socialist," in contrast to the "bourgeois" biology of others. Soon a purge was underway of the leading scientists; many were imprisoned, and several were executed. In 1948, genetics research and teaching were banned within the USSR which persisted into the 1960s.[60] The climate scientist Michael E. Mann (1965–) summarized Lysenko's legacy[61]:

> Trofim Lysenkso was a Russian agronomist, and it became Leninist doctrine to impose his views about heredity, which were crackpot theories completely

at odds with the world's scientists. Under Stalin, scientists were being jailed if they disagreed with his theories about agriculture, and Russian agriculture actually suffered. Scientists were jailed. They died in their jail cells. And potentially millions of people suffered from the disastrous agricultural policies that followed from that. So that's what happens when fringe scientific views that might support a particular ideology are allowed to trump actual mainstream science.

Trofim Lysenko has been accused of being the worst villain that science denialism has produced, as his combination of contrarian theories and political influence contributed to a famine that caused the death of millions.[62] Lysenko's exploits highlight how dangerous science denial can be once adopted by the political elite and entrenched within the state.

Relativity Denial

Science denialism had shown its toxic side in Stalinist Russia and did the same in Germany. An early relativity contrarian had been the physicist Ernst Gehrcke (1878–1960). In 1913, he received a letter from his friend the philosopher Oskar Kraus (1872–1942) who wrote sympathetically[63]:

> People are suffering from extreme fatigue, and an irritability that is due not least to the absurd theories of relativists. I have a burning desire to see the *source of error* revealed for all of the absurdities that you yourself, honored sir, have accurately characterized.

On August 24, 1920, Ernst Gehrcke had been a speaker at a notorious gathering at the Berlin Philharmonic, organized by the zealous anti-Semite Paul Weyland (1888–1972), under the name *Arbeitsgemeinschaft deutscher Naturforscher zur Erhaltung reiner Wissenschaft* ("Association of German Natural Scientists for the Preservation of Pure Science," which had a mandate[64]: "to purge German science of Jews"). While being a staunch relativity denier, Oskar Kraus, who was Jewish, had been offended by the anti-Semitism at the Berlin meeting, which he expressed in another letter to his friend Ernst Gehrcke.[65] Anti-Semitism was rampant in the early Weimar Republic, and the theory of relativity and its creator were a lightning rod for their hateful agenda.[66]

Albert Einstein (1879–1955) had sarcastically called the phoney organization behind the Berlin meeting the *Antirelativistische GmbH* ("Anti-Relativity Company").[67] Three weeks after the offensive event, Albert Einstein wrote to his friend, the mathematician Marcel Grossman (1878–1936), about his frustration with the science deniers[68]: "This world is a strange madhouse. Currently, every coachman and every waiter is debating whether relativity theory is correct. Belief in this matter depends on political party affiliation." If we substitute relativity theory with climate change, Einstein's words would be describing the major problem with climate denial.

Ernst Gehrcke and Oskar Kraus had both been members of an international relativity-denial organization, called the Academy of Nations (AoN), founded by the engineer Arvid Reuterdahl (1876–1933) and the mystic Robert Browne

(1882–1978). A letter by Arvid Reuterdahl demonstrated his zealous opposition to Albert Einstein[69]:

> There is very little danger of overdoing the fight on Einstein, I have had the pulse both here and abroad on that matter for a long time … I propose, for myself at least, to fight this charlatan until he is vanquished or I a[m] dead.

Arvid Reuterdahl wrote an article "Einsteinism, Its Fallacies and Frauds," which began by displaying his disdain for the discoverer of relativity.[70] The article had been translated into German for inclusion in *Hundred Authors Against Einstein* published in 1931, which was a collection of contrarian opinions against relativity.[71] When Albert Einstein heard about the book, he quipped[72]: "Why 100? If I were wrong, one would have been enough."

The *Hundred Authors Against Einstein* was published as part of *Deutsche Physik* (the movement known as "Aryan Physics" against the work of Albert Einstein), based on Nazi ideology. Two high-profile contrarians, Johannes Stark (1874–1957) and Philipp Lenard (1862–1947), both Nobel Prize winners in physics, were leaders of *Deutsche Physik*. Dr. Lenard had a long-standing hatred of Albert Einstein[73] and described relativity as a "Jewish fraud."[74] He was one of the first prominent scientists to join the Nazi cause and, in 1928, Adolf Hitler (1889–1945) had even visited his home in Heidelberg.[75] In January 1933, Adolph Hitler became Chancellor, and shortly thereafter Albert Einstein left Germany for the United States. Later that year, an event was organized in Leipzig by Johannes Stark to celebrate the[76]: "Commitment of the Professors at German Universities and Colleges to Adolf Hitler."

The science writer Philip Ball (1962–), in his study of physics under Hitler, commented on the power of political ideology on German science[77]:

> The case of *Deutsche Physik* … shows how the German scientists' pretensions of being 'apolitical' did not prevent politics from infecting scientific ideas themselves, and almost overwhelming them. Perhaps most importantly, the story explodes the comforting myth that science offers insulation against profound irrationality and extremism.

Deutsche Physik was a horrific example of the power of ideology which overwhelmed one of the great scientific nations in the world at the time.

Call it what you will – junk science, false paradoxes, vulgar errors, eccentric literature, pseudoscience, crank science – science denialism has a long and sordid history, existing since the beginning of science itself. Ideology is generally the driver of organized science denialism: religious ideology against heliocentricity and evolution, and political ideology against genetics and relativity.

Political Science Denialism in America

The vilification of Einstein and the maleficence of Lysenko are foreboding examples of how dangerous science denialism can be when it enters the political sphere. Martin Gardner had warned about extreme science denialism in politics[78]:

> Let us hope that Lysenko's success in Russia will serve for many generations to come as another reminder to the world how quickly and easily a science can be corrupted when ignorant political leaders deem themselves competent to arbitrate scientific disputes.

Yet, the "ignorant political leaders" have reappeared a mere two generations later[79]: "Americans should not be conned by the Trump administration's climate lysenkoism and should instead place their trust in the robust findings of climate researchers who place the public interest ahead of political ideology." When a nation turns its back on knowledge, and ideology forms the basis for decision making, there will be serious policy mistakes.

While an individual science denier can be largely ignored, the adoption of science denial by political elites is another matter, which is what has happened in America with climate denial. The new science denialism borrows some of the propaganda techniques used against Albert Einstein decades before:

- The Academy of Nations can be viewed as a prototype of modern right-wing think tanks with their contrarian "fellows" who promote climate denial.
- *Hundred Authors Against Einstein* can be viewed as a 1930s science denial "blog" as the book is a collection of denier opinions on relativity.

But even more damaging has been the involvement of political actors themselves in the censorship and state-sponsored persecution and prosecution of scientists, as demonstrated by the Republican elite's treatment of climate scientists in the new millennium (their tactics of intimidation are discussed in later chapters). The tragedy of political science denialism has come to America.

Climate Denialism

Of all the various manifestations of science denialism, climate denialism will have the most catastrophic outcome. Climate change threatens all life on the planet, as extreme weather events will only increase in frequency and severity until human-sourced greenhouse gas emissions cease. But climate denial impedes and even paralyzes our political will to take action to deal with this crisis.

The Name Game

So, what is a climate denier? A vivid description of hard-core climate deniers has been tweeted by the noted Canadian climate scientist Katharine Hayhoe (1972–) who labels them "dismissives:"[80]

> My personal definition of a dismissive is someone who, if an angel from God with brand-new tablets of stone reading 'Global Warming Is Real' in foot-high letters of flame appeared to them, would not change their minds. So why would 10000 scientific studies do so? They won't.

Below are a two more definitions that a Google search quickly picks up:

Merriam-Webster – "climate change denier" (and "climate denier")[81]: "one who denies that changes in the Earth's climate or weather patterns are caused by human activity."

Wikipedia – "climate change denier:"[82] "denial, dismissal, or unwarranted doubt that contradicts the scientific consensus on climate change, including the extent to which it is caused by humans, its effects on nature and human society, or the potential of adaptation to global warming by human actions."

And asking ChatGPT "what is a climate denier" provides the following definition:

> A climate denier is a person who rejects the scientific consensus on climate change and denies the reality of its impacts and causes. This can include denying the existence of human-caused climate change or the severity of its consequences. Climate denialism often involves spreading misinformation and false information about climate science.

The definition of climate denier that I routinely use and prefer is: "those who deny the accepted science that greenhouse gas emissions must be stopped A.S.A.P., as climate change is a present-day threat, is getting worse, and is mainly caused by us." There are many shades of climate denialism, but I have found that this definition captures most of them.

What are these common shades?[83] The following comments of climate denial can be found in the transcripts of congressional hearings and on Twitter:

i global warming does not exist or is natural
ii global warming is good for you and/or plants
iii there is no scientific consensus, or the science is too uncertain
iv scientists are corrupt
v future technologies will save us so let's just wait
vi climate policies would hurt the poor
vii China is emitting much more, or our nation is emitting only a small amount
viii there is no urgency, as it won't be that bad
ix renewable energies aren't ready to replace fossil fuels
x fossil fuels are essential to civilization and ending energy poverty.

The propaganda used by climate deniers always leads to the same comforting conclusion – no regulations are needed for fossil fuels – as each of the above illustrate.

Climate deniers shun the label, preferring the disingenuous terms "climate skeptics" or "climate realists." In a Senate hearing, for example, the contrarian scientist Roy Spencer complained about being called a denier[84]:

> that is something that continually annoys me as those of us that are called deniers, it is never actually, I think the d word was actually used by the Chairman today, it has never actually been pointed out. What is that we deny?

Climate deniers like to pretend they don't know the meaning of the term "climate denier" and will send a glib tweet, such as: "No one denies the climate" (I see these responses more than a thousand times a year on Twitter; my reply to them is, if you don't like being called a climate denier then stop being one).

In the name game, climate deniers also try to attach the term "denier" with a religious yoke, as shown by Senator Ted Cruz (1970–) (Republican from Texas)[85]:

> They call anyone who questions the science – who even points to the satellite data – they call you a, quote, 'denier.' Denier is not the language of science. Denier is the language of religion. It is heretic. You are a blasphemer. It's treated as a theology.

Some hard-core climate deniers and contrarians[86] have gone even further, invoking the Holocaust[87]; here is a sample of a few:

Judith Curry[88]: "I reserve the word 'deniers' for people that explicitly reject the history of Jewish extermination in wartime Germany … When the word 'denier' first crawled out of the political slime, I fully expected those in science and media alike to reject it, vocally and without qualification."

Tom Davis (1949–)[89]: "When so-called eco experts liken skeptics to Holocaust deniers, we have definitely gone too far."

Myron Ebell (1953–)[90]: "It's meant to be offensive … It's meant to tie … some people to Holocaust denial."

William Happer (1939–)[91]: "Anyone who objects to all of the hype gets called a denier; that is supposed to make me a Holocaust denier … the demonization of carbon dioxide is just like the demonization of the poor Jews under Hitler. Carbon dioxide is actually a benefit to the world and so were the Jews."

Marc Morano (1968–)[92]: "It remains to be seen what Gore and Moyers will have to say about proposals to make skepticism a crime comparable to Holocaust atrocities."

Roger Pielke Jr. (1968–)[93]: "The phrase 'climate change denier' is meant to be evocative of the phrase 'holocaust denier.'"

Dana Rohrabacher (1947–)[94]: "To those who disagreed … they were treated like nonentities, like they didn't exist, or were personally disparaged with labels like 'denier.' Well, you know, Holocaust denier, that's what you do. Now, how much uglier does it get?"

Roy Spencer (1955–)[95]: "They indirectly equate (1) the skeptics' view that global warming is not necessarily all manmade nor a serious problem, with (2) the denial that the Nazi's extermination of millions of Jews ever happened … I'm now going to start calling these people global warming Nazis."

This disgraceful association is made by climate deniers themselves, who exploit the Holocaust to slander those defending science against propaganda and participating in climate brawls.

The "denial" terminology had started to appear in the 1990s. At the beginning of the decade, the journalist Anuradha Vittachi tied denial to the science of climate change,[96] and in 1995, the journalist Molly Ivins wrote about the "denial of global warming."[97] In the same year, another early association of the term "denial" with those who did not accept the scientific consensus on climate change appeared in a revealing article called "The Heat is On: The Warming of the World's Climate Sparks a Blaze of Denial," by the journalist Ross Gelbspan[98]: "The skeptics assert flatly that their science is untainted by funding. Nevertheless, in this persistent and well-funded campaign of denial they have become interchangeable ornaments on the hood of a high-powered engine of disinformation." Then, in his subsequent book, Mr. Gelbspan laid out a comprehensive description of the propaganda campaign from the energy-industrial complex (the fossil fuel and related industries) and its threat to American democracy[99]:

> In the United States the mere threat of impending climate change has impelled the oil and coal industries to engine a policy of denial. While their campaign may seem at this point no more sinister than any other public relations program, it possesses a subtle antidemocratic, even totalitarian potential insofar as it curbs the free flow of information, dominates the deliberations of Congress, and obstructs all meaningful international attempts to address the gathering crisis.

All the above journalists were among the early participants in the climate brawl.

The terminology continued to evolve over the 2000s. The term "climate denier" itself showed up in 2003,[100] and the term "climate change denial" had appeared in 2000.[101] The first tweets using "global-warming denier"[102] and "climate denier"[103] both appeared in 2007, and "climate change denier," the following year.[104] Such terms also began to be heard in congressional hearings. In 2005, Richard Benedict, President, National Council for Science and the Environment, responding in writing to a question from Senator James Inhofe (1934–) (Republican from Oklahoma), had used "climate change denier,"[105] as had Professor Martin Hoffert, a year later.[106] In 2007, the term "climate deniers" appeared in a congressional hearing for the first time.[107]

"Climate denier" and related terms had become popular by the middle of the 2010s, when even the president of the United States was routinely using it (see Chapter 4). Until then, "climate skeptic" was more widely used. For example, a detailed description of climate deniers appeared in a House hearing in 2011 but was calling them "skeptics."[108] And, when I wrote *Carbon Politics and the Failure of the Kyoto Protocol* early in the 2010s, I was also still using the term "skeptic."

This was an unfortunate choice of terms, as climate skepticism is an essential aspect of science, while climate denialism is not. In 2011, Al Franken (1951–) (Democrat from Minnesota) outlined for the members of the Senate the differences between climate skeptics and climate deniers[109]:

> Now, let's make a distinction between people who are climate skeptics and people who are climate deniers. This is kind of an important distinction. There is nothing wrong with skepticism. In fact, we love skeptics. Scientists are, by nature, skeptical …

> On the other hand, a climate denier is someone who would not be convinced no matter how overwhelming the evidence. And, as I pointed out, a lot of these deniers are being paid by polluters to say what they want.

The popular astrophysicist Neil deGrasse Tyson (1958–) explained[110]: "A skeptic will question claims, then embrace the evidence. A denier will question claims, then reject the evidence." And the climate scientist Michael E. Mann tweeted[111]: "Real scientists are skeptical, but in equal-sided manner. Wholly unwarranted one-sided skepticism that rejects overwhelming evidence based on the flimsiest of arguments that don't hold up to scrutiny isn't 'skepticism'. It's denialism." Both Drs. Tyson and Mann are leading advocates of the science brawl.

The time-honoured tradition of skepticism in science should not be marred by confounding it with denialism which insults scientists and the scientific method. Moreover, the use of "climate skeptics" as a label for this anti-science group is passé. However, some habits die hard, and there are a few legitimate holdouts who still prefer the term "climate skeptics." If someone dislikes using the term "deniers," then the old-school terms "cranks," "crackpots," or "junk scientists" can be used instead – anything but "skeptics."

The Rise of Climate Denialism

Historically, propaganda from the state or church had spread science denialism to protect the absolute power of a particular social order. In America, though, propaganda was orchestrated by corporations to protect profits. While corporate greed has long existed, the organized attacks against science by business interests thrived after World War II, coinciding with the rise of the environmental and health sciences.

Whenever science threatens profits, the proliferation of denial begins. The propaganda is successful for the corporations, at least for a while. An early example of this profit-based science denial can be seen in the tobacco industry's fight against the link between smoking with cancer.[112] Coincidentally, just as the tobacco campaign was winding down, the one on climate change was ramping up, and many of the peddlers of propaganda connected with the tobacco cause switched over to fossil fuels. It didn't matter that the science for each was so different because the science didn't matter.[113] The key job requirement was being good at propaganda to keep corporate profits flowing, first for the tobacco industry and then for the energy-industrial complex.

Climate denialism was getting organized in the 1980s and was basically an American creation (and is predominant in Anglo-Saxon countries).[114] Corporate America had been emboldened by Reagan's laissez-faire politics. The end of the 1980s witnessed the rollout of the massive propaganda and lobbying offensive of the energy-industrial complex in reaction to the formation of the IPCC (Intergovernmental Panel on Climate Change), Hansen's high-profile Senate presentation, and the upcoming Earth Summit for Rio. To counter these threats to their business interests, the GCC (Global Climate Coalition) was formed, contrarian reports from right-wing think tanks were issued,[115] and contrarian testimonies before Congress against the science of climate change were initiated (discussed in later chapters).[116] The early target of the climate denial propaganda had been American politicians, but after the international buzz of the Kyoto conference, the energy-industrial complex further broadened its propaganda campaign using mainstream media to reach the public.[117] In the later 2000s, the energy-industrial complex was facing new challenges and opportunities, notably the documentary "An Inconvenient Truth," the Nobel Peace Prize awarded to Al Gore and the IPCC, the U.S. Supreme Court decision on making CO_2 a pollutant, "climategate," and the rise of social media platforms (such as Twitter) – the latter of which provided the biggest opportunity imaginable for the spread of climate denialism.

The culmination of these events, decades in the making, was an "infodemic" – a compendium of lies, doctored graphs, conspiracy theories, gaslighting, and cherry-picked information against the climate sciences. The infodemic acted like a virus spreading climate denialism across America, and especially in Washington. Even the IPCC has warned that[118]: "Vested economic and political interests have organized and financed misinformation and 'contrarian' climate change communication," and[119]: "Accurate transference of the climate science has been undermined significantly by climate change counter-movements, in both legacy and new/social media environments through misinformation." The propaganda campaign has been so successful that American politicians actually debated the validity of the peer-reviewed science at congressional hearings.

The energy-industrial complex had no trouble sowing the seeds of climate denial, which resonated especially with Republicans, by generating fear and doubt:

- Fear over economic costs from climate policies – by politicians using sound bites such as "job-killer," and "hurting families and small businesses."
- Doubt on the climate sciences – often through the official "authority" of contrarian scientists and political elites.

This political spin by conservatives has framed the debate on climate change for decades and continues to do so today. As soon as there is any focus on the science, a tsunami of propaganda drowns it out; the serious issue of climate change and the overwhelming evidence of the underlying science become lost in the distraction.

A simple way of estimating climate denialism within a society is from surveys asking if global warming is mainly caused by human activity (Table 1.1; the lower the percentage, the greater the climate denialism in that country). Using this survey, one can estimate that 59% (100%–41%) of Americans are probably climate deniers. Generally, different polls have found the degree of climate denial (those not accepting the science that global warming is mainly caused by human activity) in the United States varies between 40% and 60%.[120]

One on-going survey provides a detailed picture of climate denialism in America from which two years have been selected to highlight changes over time (Table 1.2). Over the 12-year span, considering that scientific evidence had only mounted in support of anthropogenic climate change, the fixed position among

Table 1.1 Distribution of Awareness that Global Warming Is Mostly Caused by Human Activity by Nation

Country	*2020 (%)*
United States	41
France	45
Poland	47
Czech Rep.	48
UK	52
Sweden	55
Germany	58
Italy	62
Spain	71

Source: Eichhorn, Molthof, and Nicke 2020 (for other global polls, see Flynn et al. 2021).

Table 1.2 Distribution of Awareness that Global Warming Is Mostly Caused by Human Activity by American Voters

	2008 (%)	*2020 (%)*
All	55	61
Liberal Democrats	78	92
Moderate Democrats	64	77
Moderate Republicans	52	50
Conservative Republicans	28	30

Source: Leiserowitz, Maibach, et al. 2020 (also see Leiserowitz et al. 2021; for a similar earlier poll see Anon. 2016).

Republicans is striking. One explanation is that climate denial propaganda has successfully neutralized the growing evidence from the climate sciences with Republicans. What is this propaganda? An organization – Climate Action Against Disinformation – provides a general answer[121]:

> Climate disinformation and misinformation refers to deceptive or misleading content that:
>
> - Undermines the existence or impacts of climate change, the unequivocal human influence on climate change, and the need for corresponding urgent action according to the IPCC scientific consensus and in line with the goals of the Paris Climate Agreement;
> - Misrepresents scientific data, including by omission or cherry-picking, in order to erode trust in climate science, climate-focused institutions, experts, and solutions; or
> - Falsely publicises efforts as supportive of climate goals that in fact contribute to climate warming or contravene the scientific consensus on mitigation or adaptation.

Or, as one climate policy expert has so bluntly described this propaganda[122]:

> I think it's important that the public understand that there are these powerful economic actors that have been working very hard to distort what the public thinks about this … This is a beast, or a demon, that's been unleashed by the fossil fuel industry.

The Science of Climate Denialism

Climate change is studied mainly by the physical sciences, while climate denialism is studied by the social sciences. Where does the notion of "denial" come from? Sigmund Freud (1856–1939) had introduced the concept of "denial of external reality" (*verleugnung*) back in 1924,[123] as a psychological means of coping with stress, and the concept was further elaborated in 1936, by his daughter Anna Freud (1895–1982) in *The Ego and the Mechanisms of Defense*. Since then, denial is recognized as a standard "defense mechanism."[124] An early work on psychological denial specifically in relation to climate change appeared in 2001; a conclusion of the investigation was[125]: "The most powerful zone for denial was the perceived unwillingness to abandon what appeared as personal comfort and lifestyle-selected consumption and behaviour in the name of climate change mitigation."

Climate denialism has become a formal area of research, and organizations have formed that are devoted specifically to the subject, such as the Centre for Studies of Climate Change Denialism,[126] and the Climate Psychology Alliance which stated[127]:

> Denial can also be conscious and cynical, this is sometimes called 'denialism'. This is industry-funded attempts to draw people to deny climate change. Denialism works to boost all the different forms of denial, to suit the purpose at hand, which is maintaining business as usual for the fossil fuel industry.

The noted psychologist Stephan Lewandowsky (1958–) reviewed climate denial propaganda in an article titled "Climate Change Disinformation and How to Combat It," where he stated[128]:

> Climate change presents a challenge at multiple levels: It challenges our cognitive abilities because the effect of the accumulation of emissions is difficult to understand. Climate change also challenges many people's worldview because any climate mitigation regime will have economic and political implications that are incompatible with libertarian ideals of unregulated free markets. These political implications have created an environment of rhetorical adversity in which disinformation abounds, thus compounding the challenges for climate communicators.

In a more recent review article, Matthew Hornsey and Stephan Lewandowsky also pointed out[129]:

> People's beliefs and attitudes about climate science have become a shorthand way of signalling political loyalty. This raises the possibility that people are

> not always making up their mind about climate science themselves but are rather taking their cues from elites within the political party with which they identify.

Psychological research has identified a major obstacle in dealing with climate denial[130]:

> But despite the fact that psychologists know a lot about denial, they have never had to face denial on this scale before. Millions of people share the phenomenon of climate denial. This is clearly not something that is amenable to individual or even group psychotherapy.

A NASA climate scientist, Peter Kalmus (1974–), has described his experience with this societal problem[131]:

> I even hoped that humanity would listen to scientists … after all, the scientific fundamentals are easy enough to grasp. But things have turned out worse than I dreamed possible. I underestimated the depth and intransigence of society's collective climate denial.

And on a personal note, he had tweeted[132]: "I cannot even begin to express how frustrating and devastating and terrifying it is to be a climate scientist in a society run by climate deniers."

The collective nature of the climate denial problem has attracted the attention of sociologists since the 1990s. The CCCM (climate change counter movement) had arisen in reaction to climate change and was detailed in 2000 by Aaron McCright and Riley Dunlap who presented the following[133]:

> In the past decade, global climate change became a widely accepted social problem … Not surprisingly, opponents of these efforts mobilized in recent years to mount intense opposition to calls for major international action to prevent global warming such as treaties designed to reduce carbon dioxide emissions.

Later, Robert Brulle and Riley Dunlap described auxiliary groups, including conservative think tanks, industrial trade associations, contrarian scientists, denial bloggers, social media influencers, and others, used by the CCCM. These social scientists also recommended the importance of outing the members of the counter movement[134]: "Sociological research and outreach on climate change policy obstruction has become public sociology. Exposing those responsible for climate change policy obstruction can help inoculate the public and policy-makers against their tactics."

The physical sciences can do little more to convince Republicans to act on climate change (as the table above shows their intransigence); the "solution to the politics" of the climate crisis will likely hinge more on the social sciences furthering our understanding of the workings of the climate denialism movement.

Climate Brawl

Climate denialism also has a counter movement – "climate brawl" – which became necessary because climate denial became a political force in America, and science, with all its facts and evidence, had failed to stop it. Climate brawl only targets the hard-core climate deniers and not the regular climate deniers who have been duped by these influence peddlers. There is no intention to have a scientific debate in a climate brawl; instead, the goal is to challenge the credibility of the climate denier and their sources which, in turn, discredits and weakens the propaganda. Who are these hard-core climate deniers?

In my experience dealing with hundreds of climate deniers and thousands of their comments over the years, hard-core climate deniers tend to be motivated by economic concerns and/or ideology. Those economically motivated derive financial gain from the energy-industrial complex: industry executives, employees, shareholders, investors, suppliers, contractors, consultants, think tanks, etc.; also included are residents, businesses, and politicians in the local economy where the energy-industrial complex has operations.

The other major motivation behind climate denial is political ideology, namely conservatism.[135] Climate change regulations did not fit with conservative views of individual freedom, small government, and, especially, free-market capitalism[136]; consequently, Republicans have been more inclined to reject the climate change message delivered by science given this "confirmation bias" (defined as[137]: "Once we have formed a view, we embrace information that confirms that view while ignoring, or rejecting, information that casts doubt on it"). In other words, Republicans, as easy targets for the anti-regulation PR campaigns of the energy-industrial complex, welcomed climate denialism. The political historian Matthew Dallek further explained why science denial propaganda resonated with conservatives[138]: "Still, the overwhelming misinformation on science is coming from the right … In these quarters, "freedom" — even freedom to harm fellow citizens — is more important than the public good." A special concern in Washington is that many politicians are both far right and financially connected to the energy-industrial complex, producing some of the most hard-core climate deniers, which are discussed in the fourth chapter.

There is a timeless nature to science denial and its purveyors, but their propaganda has recently exploded in proportions never seen before in history. The spread of climate denial was facilitated by new avenues for information transfer. During the new millennium, Twitter (and other social media platforms) caused a paradigm shift in communications by providing the freedom to express anything by anyone to everyone. This brave new world, however, comes with its own flaws as conspiracy theories, gaslighting, and propaganda abound, especially on climate denial.[139]

While the cranks, as they were once called, have always been there, such fringe groups had lurked mainly in the shadows of society until the internet released the "Cranken" into every corner of the world. Social media weaponized the messengers of climate denial by providing them with an unfettered global soapbox, where junk science and other propaganda travel around the world many times within seconds, especially to groups selected by the stealthy algorithms. Hard-core climate deniers and their cyber-propaganda frolic freely in this virtual world, as keyboard outlaws who scour Google for blogs, memes, doctored graphs, quote mining, and YouTube videos (pretending that these are credible sources), searching for the slightest criticism of the science. On-line climate denial thrives on Twitter, where I personally see dozens of such tweets daily. So, Twitter is a good place to start challenging climate denial.

Overall, there are more than 200 million daily active users on Twitter globally. How many of these are climate deniers? My estimate of total climate deniers on Twitter is 59 million, and half are just from the United States.[140] That is a lot, but most of them are passive tweeters. In terms of climate brawl, the climate deniers of interest are the ones who tweet propaganda aggressively, and the number of these hard-core, fanatical followers of climate denialism are only a minority. A study has found that between 2014 and January 2023, there were just 220K Twitter accounts that posted 2.5 million tweets and retweets associated with strong aspects of climate denial.[141]

Is propaganda on Twitter and other social media really a problem? Yes, it is. A multi-national survey has found that climate change is the top global threat, followed closely behind by the spread of false information online.[142] According to a group of Nobel Peace Prize winners[143]:

> We urge rights-respecting democracies to wake up to the existential threat of information ecosystems being distorted by a Big Tech business model fixated on harvesting people's data and attention, even as it undermines serious journalism and polarises debate in society and political life.

The greatest danger is that propaganda repeated, again and again, becomes the collective "truth" – called the "illusion of truth effect" or "illusory truth effect" by psychologists[144] (this is an active area of scientific research[145]) – if not challenged.[146]

Platforms such as Twitter have a huge influence on public opinion, especially by amplifying the voices of radicals and fanatics.[147] When hard-core climate deniers are tweeting, there is a flurry of likes and retweets from their loyal minions which boosts their anti-science messaging. A handful of hard-core climate deniers can have an immense reach; in a one-month period, sixteen such accounts received a total of more than half a million likes and retweets on Twitter,[148] and an analysis of these accounts concluded[149]:

> Drawing on research compiled over the past 18 months, and especially in the margins and aftermath of COP26, we have clear evidence of the challenge at hand: the failure to stem mis- and disinformation online has allowed junk science, climate delayism and attacks on climate figures to become mainstreamed. Our analysis has shown how a small but dedicated community

> of actors boast disproportionate reach and engagement across social media, reaching millions of people worldwide and bolstered by legacy print, broadcast and radio outlets.

Compounding the problem, propaganda travels more quickly than the truth on Twitter, according to MIT researchers[150]: "The data comprise ~126,000 stories tweeted by ~3 million people more than 4.5 million times … Falsehood diffused significantly farther, faster, deeper, and more broadly than the truth."

In the twitterverse, you see climate denial tweets from Twitter newbies, trolls, veteran climate deniers, and the political elite. Among politicians, Twitter has become a popular communication tool. During the 115th Congress (2017–9), for example, there were 900,000 tweets from the members, and almost 9,000 related to climate change.[151] Twitter has also become a routine communications tool for the president of the United States, including on the topic of climate change – President Trump promoted climate denial, while President Obama and President Biden joined in the climate brawl by challenging climate denial on Twitter (discussed in later chapters).

My engagement with climate denial on Twitter came about because, as a scientist, I found it appalling to witness how hard-core climate deniers rode roughshod over the climate sciences on the platform without any credible evidence. A month after joining Twitter, I began tweeting about the politics of climate skeptics (on November 28, 2012).[152] I kept on tweeting about climate skeptics until someone corrected me in mid-2016, stating[153]; "the correct term is 'denier,'" and I have since used that term in my climate brawls.

After an exceptionally tension-filled episode with a hard-core climate denier in 2018, I was quickly overwhelmed fighting a horde of his minions. Anyone on Twitter who might have helped me in a climate brawl was out of reach, as I had no way of contacting them. After more heated engagements, in February 2019, I introduced the hashtag – #ClimateBrawl – as a "Bat-Signal" distress call for assistance against aggressive climate deniers. I had never introduced a hashtag before and had no idea if it would work. To my surprise, it did – help arrived – and #ClimateBrawl has since morphed into an international community actively engaging on Twitter, under the motto be "active, civil and factual."

#ClimateBrawl is a rallying cry, then, on Twitter to challenge the propaganda of climate deniers, which many have done – known as the "friends of #ClimateBrawl." Can anyone join the friends of #ClimateBrawl? Yes, you don't have to be a scientist to challenge the average climate denier but be prepared. Those who engage the agents of denialism are often met by a barrage of questions, phoney science, and angry insults by swarms of aggressive hard-core climate deniers. These encounters are stressful and exhausting, and many may be, understandably, reluctant to volunteer for such a toxic engagement. However, thanks to #ClimateBrawl, there will be plenty of help. Along with this support group, another advantage would be knowing the games of deception that are played by climate deniers which are discussed next.

The Climate Brawl Handbook, Twitter Edition

Climate denialism is a counter movement founded on ideology and propaganda. Since it is almost impossible to change a person's ideology, we must challenge the propaganda; in other words, there needs to be a climate brawl. A good place to start would be Twitter, as one study concluded that social media[154]: "offers a novel opportunity to study previously inaccessible aspects of social interaction about climate change." Twitter, in particular[155]: "attracts a large number of influencers on political and climate issues" and[156]:

> is the ideal platform for studying climate communications because it is widely used by politicians and journalists, has broad social and cultural influence … many studies highlight the importance of Twitter (and social media in general) as a critical tool for studying climate communication, political polarization and misinformation.

Twitter has one of the largest depositories of climate denial in the world. You don't even have to search it; just tweet a few choice words about the science of climate change, and the climate deniers will find you.

The "Climate Brawl handbook" is a guide[157] against the propaganda of climate denial based on my personal experience. On Twitter, you learn the strategies, tactics, and vulnerabilities of the propaganda of climate denialism. Over the past decade, I noticed patterns in the types of tweets repeated over and over again. Below are some helpful tips for replies, based on my engagements in the Twitter trenches; I discovered these practical strategies by trial and error, after over one hundred and eighty thousand tweets.

The Climate Denial Game

Climate deniers see their activities as a game of chess, where they are planning three to four tweets ahead, waiting for gotcha moments to pounce on supporters of science and ensnare them in futile debates. This sort of insidious gamesmanship happens in Congress, in mainstream media, but nowhere more than Twitter.

In short, there are two basic approaches in a climate brawl for challenging their propaganda:

Defensive – presenting peer-reviewed studies to counter the claims of the climate denier

Offensive – demanding peer-reviewed studies that justify the claims of the climate denier.

I began with the traditional defensive strategy, posting on Twitter evidence-based information with links to scientific sources to counter the shoddy evidence used by climate deniers. With this strategy, it soon became clear to me that you win every battle but lose the war. Though the defensive strategy is a logical way to reply, the climate denier has expected it, as the thread usually goes as follows:

DENIER: "The science is wrong."
ME: "Here is peer-reviewed evidence that it is not" (add link).
DENIER: "The evidence is wrong."
ME: "Here is more peer-reviewed evidence" (add link).
DENIER: "That evidence is also wrong."

The engagement then deteriorates into an endless feedback loop, as designed by the climate denier, since you have given them the opportunity to tweet one piece of phoney science after another. They have a good supply of junk science, and when that runs out, the climate deniers just repeat everything again which is further endorsed by their zealous followers. Climate deniers do not want to learn about the climate sciences; they just want to deny.

What are climate deniers hoping to accomplish by sharing their propaganda and junk science? As always, they are implying that a scientific debate exists, when it does not. The science has already been debated by experts over decades; so, there is no reason for the non-expert to debate it. Replying to them with a science-based rebuttal is all that climate deniers want and then their game is on. A more effective strategy, therefore, is not to show them more science (they don't care) but make them defend the sources of their "science" (they can't).

So, I switched over to the offensive strategy, which works better as the following generic thread illustrates:

DENIER: "The science is wrong."
ME: "Share the peer-reviewed evidence that states this."
DENIER: "Well, this meme shows that the science is wrong."
ME: "Still waiting for your peer-reviewed evidence."
DENIER: "Peer review is just pal review."
ME: "Your tweets betray your character and credentials; no need to engage further."

One benefit of the offensive strategy is that you have taken control of the Twitter engagement and have placed the climate denier on the defensive. Instead of debating science, you are now challenging the credibility of the climate denier and their sources. By asking for peer-reviewed studies[158] that support the junk science tweeted by them, the argument of the climate denier collapses. The offensive climate brawl approach works on Twitter, but it should also work in Congress against contrarians, members of the denial cabal, and the energy-industrial complex.

When climate deniers realize that they are losing the argument, they begin another strategy which is deflecting to a new topic. A couple of typical examples would be when there are discussions about how greenhouse gas emissions need to be reduced as soon as possible, and those supporting fossil fuel interests slyly flip the focus over to people freezing to death, energy poverty, or massive brownouts, but no one is proposing to reduce greenhouse gas emissions by risking energy reliability; also, climate deniers will suddenly throw out a strawman comment about a celebrity climate advocate who is flying in a private plane. Ignore all deflections and keep the conversation on track.

Below, we will see how to fight back on specific tweets popular among climate deniers and learn more about how the propaganda of climate denial works.

Credentials Don't Matter

When confronted with real science, such as from the IPCC or NASA, a climate denier may accuse the person of "appealing to authority" which is amusing, since this is an entirely appropriate thing to do. Of course, they themselves never appeal to any authority, as they have none and, in any case, believe that they are the smartest ones in the Twitter room. They will claim to know more because of their common sense and boast about being independent thinkers. When I challenge them on their lack of scientific knowledge, climate deniers will counter that I also lack training in the climate sciences, which is true – however, you don't need to know the science to accept it, but you do to refute it. Another common rebuttal is the use of dismissive gestures, such as the science is a hoax or scam, or the scientists/IPCC/NASA are corrupt or communist. Suggested Twitter replies

> Just because you don't understand the science does not make it wrong.
> Listen to the science; it knows a lot more than you do.

Uncertainty

Uncertainty, often exploited by climate deniers, is not the Achilles' heel of the climate sciences but a backbone of all science. Back in 1874, William Stanley Jevons (1835–82) had described the situation[159]: "Perfect knowledge alone can give certainty, and in nature perfect knowledge would be infinite knowledge, which is clearly beyond our capacities. We have, therefore, to content ourselves with partial knowledge—knowledge mingled with ignorance, producing doubt." Scientific investigations acknowledge uncertainty in order to better identify the parameters of the research and the directions for future study, which climate deniers exploit as a weakness. In the case of climate change, the body of knowledge is too great to ignore and waiting for science to reduce uncertainty further before acting is a recipe for disaster, as we are already witnessing.

The humbling principle of uncertainty is related to proof generally lacking for the theories of science which are never settled – in other words, according to climate deniers, how can there be a scientific consensus if knowledge is always evolving? The answer is that a theory is accepted when there is general agreement in the peer-reviewed literature (forming a consensus until, and if, new evidence requires modification or rejection of the theory). Suggested Twitter replies:

Uncertainty is a fundamental principle of all science.
If we wait for certainty, it will be too late to react.
Much more important than the uncertainty is what is known by science.

The Climate Model Deniers

Computer models are important to many sciences, including climate change.[160] While the science used by the models to project future climate conditions is robust, unknown inputs must be factored in, such as population growth, economic growth, rate of adoption of low-carbon energy sources, political will to reduce greenhouse gas emissions, etc.; so, scenarios with various socioeconomic assumptions are added to the models to account for human behaviour. Yet, climate deniers claim that the models are always wrong (and are not science) at congressional hearings and routinely on Twitter, but the projections of the climate models are consistent with observations.[161] Suggested Twitter replies:

Science has confirmed the models are accurate (add a link to the reference).
Strange that if climate modelling was wrong, the ground-breaking work was awarded the Nobel Prize in Physics in 2021.
Please share your peer-reviewed studies that demonstrate the models are wrong.

Weather

One of the most tiresome arguments from climate deniers is their deliberate confusion between weather and climate. An article from 1889 (with the surprising title "Is Our Climate Changing") from the pioneering American meteorologist Cleveland Abbe (1838–1916) explained the difference[162]: "Weather is the condition of the atmosphere at any moment of time; climate is the general or prevailing condition of the important features of the weather during a considerable period." Yet, climate deniers discredit climate change during unseasonably cold spells [and dispute the role of global warming in extreme weather events (and wildfires)[163]]. Suggested Twitter replies:

Climate change impacts all weather events.
Science knows that global warming will increase the frequency and severity of extreme weather events.

The Pollution Deniers

Climate deniers argue that carbon dioxide is essential to the growth of plants, and therefore, should not be regarded as pollution (but a benefit). Pollution is defined as[164]: "The introduction of harmful substances or products into the environment," and the carbon dioxide emitted from the burning of fossil fuels has been found to be harmful by causing climate change (and the U.S. Supreme Court agreed, declaring carbon dioxide emissions to be a "pollutant"[165]). Suggested Twitter replies:

Of course, the massive emissions of CO_2 from the burning of fossil fuels are pollution (provide a link to the definition of pollution and/or the Supreme Court announcement).
Manure is good for plants too, but I don't want you dumping it in my backyard.
There is no net benefit to adding CO_2 to the atmosphere.

The Google Moment

A common tweet from climate deniers is sharing a piece of evidence from their Google search (which every scientist has somehow missed) demonstrating that the science of climate change is wrong – the "Google moment." Climate deniers scour the internet for memes, doctored graphs, and YouTube videos looking for that Google moment when they find something that they believe challenges the science of climate change but is no more than propaganda, lies, half-truths, quote mining, and cherry-picked data, often just copied from a climate denial blog site. Despite their in-depth search of Google, climate deniers have somehow missed the thousands of peer-reviewed studies which can be readily found that have contributed to our understanding of climate change.

One such Google moment for many climate deniers is that the sun is the cause of modern climate change. In fact, solar activity has decreased and is slightly decreasing the temperature of the Earth, which is being overwhelmed by global warming.[166] However, climate deniers will resolutely hang on to this fallacy, even though no peer-reviewed literature supports the sun theory of modern global warming. A game played by climate deniers is stating that the sun warms the atmosphere or is responsible for the climate (which is true), but the sun is not causing modern climate change.

Another trick is showing plots of carbon dioxide levels and temperatures going back millions of years, illustrating that warming eons ago had been caused by other forcings, such as the sun, but the fact that the sun caused changes in the climate before does not mean that it is the cause this time. Climate deniers may be surprised to learn that climate scientists are already aware of the influence of the sun on climate and climate change, in the past and now. No one can do a Google search and find evidence that would actually discredit accepted science because it does not exist. Suggested Twitter replies:

A Google search gives you better knowledge than a climate scientist?
In what comic book did you find that cartoon?
You seem to know a lot about science, what degrees do you have?

"Sealioning"

Climate deniers ask questions, but they do not listen to the answers. At the start of a Twitter engagement, a climate denier may appear genuinely interested in the science of climate change; however, sincere requests to learn about the science of climate change became extinct on Twitter a long time ago (there may be the odd exception). Upon answering their question, they then ask another question, then another and another. The tone of the tweets will also change, becoming confrontational. The geoscientist Pete Akers has explained this strategy[167]:

> people who troll online by pretending to ask sincere questions, but just keep feigning ignorance and repeating 'polite' follow ups until someone gets fed up. That way, they can cast their opponents as attacking them and being unreasonable. It's pretty common on comment sections of weather blogs re: climate change.

They hope to rattle those defending science until they become testy and look petty and unprofessional. The questions are only a trap, so don't take the bait. Suggested Twitter replies:

> GIYF (Google is Your Friend; in other words, look it up yourself)
> You are just sealioning (provide a link to a definition of sealioning).
> Why are you asking questions when you won't listen to the answers?

The Last Resort

When all avenues have been exhausted and you have outmaneuvered the climate deniers in all their games, their last act of defiance will be to

- Use foul language and insult you
- Block you.

These common outcomes can be seen as a sign of victory in the twitterverse for the friends of #ClimateBrawl. I will often take a screenshot of such a tweet (which limits the exposure of the denier on Twitter) and post it under the banner that their tweet is "proof that climate denial is a lost cause." In other words, I share pictures of the tweets by climate deniers when they discredit themselves. Suggested Twitter reply:

> Your tweets betray your character and credentials. No need to engage further. Have a nice day.

Trolls and Bots

Besides the typical climate denier, there are two more tenacious types of predatory accounts on Twitter that relentlessly promote toxic propaganda and/or targeted harassment: automated accounts called "bots" and human trolls. Both are propaganda tweeting machines that amplify the messages of climate denial. I don't see many bots in my Twitter engagements (at least, I don't think so), but there are many trolls, as dozens of these on-line bullies harass me every day. In frustration over the widespread influence of bots and trolls on Twitter, I complained in an interview[168]: "It's no longer social media, it's bot media … It's difficult to know the difference."

A study found that about one-quarter of all tweets on the climate crisis originated from bots,[169] and climate denialism relies on more bots than other kinds of propaganda.[170] Bots, though, often hibernate, coming out on Twitter mainly for high-profile events, such as national elections or major conferences on climate change. Twitter does tackle bots, explaining that they[171]: "permanently suspend millions of accounts every month that are automated or spammy, and we do this before they ever reach an eyeball in a Twitter Timeline or Search." Malicious bots are included in their "platform manipulation" statistics which recorded over ten million cases in 2021.[172]

But the distinction between bots and trolls does not really matter, as both seek to manipulate the twitterverse. Bots, of course, are indifferent to science and the truth, but so are the trolls. Trolls can tweet toxic propaganda with impunity, as their identity can remain anonymous and disguised, so their friends, family and co-workers never know of their secret lives. Not all trolls are anonymous, but the nastiest usually are. A particularly awful sort is what I call a stalker troll who attacks repeatedly with doxing and personally insulting tweets meant to discredit you with your followers and others in the twitterverse or to bait you into replying in an equally foul manner. The best strategy is to completely ignore these toxic tweets and consider them a badge of honour, because the worst attacks are reserved only for high-profile members of the climate brawl. While these trolls can be reported to Twitter, in my experience, this seldom leads to any punitive actions by the platform.

The main argument against increased activism against trolls (and climate deniers generally) is that by replying you are giving them the attention they desperately seek and raising their exposure, hence the internet adage, "don't feed the trolls." I understand this reasoning but preventing the harm caused by their unchecked propaganda (discussed earlier) needs to be addressed. Therefore, I have chosen to challenge their tweets and, unlike many on Twitter, I never block or mute climate deniers for that will not silence their propaganda. But I restrict my engagements to mostly hard-core climate deniers (the ones with the greatest influence) and discredit them quickly (to minimize their exposure); hence, a better saying might be, "don't overfeed the trolls." The Twitter trenches must not be abandoned to the propaganda of climate denial.

Climate Denial is a Lost Cause

In Twitter

Strong evidence that climate denial is waning on Twitter is that the messaging has become stale. I see thousands of tweets from climate deniers every year but seldom see any new information or ideas. By now, after many years, more variety in the material tweeted should have appeared, but the same debunked memes are simply recycled over and over again.

Has #ClimateBrawl had an influence on climate denial on Twitter? There is evidence of major changes in the twitterverse. Gone are the days of a solitary advocate of science being swarmed by followers of hard-core climate deniers, without others coming to help – something #ClimateBrawl was designed to do. According to a global ranking of Twitter influencers,[173] during the first half of 2020, four hard-core climate deniers had been in the top 50 influencers for climate change, but by the end of the year, there was only one left (who barely made it at number 49). #ClimateBrawl has, hopefully, helped mute climate denial on Twitter.

An abrupt surge in climate denial on Twitter was observed in the second half of 2022, after Elon Musk had taken over the social platform; during 2022, climate denial tweets had averaged 20K per month from January to June but jumped to 140K per month from July to December.[174] Mr. Musk had removed the timid controls that Twitter had in place to limit propaganda, and, more importantly, he removed the suspensions of those previously banned, including several of the most high-profile climate deniers, such as psychologist Jordan Peterson and Donald Trump (although the latter has not yet returned to the platform, the removal of Trump's suspension has likely served to excite his base into action). Meanwhile, some climate scientists and climate activists have left Twitter and turned to other sites (such as Mastodon).[175]

In America

A detailed survey of American attitudes on climate change organized the results into six categories described as[176]:

> The *Alarmed* are the most engaged with global warming: they are convinced it is happening and human-caused, are very worried about it, and strongly support climate action. The *Concerned* are also convinced global warming is happening and human-caused, but they worry about it less and are less motivated to take action. The *Cautious* are uncertain about whether or not global warming is happening and human-caused, and are not very worried about it, so they are less motivated to act. The *Disengaged* are largely unaware of global warming. The *Doubtful* question whether global warming is happening or human-caused, and perceive it as a low risk, so they are among the least motivated to act. The *Dismissive* reject the idea that global warming is happening and human-caused, do not view it as a threat, and tend to strongly oppose climate policies.

How many Americans, then, can be estimated to be climate deniers? In my opinion, the latter four categories can be taken to represent different shades of climate denialism, and the percentage of these four categories combined was 46% in 2022, which roughly translates into 110 million adult Americans who are climate deniers.[177] According to the survey, the percentages for these four categories combined have fallen from 59% in 2012 to 46% in 2022, a drop of 22% overall (mainly from those categorized as Cautious). While declining, there are still significant numbers of climate deniers in America.

The extreme "Dismissive" category, those who reject that global warming is even happening, overlaps best with what I refer to as hard-core climate deniers. This group has remained steady in its numbers for the past decade (except for a few brief dips); Dismissives returned to their historic levels at 11% by the end of 2022 which is an estimated 27 million Americans who can be considered hard-core climate deniers. This recalcitrant group dedicated to extreme climate denial is unlikely to ever accept the science of climate change.

In Congress

In Congress, there are 535 members, and the number of climate deniers there have also been declining:

115th Congress (2017) = 180 members
116th Congress (2019) = 150 members
117th Congress (2021) = 139 members.

And the conclusion of the congressional survey was[178]:

> The reality that so many members of Congress and senators—including many newly elected officials—deny basic science … the agents of doubt peddling climate change misinformation will have succeeded in darkening not just the future but also the shape of the present as well.

The overall decrease between the surveys in 2017 and 2021 was 23%, which is consistent with the reduction in climate denialism observed in the general American population.

Censure

Climate denialism has been harshly condemned across the globe; below are some quotes from famous people which can be regarded as part of the climate brawl:

- Scholar, Noam Chomsky (1928–)[179]: "The problem is the great mass of the population who are being deluded by constant propaganda."
- Economist, Paul Krugman (1953–) had called climate change denial "treason against the planet,"[180] and wrote[181]: "But there are almost no good-faith climate-change deniers. And denying science for profit, political advantage or ego satisfaction is not OK; when failure to act on the science may have terrible consequences, denial is, as I said, depraved."
- Scientist, Michael E. Mann (1965–)[182]: "Every time we seem to have gotten there [to do something about climate change], the forces of denial have simply doubled down."

- King Charles III (1948–)[183]: "All of a sudden, and with a barrage of sheer intimidation, we are told by powerful groups of deniers that the scientists are wrong and we must abandon all our faith in so much overwhelming scientific evidence."
- Pope Francis (1936–)[184]: "Anyone who denies [climate change] should go to the scientists and ask them. They speak very clearly … climate change is having an effect, and scientists are telling us which path to follow."
- Former president of Ireland, Mary Robinson (1944–)[185]: "Climate change denial is not just ignorant, it is malign, it is evil, and it amounts to an attempt to deny human rights to some of the most vulnerable people on the planet."
- President of the United States Barack Obama (1961–)[186]: "So unfortunately, inside of Washington we've still got some climate deniers who shout loud, but they're wasting everybody's time on a settled debate."

The conclusions to be drawn from the global censure and the trends in Twitter, American society, and Congress are that climate denial is increasingly being recognized as a lost cause, and climate deniers are becoming a dying breed. While their numbers and influence are declining, there is still an influential group of hard-core climate deniers, including in Washington, as discussed later in the book.

Notes

1 Bulletin of the Atomic Scientists 2020; also see Lin 2019.
2 See Hardiman undated; Johnson 2006, ch. 4.
3 Browne 1672, First Book, ch. 10.
4 De Morgan 2007, Introductory 2–4.
5 Fiske 1899, p. 436.
6 Lafleur 1951, p. 284.
7 Gardner 1957, pp. 7, 11; for a more recent discussion of cranks, see Bernstein 1993.
8 Andrews 1796, p. 87.
9 Holmes 2019; for a recent connection of pseudo-science to science denial, see Hansson 2017.
10 Quigley 2017, p. H2633.
11 Burrowes 1903, p. 18.
12 United States House of Representatives 1984, p. 363.
13 Wirth 1990, p. 7475; for the term "junk scientists," see United States House of Representatives 1996, p. 333. In 2004, Robert F. Kennedy Jr. wrote an article titled "The Junk Science of George W. Bush (Kennedy 2004).
14 Pliny 35.85; see Rackham 1995, p. 324.
15 See Lawler 2012.
16 Diogenes Laertius 2.31; see Hicks 2000, p. 161.
17 Plato, Republic 506C; see Shorey 1935, p. 93.
18 Asimov 1980, p. 19.
19 Asimov 1983, p. 26.
20 Nichols 2014; also see Nichols 2017.
21 Kenyon 2016.
22 See Azarian 2018.
23 Kruger and Dunning 2009, pp. 44–45.
24 Johnson 2020; also see, for example, Collins 2012.

25 Fiske 1899, p. 453; also see Gardner 1957, pp. 13, 86.
26 Westman 2011, p. 196.
27 See Le Stark 2016, ch. 2.
28 Finocchario 1989.
29 See Gould 2001, p. 46, also see p. 38. The speech was given by Simplicio, who was defending Ptolemy and Aristotle (against Copernicus).
30 History.com Editors 2020.
31 Drake 1978, pp. 356–357.
32 De Morgan 2007, 296–297.
33 De Morgan 2007, 242.
34 De Morgan 2007, 246.
35 De Morgan 2007, 262–263.
36 Fiske 1899, p. 418.
37 Darwin was also mentioned in De Morgan 2007, 344–345.
38 See Martin 2016, Part V, section 3.
39 See Priestman 2020.
40 Darwin 1876, p. 2.
41 See Lucas 1979, p. 318.
42 See Lucas 1979, p. 321.
43 See Lucas 1979, p. 324; note that Lucas was generally critical of Huxley and sympathetic towards Wilberforce.
44 See Barton 1998, pp. 411, 437; the article reviews the formation of the X Club; also see Barton 2006.
45 Bryan undated, p. 15; see pp. 12–15 for more comments opposing evolution. The pamphlet was based on a speech Bryan gave in 1904 (see Kazin 2006, p. 137).
46 See Kazin 2006, p. 287.
47 See Kazin 2006, p. 294. Tragically, Bryan died five days after the trial.
48 Hatch 1979, p. 15474.
49 Brenan 2019; also see, for example, Rennie 2002; Masci 2019.
50 See Associated Press 2012.
51 Sullivan 2019.
52 Funk 2014; while 60% believed that humans evolved over time, 24% still attributed this to being guided by a supreme being; also see Funk 2019.
53 Brenan 2019; also see, for example, Rennie 2002; Masci 2019.
54 Fairbanks 2020.
55 Kean 2017.
56 See Gratzer 2000, p. 186.
57 See Gardner 1957, pp. 146–147; Kean 2017.
58 Gardner 1957, p. 146.
59 Kean 2017.
60 Gardner 1957, pp. 144–145; Graham 1993, pp. 129–134.
61 United States House of Representatives 2017, p. 98.
62 See Kean 2017.
63 Wazeck 2014, p. 1.
64 See Grundmann 2005, p. 99.
65 Wazeck 2014, p. 241.
66 See Grundmann 2005, pp. 91–110.
67 Kleinert undated; also see Grundmann 2005, pp. 98–99; Ball 2014, p. 87; for other criticisms of relativity see Gardner 1957, ch. 7.
68 See Van Dongen 2010, p. 78.
69 Wazeck 2014, p. 253; the letter was to the American astronomer Thomas Jefferson Jackson See.
70 University of St. Thomas undated.

71 Israel, Ruckhaber, and Weinmann 1931.
72 Robinson 2015, p. 51; also see Ball 2014, p. 94.
73 One month after the meeting in Berlin, on September 23, 1920, he had debated Dr. Einstein at a session of the Society of German Natural Scientists and Physicians in Bad Nauheim (Kleinert undated; Ball 2014, pp. 87–88).
74 Ball 2014, p. 85.
75 Grundmann 2005, p. 99.
76 Ball 2014, p. 72.
77 Ball 2014, p. 83.
78 Gardner 1957, p. 151.
79 Mann and Ward 2019.
80 Hayhoe 2020.
81 Merriam-Webster 2020.
82 Wikipedia 2020.
83 For an excellent analysis of climate denialism and review of the literature see Coan et al. 2021.
84 United States Senate 2013, p. 315.
85 Cruz 2015.
86 For a discussion on the term "contrarian" see Jacques 2012, pp. 9–10.
87 For a general discussion and controversy over the use of the term "climate denier" see Jacques 2012; Wihbey 2012; McKewon 2014.
88 Curry 2009.
89 United States House of Representatives 2007a, p. 9.
90 Dykstra 2017.
91 Happer 2014; for more on Happer see Anon. (undated); Goldenberg 2015.
92 Morano 2006.
93 Morano 2006.
94 Rohrabacher 2011, p. H8311.
95 Spencer 2014.
96 Vittachi 1990.
97 Ivins 1995.
98 Gelbspan 1995.
99 Gelbspan 1998, p. 154. In 1996, the book "The Politics of Denial" briefly discussed climate change (Milburn and Conrad 1996, pp. 214–216).
100 Marshall and Lynas 2003. The article goes on to list "climate-change deniers," including Richard Lindzen and Bjorn Lomborg.
101 Müller 2000, p. 2.
102 Plait 2007.
103 Maheux 2007.
104 Grist 2008; Patel 2008.
105 United States Senate 2005, p. 51.
106 United States House of Representatives 2006, pp. 47, 48, 50.
107 United States House of Representatives 2007b, p. 78.
108 United States House of Representatives 2011, pp. 43–47.
109 Franken 2011, p. S8591.
110 Tyson 2016; also see Jacques 2012, pp. 9–10; for a review, see Weart 2011.
111 Mann 2022.
112 Oreskes and Conway 2010, ch. 1, "Doubt is Our Product."
113 Oreskes and Conway 2010, p. 168.
114 Hickman 2011; Björnberg et al. 2017.
115 Franta 2021.
116 United States House of Representatives 1989, pp. 56–60, 78–86.
117 Walker 1998.

118 IPCC 2022, p. 14.
119 IPCC 2023, p. TS-11.
120 Eichhorn, Molthof, and Nicke 2020; Leiserowitz, Maibach, et al. 2020; Leiserowitz, Marlon, et al. 2020; Tyson and Kennedy 2020; Leiserowitz et al. 2021; also see Monmouth University 2018; Saad 2019, 2021.
121 Climate Action Against Disinformation undated.
122 See Crowe 2019.
123 See Vaillant 1992, p. 10.
124 See Whitbourne 2011.
125 Stoll-Kleemann, O'Riordan, and Jaeger 2001, p. 113; also see American Psychological Association 2009.
126 Hultman 2018.
127 Climate Psychology Alliance 2022, p. 10.
128 Lewandowsky 2020.
129 Hornsey and Lewandowsky 2022, p. 1456; for another review of the psychology of climate denial see Wong-Parodi and Feygina 2020.
130 Gorman and Gorman 2019.
131 Kalmus 2022b.
132 Kalmus 2022a.
133 McCright and Dunlap 2000, p. 499; also see McCright and Dunlap 2011.
134 Brulle and Dunlap 2021; for other references on climate change and the social sciences, see Washington and Cook 2011; Brulle 2014; Crowe 2019; Klinenberg, Araos, and Koslov 2020; Ekberg et al. 2023.
135 See, for example, Lewandowsky 2020, pp. 2–4.
136 See, for example, Collomb 2014.
137 Heshmat 2015.
138 Dallek 2020.
139 For studies on climate change and climate denialism on Twitter, see Cody et al. 2015.
140 Climate deniers – global = 29% [medium of global survey which found climate change was a minor or not a global threat (Fagan and Huang 2019; for more on international surveys see Milman and Harvey 2019; Eichhorn, Molthof, and Nicke 2020; Flynn et al. 2021)] × 206 million [on Twitter (Statista 2022; also see Odabas 2022)] = 59 million; climate deniers – U.S. = 39% × 76.9 million = 30 million.
141 Falkenberg and Baronchelli 2023.
142 Poushter, Fagan, and Gubbala 2022.
143 Milmo 2022.
144 See Stafford 2016; Lewandowsky et al. 2020, p. 5.
145 See Björnberg et al. 2017; Harvey et al. 2018; Anon. 2019; Schmid and Betsch 2019; Starbird 2019; Van der Linden 2019, p. 889; Bulletin of the Atomic Scientists 2020; Lewandowsky et al. 2020, p. 14; Treen, Williams, and O'Neill 2020; Hiar 2021a; Mann 2021, p. 42, also see p. 145; Sinatra and Hofer 2021.
146 See Wilhelm 2019; Hyman 2021.
147 See, for example, Ecker et al. 2022.
148 King, Janulewicz, and Arcostanzo 2022, p. 6; for further information see King 2023.
149 King, Janulewicz, and Arcostanzo 2022, p. 2.
150 Vosoughi, Roy, and Aral 2018.
151 For studies on politicians tweeting about climate change see Yu, Margolin, and Allred 2021.
152 Kutney 2012.
153 Bertolus 2016.
154 Williams et al. 2015, p. 126.
155 Escalón 2023.
156 Falkenberg et al. 2022.

157 For a more scientific-based handbook see Lewandowsky et al. 2020.
158 Among the thousands of peer-reviewed studies on climate change, there are but a handful from contrarians (a 2015 study identified 38), and even these few have been found to be flawed, see Benestad et al. 2016 (this paper provides a detailed analysis of the faults in the contrarian papers).
159 Jevons 1874, p. 224; for a discussion on climate denialism and uncertainty see Lewandowsky et al. 2015.
160 NOAA undated.
161 See, for example, McSweeney and Hausfather 2018; Harvey 2019; Buis 2020; Skeptical Science 2020a.
162 Abbe 1889, p. 679.
163 See, for example, National Academy of Sciences, Engineering and Medicine 2016; Skeptical Science 2020b; WMO 2020.
164 Dictionary.com 2022.
165 United States Supreme Court 2007.
166 NASA 2020.
167 See Shepherd 2019.
168 See Lavelle 2019.
169 Milman 2020; Hiar 2021b.
170 Escalón 2023.
171 Roth and Pickles 2020.
172 Twitter 2022.
173 SustMeme 2020, which tracked the "Global ranking of Top 500 influencers and players active on Twitter in Climate Science & Forecast;" on February 9, 2021, SustMeme changed their analysis from "Top 500 influencers" to "Top 500 positive influencers" and banned the four climate deniers. The ranking ended in April 2023.
174 Falkenberg and Baronchelli 2023.
175 Milman 2022; Waldman 2022; Guynn 2023; King 2023; also see Escalón 2023.
176 Leiserowitz et al. 2023.
177 Climate deniers = 46% [17% ("Cautious") + 7% ("Disengaged") + 11% ("Doubtful") + 11% ("Dismissive"), Leiserowitz et al. 2023] × 246.57 million [Americans over the age of 19 (Statista 2020)] = 113 million. Hard-core climate deniers = 11% ("Dismissive")] × 246.57 million = 27 million.
178 Drennen and Hardin 2021, and references therein.
179 See Roberts 2020.
180 Krugman 2009.
181 Krugman 2018.
182 Mann 2019.
183 See Quinn 2014.
184 See Burton 2017.
185 See Carrington 2019.
186 Obama 2014.

References

Abbe, Cleveland (1889). "Is Our Climate Changing?" *The Forum*, Vol. VI, p. 678. New York: The Forum Publishing Co.; https://books.google.ca/books?id=3aLPAAAAMAAJ&dq=%22is+our+climate+changing%22,+abbe&source=gbs_navlinks_s

American Psychological Association (2009). *Psychology and Global Climate Change: Addressing a Multi-Faceted Phenomenon and Set of Challenges*. Report of the American Psychological Association Task Force on the Interface Between Psychology and Global Climate Change; https://www.apa.org/science/about/publications/climate-change

Andrews, James Pettit (1796). *History of Great Britain*, Vol. II. London: T. Cadell Jun. and W. Davies; https://books.google.ca/books?id=WtdUAAAAcAAJ&source=gbs_navlinks_s

Anon. (undated). *Professor Denies Global Warming Theory*. The Daily Princetonian; https://www.dailyprincetonian.com/article/2009/01/professor-denies-global-warming-theory

Anon. (2019). "Debunking Science Denialism." *Nature Human Behaviour* 3, p. 887, September 12; https://www.nature.com/articles/s41562-019-0746-8#citeas

Asimov, Isaac (1980). "A Cult of Ignorance." *Newsweek*, p. 19; January 24; https://aphelis.net/wp-content/uploads/2012/04/ASIMOV_1980_Cult_of_Ignorance.pdf

Asimov, Isaac (1983). *The Roving Mind*. Buffalo: Prometheus Books; https://pdfcoffee.com/asimov-isaac-the-roving-mind-pdf-free.html

Associated Press (2012). "Pol: Evolution Lie from 'Hell,'" *Politico*, October 6; https://www.politico.com/story/2012/10/pol-evolution-lie-from-hell-082108

Azarian, Bobby (2018). "The Dunning-Kruger Effect May Help Explain Trump's Support." *Psychology Today*, August 22; https://www.psychologytoday.com/ca/blog/mind-in-the-machine/201808/the-dunning-kruger-effect-may-help-explain-trumps-support

Ball, Philip (2014). *Serving the Reich: The Struggle for the Soul of Physics Under Hitler*. Chicago, IL: The University of Chicago Press; https://books.google.ca/books?id=G9BjBAAAQBAJ&dq=Serving+the+Reich:+The+Struggle+for+the+Soul+of+Physics+under+Hitler&source=gbs_navlinks_s; also see "How 2 Pro-Nazi Nobelists Attacked Einstein's 'Jewish Science' [Excerpt]." *Scientific American*, February 13; https://www.scientificamerican.com/article/how-2-pro-nazi-nobelists-attacked-einstein-s-jewish-science-excerpt1/

Barton, Ruth (1998). "'Huxley, Lubbock, and Half a Dozen Others': Professionals and Gentlemen in the Formation of the X Club, 1851–1864." *Isis* 89 (3), p. 410; https://www.jstor.org/stable/237141?seq=1

Barton, Ruth (2006). "X Club." *Oxford Dictionary of National Biography*, September 28; https://www.oxforddnb.com/view/10.1093/ref:odnb/9780198614128.001.0001/odnb-9780198614128-e-92539;jsessionid=DBFD0DAD5A6E96039AA2BDFB8C447A99?back=4168

Benestad, Rasmus E., Nuccitelli, Dana, Lewandowsky, Stephan, Hayhoe, Katharine, Hygen, Hans O., van Dorland, Rob, and Cook, John (2016). "Learning from Mistakes in Climate Research." *Theoretical and Applied Climatology* 126, p. 699, August 20; https://link.springer.com/article/10.1007/s00704-015-1597-5

Bernstein, Jeremy (1993). *Cranks, Quarks and the Cosmos*. New York: Basic Books; https://archive.org/details/cranksquarkscosm00bern/page/n9/mode/2up

Bertolus, Jack (2016). Tweet, Twitter, August 17; https://twitter.com/jackcb1991/status/765888665700532224

Björnberg, Karin E., Karlsson, Mikael, Gilek, Michael, and Hansson, Sven O. (2017). "Climate and Environmental Science Denial: A Review of the Scientific Literature Published in 1990–2015." *Journal of Cleaner Production* 167, p. 229, November 20; https://www.sciencedirect.com/science/article/pii/S0959652617317821?via%3Dihub

Brenan, Megan (2019). *40% of Americans Believe in Creationism*. Gallup, July 26; https://news.gallup.com/poll/261680/americans-believe-creationism.aspx

Browne, Thomas (1672). *Pseudodoxia Epidemica*. London; http://penelope.uchicago.edu/pseudodoxia/pseudodoxia.shtml

Brulle, Robert J. (2014). "Institutionalizing Delay: Foundation Funding and the Creation of U.S. Climate Change Counter-Movement Organizations." *Climatic Change* 122, p. 681; https://link.springer.com/article/10.1007/s10584-013-1018-7#page-1

Brulle, Robert J., and Dunlap, Riley E. (2021). "A Sociological View of the Effort to Obstruct Action on Climate Change." *Footnotes* 49 (3), summer; https://www.asanet.org/sociological-view-effort-obstruct-action-climate-change

Bryan, William Jennings (undated). *The Prince of Peace*. New York: Fleming H. Revell Company; https://archive.org/stream/princeofpeace00brya#page/5/mode/1up/search/Darwinian

Buis, Alan (2020). *Study Confirms Climate Models Are Getting Future Warming Projections Right*. NASA Global Climate Change, January 9; https://climate.nasa.gov/news/2943/study-confirms-climate-models-are-getting-future-warming-projections-right/

Bulletin of the Atomic Scientists (2020). *Closer Than Ever: It Is 100 Seconds to Midnight, 2020 Doomsday Clock Statement*. January 23; https://thebulletin.org/doomsday-clock/2020-doomsday-clock-statement/

Burrowes, Peter E. (1903). *Revolutionary Essays in Socialist Faith and Fancy*. New York: The Comrade Publishing Co.; https://babel.hathitrust.org/cgi/pt?id=nyp.33433075936223&view=1up&seq=9&q1=%22junk%20science%22

Burton, Tara I. (2017). "Pope Francis Warns 'History Will Judge' Climate Change Deniers." *Vox*, September 11; https://www.vox.com/identities/2017/9/11/16290546/pope-francis-climate-change-deniers-daca

Carrington, Damian (2019). "Climate Change Denial Is Evil, Says Mary Robinson." *Guardian*, March 26; https://www.theguardian.com/environment/2019/mar/26/climate-change-denial-is-evil-says-mary-robinson

Climate Action Against Disinformation (undated). *What Is Climate Mis/Disinformation?* https://caad.info/what-is-climate-disinformation/

Climate Psychology Alliance (2022). "Climate Change Denial." In: *Handbook of Climate Psychology*. August 29; https://www.climatepsychologyalliance.org/index.php/component/content/article/climate-psychology-handbook?catid=15&Itemid=101

Coan, Travis G., Boussalis, Constantine, Cook, John, and Nanko, Mirjam O. (2021). "Computer-Assisted Classification of Contrarian Claims about Climate Change." *Scientific Reports* 11, p. 22320; https://www.nature.com/articles/s41598-021-01714-4#citeas

Cody, Emily M., Reagan, Andrew J., Mitchell, Lewis, Dodds, Peter S., and Danforth, Christopher M. (2015). "Climate Change Sentiment on Twitter: An Unsolicited Public Opinion Poll." *PLoS One*, August 20; https://journals.plos.org/plosone/article?id=10.1371/journal.pone.0136092

Collins, Loren (2012). *Bullspotting: Finding Facts in the Age of Misinformation*. Amherst: Prometheus Books; https://books.google.ca/books?id=8gXvP3MCAqcC&source=gbs_navlinks_s

Collomb, Jean-Daniel (2014). "The Ideology of Climate Change Denial in the United States." *European Journal of American Studies*, 9 (1); https://journals.openedition.org/ejas/10305#bodyftn11

Crowe, Kelly (2019). "How 'Organized Climate Change Denial' Shapes Public Opinion on Global Warming." *CBC News*, September 27; https://www.cbc.ca/news/technology/climate-change-denial-fossil-fuel-think-tank-sceptic-misinformation-1.5297236

Cruz, Ted (2015). "Scientific Evidence Doesn't Support Global Warming, Sen. Ted Cruz Says." *NPR*, December 9; https://www.npr.org/2015/12/09/459026242/scientific-evidence-doesn-t-support-global-warming-sen-ted-cruz-says

Curry, Judith (2009). *The Curry Letter: A Word about "Deniers"* ... Watts Up With That, November 28; https://wattsupwiththat.com/2009/11/28/the-curry-letter-a-word-about-deniers/

Dallek, Matthew (2020). "The GOP Has a Long History of Ignoring Science. Trump Turned It into Policy." *Washington Post*, October 9; https://www.washingtonpost.com/outlook/the-gop-has-a-long-history-of-ignoring-science-trump-turned-it-into-policy/2020/10/09/53574602-0917-11eb-859b-f9c27abe638d_story.html

Darwin, Charles (1876). *The Descent of Man, and Selection in the Relationship to Sex*. New York: D. Appleton and Co.; https://books.google.ca/books?id=MkJKAAAAYAAJ&vq=

climate&dq=%E2%80%9CThis+discovery+of+fire,+probably+the+greatest+ever+made+by+man,+excepting+language,+dates+from+before+the+dawn+of+history.%E2%80%9D&source=gbs_navlinks_s

De Morgan, Augustus (2007). *A Budget of Paradoxes*. Ebook, Vol. 1. New York: Dover Publications; https://www.gutenberg.org/files/23100/23100-h/23100-h.htm

Dictionary.com (2022). *Pollution*. https://www.dictionary.com/browse/pollution

Drake, Stillman (1978). *Galileo at Work*. Mineola: Dover Publications; https://books.google.ca/books?id=OwOlRPbrZeQC&vq=%22yet+it+moves%27&source=gbs_navlinks_s

Drennen, Ari, and Hardin, Sally (2021). *Climate Deniers in the 117th Congress*. Center for American Progress, March 30; https://www.americanprogress.org/issues/green/news/2021/03/30/497685/climate-deniers-117th-congress/

Dykstra, Peter (2017). "Climate Deniers, You're Climate Deniers – Deal with It." *Scientific American*, March 7; https://blogs.scientificamerican.com/guest-blog/climate-deniers-youre-climate-deniers-deal-with-it/

Ecker, Ullrich K. H., Lewandowsky, Stephan, Cook, John, Schmid, Philipp, Fazio, Lisa K., Brashier, Nadia, Kendeou, Panayiota, Vraga, Emily K., and Amazeen, Michelle A. (2022). "The Psychological Drivers of Misinformation Belief and Its Resistance to Correction." *Nature Reviews Psychology* 1, p. 13; https://www.nature.com/articles/s44159-021-00006-y

Eichhorn, Jan, Molthof, Luuk, and Nicke, Sascha (2020). *From Climate Change Awareness to Climate Crisis Action*. Open Society, European Policy Institute; https://dpart.org/wp-content/uploads/2020/11/Comparative_report.pdf

Ekberg, Kristoffer, Forchtner, Bernhard, Hultman, Martin, and Jylhä, Kirsti M. (2023). *Climate Obstruction How Denial, Delay and Inaction Are Heating the Planet*. London: Routledge; https://www.routledge.com/Climate-Obstruction-How-Denial-Delay-and-Inaction-are-Heating-the-Planet/Ekberg-Forchtner-Hultman-Jylha/p/book/9781032019475

Escalón, Sebastián (2023). "Investigating Climate Sceptics' Disinformation Strategy on Twitter." *CNRS News*, March 13; https://news.cnrs.fr/articles/investigating-climate-sceptics-disinformation-strategy-on-twitter

Fagan, Moira, and Huang, Christine (2019). *A Look at How People Around the World View Climate Change*. Pew Research Center, April 18; https://www.pewresearch.org/fact-tank/2019/04/18/a-look-at-how-people-around-the-world-view-climate-change/

Fairbanks, Daniel J. (2020). "Mendel and Darwin: Untangling a Persistent Enigma." *Heredity* 124, p. 263; https://www.nature.com/articles/s41437-019-0289-9

Falkenberg, Max, and Baronchelli, Andrea (2023). *Figures on Twitter Climate Scepticism*. Private Communications, April 2.

Falkenberg, Max, Galeazzi, Alessandro, Torricelli, Maddalena, et al. (2022). "Growing Polarization around Climate Change on Social Media." *Nature Climate Change* 12, p. 1114; https://www.nature.com/articles/s41558-022-01527-x

Finocchario, Maurice A. (trans.) (1989). *The Galileo Affair: A Documentary History*. See Gagné, Marc (undated). *Texts from the Galileo Affair*; https://web.archive.org/web/20070930013053/http://astro.wcupa.edu/mgagne/ess362/resources/finocchiaro.html#conreport

Fiske, John (1899). *A Century of Science and Other Essays*. Boston, MA: Houghton, Mifflin and Company; https://books.google.ca/books?id=2LoyAQAAMAAJ&dq=paradoxer&source=gbs_navlinks_s

Flynn, Cassie, Yamasumi, Eri, Fisher, Stephen, Snow, Dan, Grant, Zack, Kirby, Martha, Browning, Peter, Rommerskirchen, Moritz, and Russell, Inigo (2021). *People's Climate Vote*. UNDP and the University of Oxford, January 26; https://www.undp.org/content/

undp/en/home/librarypage/climate-and-disaster-resilience-/The-Peoples-Climate-Vote-Results.html

Franken, Al (2011). *Climate Change*. Congressional Record – Senate, p. S8589, December 14; https://www.govinfo.gov/content/pkg/CREC-2011-12-14/pdf/CREC-2011-12-14-pt1-PgS8589.pdf

Franta, Benjamin (2021). "Early Oil Disinformation on Global Warming." *Environmental Politics* 30 (4), 663; https://www.tandfonline.com/doi/full/10.1080/09644016.2020.1863703#

Funk, Cary (2014). *Republicans' Views on Evolution*. Pew Research Center, January 3; https://www.pewresearch.org/fact-tank/2014/01/03/republican-views-on-evolution-tracking-how-its-changed/

Funk, Cary (2019). *How Highly Religious Americans View Evolution Depends on How They're Asked about It*. Pew Research Center, February 6; https://www.pewresearch.org/fact-tank/2019/02/06/how-highly-religious-americans-view-evolution-depends-on-how-theyre-asked-about-it/

Gardner, Martin (1957). *Fads and Fallacies in the Name of Science*. New York: Dover Publications; https://books.google.ca/books?id=TwP3SGAUsnkC&vq=gardner+fads+and+fallacies&source=gbs_navlinks_s

Gelbspan, Ross (1995). "The Heat Is On: The Warming of the World's Climate Sparks a Blaze of Denial." *Harper's Magazine*, December; https://web.archive.org/web/20160307042257/http://www.sorryaboutthat.net/gelbspan.html

Gelbspan, Ross (1998). *The Heat Is On: The Climate Crisis, the Cover-Up, the Prescription*. Cambridge: Perseus Books; https://archive.org/details/heatisonclim00gelb/page/183/mode/1up?q=denial

Goldenberg, Suzanne (2015). "Greenpeace Exposes Sceptics Hired to Cast Doubt on Climate Science." *Guardian*, December 8; https://www.theguardian.com/environment/2015/dec/08/greenpeace-exposes-sceptics-cast-doubt-climate-science

Gorman, Sara, and Gorman, Jack M. (2019). "Climate Change Denial." *Psychology Today*, January 12; https://www.psychologytoday.com/us/blog/denying-the-grave/201901/climate-change-denial

Gould, Stephen Jay (ed.), and Drake, Stillman (trans.) (2001). *Galileo Galilei, Dialogue Concerning the Two Chief World Systems: Ptolemaic and Copernican*. New York: The Modern Library; https://books.google.ca/books?id=c-nIrKjBqOwC&dq=galileo,+%22Dialogue+Concerning+the+Two+Chief+World+Systems%22,+%22denying+scientific+principles%22&source=gbs_navlinks_s

Graham, Loren R. (1993). *Science in Russia and the Soviet Union, A Short History*. Cambridge: Cambridge University Press; https://books.google.ca/books?id=m_wPpj64GqMC&source=gbs_navlinks_s

Gratzer, Walter (2000). *The Undergrowth of Science*. New York: Oxford University Press; https://books.google.ca/books?id=quixwL-N41UC&printsec=frontcover&source=gbs_ge_summary_r&cad=0#v=onepage&q&f=false

Grist (2008). Tweet, Twitter, April 4; https://twitter.com/grist/status/783014681

Grundmann, Siegfried (Hentschel, Ann M., trans.) (2005). *The Einstein Dossiers*. Berlin: Springer-Verlag; https://books.google.ca/books?id=1bxYPMHPhGcC&source=gbs_navlinks_s

Guynn, Jessica (2023). "Twitter Has Always Been a Hotspot for Climate Change Misinformation. On Musk's Watch, It's Heating Up." *USA Today*, January 10; https://www.usatoday.com/story/tech/2023/01/10/twitter-climate-change-misinformation-surge-elon-musk/11002993002/

Hansson, Sven Ove (2017). "Science Denial as a Form of Pseudoscience." *Studies in History and Philosophy of Science, Part A*, 63, p. 39; https://www.sciencedirect.com/science/article/abs/pii/S0039368116300681?via%3Dihub

Happer, William (2014). "Princeton Prof: 'Shut Up' Over Climate Change." *CNBC*, Squawk Box, YouTube, July 14; https://www.cnbc.com/video/2014/07/14/princeton-prof-shut-up-over-climate-change.html

Hardiman, Rachel (undated). *Welcome to the Paradoxography Website*. Paradoxography; https://sites.google.com/site/paradoxography/Home

Harvey, Chelsea (2019). "Climate Models Got It Right on Global Warming." *Scientific American, E&E News*, December 5; https://www.scientificamerican.com/article/climate-models-got-it-right-on-global-warming/

Harvey, Jeffrey A., Van den Berg, Daphne, Ellers, Jacintha, et al. (2018). "Internet Blogs, Polar Bears, and Climate-Change Denial by Proxy." *BioScience* 68 (4), p. 281; https://academic.oup.com/bioscience/article/68/4/281/4644513?login=false

Hatch, Orrin (1979). *Equal Time in Our Schools*. Congressional Record–Senate, p. 15474, June 19; https://www.govinfo.gov/content/pkg/GPO-CRECB-1979-pt12/pdf/GPO-CRECB-1979-pt12-5-1.pdf

Hayhoe, Katharine (2020). Tweet, Twitter, March 4; https://twitter.com/KHayhoe/status/1235204612354867201

Heshmat, Shahram (2015). "What Is Confirmation Bias?" *Psychology Today*, April 23; https://www.psychologytoday.com/ca/blog/science-choice/201504/what-is-confirmation-bias

Hiar, Corbin (2021a). "Greens Track Deniers Ahead of Biden Climate Push." *E&E News*, January 4; https://www.eenews.net/stories/1063721645

Hiar, Corbin (2021b). "Twitter Bots Are a Major Source of Climate Disinformation." *Scientific American, E&E News*, January 22; https://www.scientificamerican.com/article/twitter-bots-are-a-major-source-of-climate-disinformation

Hickman, Leo (2011). "Is Climate Skepticism a Largely Anglo-Saxon Phenomenon?" *Guardian*, November 11; https://www.theguardian.com/environment/blog/2011/nov/11/climate-change-scienceofclimatechange#comment-13244224

Hicks, R. D. (trans.) (2000). *Diogenes Laertius Lives of Eminent Philosophers*. Vol. I. Cambridge: Harvard University Press.

History.com Editors (2020). *Galileo Is Accused of Heresy*. History, This Day in History, April 12; https://www.history.com/this-day-in-history/galileo-is-accused-of-heresy

Holmes, Oliver Wendell (2019). *The Professor at the Breakfast-Table*. Good Press; https://books.google.ca/books?id=ptTCDwAAQBAJ&source=gbs_navlinks_s

Hornsey, Matthew J., and Lewandowsky, Stephan (2022). "A Toolkit for Understanding and Addressing Climate Scepticism." *Nature Human Behaviour* 6, p. 1454; https://www.nature.com/articles/s41562-022-01463-y

Hultman, Martin (2018). *Centre for Studies of Climate Change Denialism* (CEFORCED). Chalmers University of Technology, November 7; https://www.chalmers.se/en/departments/tme/centres/ceforced/Pages/default.aspx

Hyman, Ira (2021). "Why Disinformation Campaigns Are Dangerous." *Psychology Today*, January 15; https://www.psychologytoday.com/us/blog/mental-mishaps/202101/why-disinformation-campaigns-are-dangerous?amp&__twitter_impression=true

IPCC (2022). *Climate Change 2022: Impacts, Adaptation and Vulnerability.* IPCC, WG II; https://www.ipcc.ch/report/ar6/wg2/

IPCC (2023). *Climate Change 2022: Mitigation of Climate Change.* IPCC, WG III; https://report.ipcc.ch/ar6/wg3/IPCC_AR6_WGIII_Full_Report.pdf

Israel, Hans, Ruckhaber, Erich, and Weinmann, Rudolf (1931). *Hundert Autoren Gegen Einstein.* Leipzig: R. Voigtlander's Verlag; https://archive.org/details/HundertAutoren GegenEinstein

Ivins, Molly (1995). "Molly Ivins on Climate Change Deniers." *Texas Observer,* September 29; https://www.texasobserver.org/molly-ivins-on-climate-change-deniers/

Jacques, Peter J. (2012). "A General Theory of Climate Denial." *Global Environmental Politics* 12 (2), p. 7; https://direct.mit.edu/glep/article/12/2/9/14558/A-General-Theory-of-Climate-Denial

Jevons, William Stanley (1874). *The Principles of Science*. Special American Edition, Vol. 1. New York: Macmillan and Co.; https://books.google.ca/books?id=T8sHAAAAIAAJ&vq=ignorance&source=gbs_navlinks_s

Johnson, David Kyle (2020). "The Galileo Gambit and Appealing to Ignorance." *Psychology Today,* June 21; https://www.psychologytoday.com/ca/blog/logical-take/202006/the-galileo-gambit-and-appealing-ignorance

Johnson, Scott, F. (2006). *The Life and Miracles of Thekla: A Literary Study*. Hellenic Studies Series 13. Washington, DC: Center for Hellenic Studies; https://chs.harvard.edu/chapter/preface-37/

Kalmus, Peter (2022a). Tweet, Twitter, July 1; https://twitter.com/ClimateHuman/status/1543056729394384896

Kalmus, Peter (2022b). "To Understand the Scale of the Climate Emergency, Look at Hurricanes." *Guardian,* October 1; https://www.theguardian.com/commentisfree/2022/oct/01/climate-emergency-hurricane-ian-crisis

Kazin, Michael (2006). *A Godly Hero, the Life of William Jennings Bryan*. New York: Alfred A. Knopf; https://books.google.ca/books/about/A_Godly_Hero.html?id=GUzEO18oGYUC&redir_esc=y

Kean, Sam (2017). "The Soviet Era's Deadliest Scientist Is Regaining Popularity in Russia." *The Atlantic*, December 19; https://www.theatlantic.com/science/archive/2017/12/trofim-lysenko-soviet-union-russia/548786/

Kennedy Jr., Robert F. (2004). "The Junk Science of George W. Bush." *The Nation*, February 19; https://www.thenation.com/article/archive/junk-science-george-w-bush/

Kenyon, Georgina (2016). "The Man Who Studies the Spread of Ignorance." *BBC News*, January 6; https://www.bbc.com/future/article/20160105-the-man-who-studies-the-spread-of-ignorance?referer=https%3A%2F%2Fen.m.wikipedia.org%2F

King, Jennie (2023). *Deny, Deceive, Delay Vol. 2: Exposing New Trends in Climate Mis- and Disinformation at COP27*. Institute for Strategic Dialogue, January 19; https://www.isdglobal.org/isd-publications/deny-deceive-delay-vol-2-exposing-new-trends-in-climate-mis-and-disinformation-at-cop27/

King, Jennie, Janulewicz, Lukasz, and Arcostanzo, Francesca (2022). *Deny, Deceive, Delay: Documenting and Responding to Climate Disinformation at COP26 & Beyond – Executive Summary*. Institute for Strategic Dialogue, June 9; https://www.isdglobal.org/isd-publications/deny-deceive-delay-documenting-and-responding-to-climate-disinformation-at-cop26-and-beyond/

Kleinert, Andreas (undated). *Paul Weyland, the Einstein-Killer from Berlin*. Martin-Luther-Universitat-Halle-Wittenberg; http://www.physik.uni-halle.de/Fachgruppen/history/weyland.htm

Klinenberg, Eric, Araos, Malcolm, and Koslov, Liz (2020). "Sociology and the Climate Crisis." *Annual Review of Sociology* 46, p. 649; https://www.annualreviews.org/doi/full/10.1146/annurev-soc-121919-054750?casa_token=03eInjK7ltIAAAAA:1os6QYg-NG1gRqAeXRW3i7-zC7HLrS3j4DzIkz_IYyOjXkTnguMeFws-Bsi3H0TX_aidTVxdvPmhhA

Kruger, Justin, and Dunning, David (2009). "Unskilled and Unaware of It: How Difficulties in Recognizing One's Own Incompetence Lead to Inflated Self-Assessments." *Psychology* 1, p. 30; https://www.researchgate.net/publication/12688660_Unskilled_and_Unaware_of_It_How_Difficulties_in_Recognizing_One's_Own_Incompetence_Lead_to_Inflated_Self-Assessments

Krugman, Paul (2009). "Betraying the Planet." *New York Times*, June 28; https://www.nytimes.com/2009/06/29/opinion/29krugman.html

Krugman, Paul (2018). "The Depravity of Climate-Change Denial." *New York Times*, November 26; https://www.nytimes.com/2018/11/26/opinion/climate-change-denial-republican.html

Kutney, Gerald (2012). Tweet, Twitter, November 28; https://twitter.com/GeraldKutney/status/273832668205043713

Lafleur, Laurence J. (1951). "Cranks and Scientists." *The Scientific Monthly* 73 (5), p. 284; https://www.jstor.org/stable/20436?seq=1

Lavelle, Marianne (2019). "'Trollbots' Swarm Twitter with Attacks on Climate Science Ahead of UN Summit." *Inside Climate News*, September16; https://insideclimatenews.org/news/16092019/trollbot-twitter-climate-change-attacks-disinformation-campaign-mann-mckenna-greta-targeted

Lawler, Susan (2012). *Belief and Scepticism: Creating Nonsense by Mislabelling Scientists and Deniers*. The Conversation, May 1; https://theconversation.com/belief-and-scepticism-creating-nonsense-by-mislabelling-scientists-and-deniers-6790

Le Stark, Michael (2016). *The Deep Sky Chronicles*. Lulu Publishing Services; https://books.google.ca/books?id=16UZDQAAQBAJ&dq=%E2%80%9CThere+may+be+babblers,+wholly+ignorant+of+mathematics%22,+copernicus&source=gbs_navlinks_s

Leiserowitz, Anthony, Maibach, Edward, Rosenthal, Seth, Kotcher, John, Ballew, Matthew, Bergquist, Parrish, Gustafson, Abel, Goldberg, Matthew, and Wang, Xinran (2020). *Politics & Global Warming, April 2020*. Yale University and George Mason University. New Haven, CT: Yale Program on Climate Change Communication, June 4; https://climatecommunication.yale.edu/publications/politics-global-warming-april-2020/

Leiserowitz, Anthony, Maibach, Edward, Rosenthal, Seth, Kotcher, John, Ballew, Matthew, Marion, Jennifer, Carman, Jennifer, Verner, Marija, Lee, Sanguk, Myers, Teresa, and Goldberg, Matthew (2023). *Global Warming's Six Americas, December 2022*. Yale University and George Mason University. New Haven, CT: Yale Program on Climate Change Communication, March 14; https://climatecommunication.yale.edu/publications/global-warmings-six-americas-december-2022/

Leiserowitz, Anthony, Maibach, Edward, Rosenthal, Seth, Kotcher, John, Carman, Jennifer, Wang, Xinran, Goldberg, Matthew, Lacroix, Karine, and Marlon, Jennifer (2021). *Politics & Global Warming, December 2020*. Yale University and George Mason University. New Haven, CT: Yale Program on Climate Change Communication, January 14; https://climatecommunication.yale.edu/publications/politics-global-warming-december-2020/

Leiserowitz, Anthony, Marlon, Jennifer, Wang, Xinran, Bergquist, Parrish, Goldberg, Matthew, Kotcher, John, Maibach, Edward, and Rosenthal, Seth (2020). *Global Warming's Six Americas in 2020*. New Haven, CT: Yale Program on Climate Change Communication, October 10; https://climatecommunication.yale.edu/publications/global-warmings-six-americas-in-2020/

Lewandowsky, Stephan (2020). "Climate Change, Disinformation, and How to Combat It." *Annual Review of Public Health, Forthcoming*, September 16; https://papers.ssrn.com/sol3/papers.cfm?abstract_id=3693773

Lewandowsky, Stephan, Cook, John, Ecker, Ullrich K. H., et al. (2020). *The Debunking Handbook 2020*. https://www.climatechangecommunication.org/wp-content/uploads/2020/10/DebunkingHandbook2020.pdf

Lewandowsky, Stephan, Oreskes, Naomi, Risbey, James S., Newell, Ben R., and Smithson, Michael (2015). "Seepage: Climate Change Denial and Its Effects on the Scientific Community." *Global Environmental Change*, 33 (July), p. 1; https://www.sciencedirect.com/science/article/pii/S0959378015000515

Lin, Herbert (2019). "The Existential Threat from Cyber-enabled Information Warfare." *Bulletin of the Atomic Scientists* 75 (4), p. 187; https://www.tandfonline.com/doi/abs/10.1080/00963402.2019.1629574?journalCode=rbul20

Lucas, J. R. (1979). "Wilberforce and Huxley: A Legendary Encounter." *The Historical Journal* 22 (2), p. 313; https://www.jstor.org/stable/2638867

Maheux, Ged (2007). Tweet, Twitter, November 8; https://twitter.com/gedeon/status/400028802

Mann, Michael E. (2019). Tweet, Twitter, February 7; https://twitter.com/MichaelEMann/status/1093704873764372480

Mann, Michael E. (2021). *The New Climate War*. New York: Public Affairs.

Mann, Michael E. (2022). Tweet, Twitter, November 7; https://twitter.com/michaelemann/status/1589652192411410432

Mann, Michael, E., and Ward, Bob (2019). "Donald Trump Is Using Stalinist Tactics to Discredit Climate Science." *Guardian*, March 20; https://www.theguardian.com/commentisfree/2019/mar/20/donald-trump-stalinist-techniques-climate-science

Marshall, George, and Lynas, Mark (2003). "Why We Don't Give a Damn." *New Statesman*, December 1; https://www.newstatesman.com/node/191306 (link broken).

Martin, Gustavo Rodriguez (ed.) (2016). *The Critical Shaw on Literature*. RosettaBooks; https://books.google.ca/books?id=lGzVCwAAQBAJ&dq=george+bernard+shaw,+%22darwin+denounced%22&source=gbs_navlinks_s

Masci, David (2019). *Darwin in America*. Pew Research Centre, February 6; https://www.pewforum.org/essay/darwin-in-america/

McCright, Aaron M., and Dunlap, Riley E. (2000). "Challenging Global Warming as a Social Problem: An Analysis of the Conservative Movement's Counter-Claims." *Social Problems* 47 (4), p. 499; https://www.jstor.org/stable/3097132

McCright, Aaron M., and Dunlap, Riley E. (2011). "The Politicization of Climate Change and Polarization in the American Public's Views of Global Warming, 2001–2010." *The Sociological Quarterly* 52 (2), p. 155; https://www.tandfonline.com/doi/abs/10.1111/j.1533-8525.2011.01198.x?journalCode=utsq20

McKewon, Elaine (2014). *The Journal That Gave in to Climate Deniers' Intimidation*. Yale Climate Connections, April 17; https://yaleclimateconnections.org/2014/04/the-journal-that-gave-in-to-climate-deniers-intimidation/

McSweeney, Robert, and Hausfather, Zeke (2018). "Climate Modelling, Q&A: How Do Climate Models Work?" *CarbonBrief*, January 15; https://www.carbonbrief.org/qa-how-do-climate-models-work

Merriam-Webster (2020). *Climate Change Denier*. Merriam-Webster.com Dictionary; https://www.merriam-webster.com/dictionary/climate%20change%20denier

Milburn, Michael A., and Conrad, Sheree D. (1996). *The Politics of Denial*. Cambridge: The MIT Press; https://books.google.ca/books?id=ntVE1n3g51wC&source=gbs_navlinks_s

Milman, Oliver (2020). "Revealed: Quarter of All Tweets about Climate Crisis Produced by Bots." *Guardian*, February 21; https://www.theguardian.com/technology/2020/feb/21/climate-tweets-twitter-bots-analysis

Milman, Oliver (2022). "#ClimateScam: Denialism Claims Flooding Twitter have Scientists Worried." *Guardian*, December 2; https://www.theguardian.com/technology/2022/dec/02/climate-change-denialism-flooding-twitter-scientists

Milman, Oliver, and Harvey, Fiona (2019). "US Is Hotbed of Climate Change Denial, Major Global Survey Finds." *Guardian*, May 8; https://www.theguardian.com/environment/2019/may/07/us-hotbed-climate-change-denial-international-poll

Milmo, Dan (2022). "Nobel Peace Prize Winners Call for Action on Online Disinformation." *Guardian*, September 2; https://www.theguardian.com/media/2022/sep/02/nobel-peace-prize-winners-call-for-action-on-online-disinformation

Monmouth University (2018). *Climate Concerns Increase; Most Republicans Now Acknowledge Change*. Monmouth University Polling, November 29; https://www.monmouth.edu/polling-institute/reports/monmouthpoll_us_112918/

Morano, Marc (2006). *Nuremberg-Style Trials Proposed for Global Warming Skeptics*. U.S. Senate Committee on Environment and Public Works, Newsroom, Press Releases, October 11; https://www.epw.senate.gov/public/index.cfm/press-releases-all?ID=A4017645-DE27-43D7-8C37-8FF923FD73F8

Müller, Benito (2000). *The Hague Climate Conference*. Oxford Climate Policy, December 7; https://oxfordclimatepolicy.org/sites/default/files/hague_0.pdf

NASA (2020). *Is the Sun Causing Global Warming?* NASA Global Climate Change; https://climate.nasa.gov/faq/14/is-the-sun-causing-global-warming/

National Academy of Sciences, Engineering and Medicine (2016). *Attribution of Extreme Weather Events in the Context of Climate Change*. Washington, DC: The National Academies Press; https://www.nap.edu/catalog/21852/attribution-of-extreme-weather-events-in-the-context-of-climate-change

Nichols, Tom (2014). *The Death of Expertise*. The Federalist, January 17; https://thefederalist.com/2014/01/17/the-death-of-expertise/

Nichols, Tom (2017). *The Death of Expertise*. New York: Oxford University Press; https://books.google.ca/books?id=x3TYDQAAQBAJ&vq=science&source=gbs_navlinks_s

NOAA (undated). *Climate Models*. NOAA Climate.gov; https://www.climate.gov/maps-data/primer/climate-models

Obama, Barack (2014). *President Obama Speaks on American Energy*. The White House, President Barack Obama, May 9; https://obamawhitehouse.archives.gov/photos-and-video/video/2014/05/09/president-obama-speaks-american-energy#transcript

Odabas, Meltem (2022). *10 Facts about Americans and Twitter.* Pew Research Center, May 5; https://www.pewresearch.org/fact-tank/2019/08/02/10-facts-about-americans-and-twitter/

Oreskes, Naomi, and Conway, Erik M. (2010). *Merchants of Doubt*. New York: Bloomsbury Press.

Patel, Raj (2008). Tweet, Twitter, April 2; https://twitter.com/Rajio/status/781409291

Plait, Phil (2007). Tweet, Twitter, June 6; https://twitter.com/BadAstronomer/status/93564192

Poushter, Jacob, Fagan, Moira, and Gubbala, Sneha (2022). *Climate Change Remains Top Global Threat across 19-Country Survey*. Pew Research Center, August 31; https://www.pewresearch.org/global/2022/08/31/climate-change-remains-top-global-threat-across-19-country-survey/

Priestman, Martin (2020). "The Other Darwin's Plots: Evolution as Literature in Erasmus Darwin, Samuel Butler and George Bernard Shaw." *Érudit*, June 10; https://www.erudit.org/en/journals/ravon/1900-v1-n1-ravon05340/1069962ar/

Quigley, Mike (2017). *Climate Change*. Congressional Record – House, p. H2633, April 4; https://www.govinfo.gov/content/pkg/CREC-2017-04-04/pdf/CREC-2017-04-04-pt1-PgH2633-5.pdf

Quinn, Ben (2014). "Climate Change Sceptics Are 'Headless Chickens,' Says Prince Charles." *Guardian*, January 31; https://www.theguardian.com/environment/2014/jan/31/climate-change-sceptics-headless-chickens-prince-charles

Rackham, H. (trans.) (1995). *Pliny Natural History Books XXXIII–XXXV*. Cambridge: Harvard University Press.

Rennie, John (2002). "15 Answers to Creationist Nonsense." *Scientific American* 287 (1), July 1; https://www.scientificamerican.com/article/15-answers-to-creationist/

Roberts, David (2020). "Noam Chomsky's Green New Deal." *Vox*, September 21; https://www.vox.com/energy-and-environment/21446383/noam-chomsky-robert-pollin-climate-change-book-green-new-deal

Robinson, Andrew (2015). *Einstein, a Hundred Years of Relativity*. Princeton, NJ: Princeton University Press; https://books.google.ca/books?id=Px4_CQAAQBAJ&source=gbs_navlinks_s

Rohrabacher, Dana (2011). *The Specter of Global Governance*. Congressional Record – House, p. H8309, December 8; https://www.govinfo.gov/content/pkg/CREC-2011-12-08/pdf/CREC-2011-12-08-pt1-PgH8309.pdf

Roth, Yoel, and Pickles, Nick (2020). *Bot or Not? The Facts about Platform Manipulation on Twitter*. Twitter Blog, May 18; https://blog.twitter.com/en_us/topics/company/2020/bot-or-not.html

Saad, Lydia (2019). *Americans as Concerned as Ever about Global Warming*. Gallup, March 25; https://news.gallup.com/poll/248027/americans-concerned-ever-global-warming.aspx

Saad, Lydia (2021). *Global Warming Attitudes Frozen Since 2016*. Gallup, April 5; https://news.gallup.com/poll/343025/global-warming-attitudes-frozen-2016.aspx

Schmid, Philipp, and Betsch, Cornelia (2019). "Effective Strategies for Rebutting Science Denialism in Public Discussions." *Nature Human Behaviour* 3, p. 931, September; https://www.nature.com/articles/s41562-019-0632-4

Shepherd, Marshall (2019). "'Sealioning' Is a Common Trolling Tactic on Social Media -- What Is It?" *Forbes*, March 7; https://www.forbes.com/sites/marshallshepherd/2019/03/07/sealioning-is-a-common-trolling-tactic-on-social-media-what-is-it/#1d3fc2ab7a41

Shorey, Paul (trans.) (1935). *Plato VI, Republic II*. Cambridge: Harvard University Press.

Sinatra, Gale M., and Hofer, Barbara K. (2021). *Science Denial: Why It Happens and What to Do about It*. New York: Oxford University Press.

Skeptical Science (2020a). *How Reliable Are Climate Models?*; https://skepticalscience.com/climate-models.htm

Skeptical Science (2020b). *Is Extreme Weather Caused by Global Warming?*; https://www.skepticalscience.com/extreme-weather-global-warming.htm

Spencer, Roy W. (2014). *Time to Push Back against the Global Warming Nazis*. Roy Spencer Blog, February 20; http://www.drroyspencer.com/2014/02/time-to-push-back-against-the-global-warming-nazis/

Stafford, Tom (2016). "How Liars Create the 'Illusion of Truth.'" *BBC*, October 26; https://www.bbc.com/future/article/20161026-how-liars-create-the-illusion-of-truth#:~:text=%E2%80%9CRepeat%20a%20lie%20often%20enough,%22illusion%20of%20truth%22%20effect

Starbird, Kate (2019). "Disinformation's Spread: Bots, Trolls and All of Us." *Nature* 571, p. 449, July 24; https://www.nature.com/articles/d41586-019-02235-x

Statista (2020). *Resident Population of the United States by Sex and Age as of July 1, 2019*; November 5; https://www.statista.com/statistics/241488/population-of-the-us-by-sex-and-age/#:~:text=The%20estimated%20population%20of%20the,and%20around%2011.5%20million%20females

Statista (2022). *Leading Countries Based on Number of Twitter Users as of January 2022*; May 22; https://www.statista.com/statistics/242606/number-of-active-twitter-users-in-selected-countries/

Stoll-Kleemann, S., O'Riordan, Tim, and Jaeger, Carlo C. (2001). "The Psychology of Denial Concerning Climate Mitigation Measures: Evidence from Swiss Focus Groups." *Global Environmental Change* 11 (2), p. 107, July; https://www.sciencedirect.com/science/article/abs/pii/S0959378000000613

Sullivan, Megan Elizabeth (2019). "'Alternative Facts' in the Classroom: Creationist Educational Policy and the Trump Administration." *National Law Review*, IX (70); https://www.natlawreview.com/article/alternative-facts-classroom-creationist-educational-policy-and-trump-administration

SustMeme (2020). *SustMeme Climate & Energy Top 500*. https://www.rise.global/sustmeme-climate-energy/hof

Treen, Kathie M. d'I., Williams, Hywel T. P., and O'Neill, Saffron J. (2020). "Online Misinformation about Climate Change." *Wiley Interdisciplinary Reviews Climate Change* 11 (5), June 18; https://wires.onlinelibrary.wiley.com/doi/full/10.1002/wcc.665

Twitter (2022). *Transparency, Platform Manipulation*, July 28; https://transparency.twitter.com/en/reports/platform-manipulation.html#2021-jul-dec

Tyson, Alec, and Kennedy, Brian (2020). *Two-Thirds of Americans Think Government Should Do More on Climate*. Pew Research Center, June 23; https://www.pewresearch.org/science/2020/06/23/two-thirds-of-americans-think-government-should-do-more-on-climate/

Tyson, Neil deGrasse (2016). Tweet, Twitter, May 16; https://twitter.com/neiltyson/status/732191219883900928

United States House of Representatives (1984). *Superfund Reauthorization*. Hearings before the Subcommittee on Commerce, Transportation, and Tourism of the Committee on Energy and Commerce, House of Representatives, 98th Congress, Second Session, February–March; https://babel.hathitrust.org/cgi/pt?id=mdp.39015012903111&view=1up&seq=1&size=125&q1=%22junk%20science%22

United States House of Representatives (1989). *Global Warming*. Hearings before the Subcommittee on Energy and Power of the Committee on Energy and Commerce, House of Representatives, 101st Congress, First Session, February 21 and May 4; https://babel.hathitrust.org/cgi/pt?id=uc1.31210007429986;view=1up;seq=1

United States House of Representatives (1996). *U.S. Global Change Research Programs: Data Collection and Scientific Priorities*. Hearing before the Committee on Science, U.S. House of Representatives, 104th Congress, Second Session, March 6; https://babel.hathitrust.org/cgi/pt?id=pst.000025977899&view=1up&seq=3

United States House of Representatives (2006). *Department of Energy's Plan for Climate Change Technology Programs*. Hearing before the Subcommittee on Energy, Committee on Science, House of Representatives, 109th Congress, Second Session, September 20; https://babel.hathitrust.org/cgi/pt?id=pst.000058944103&view=1up&seq=1&skin=2021&q1=deniers

United States House of Representatives (2007a). *Allegations of Political Interference with the Work of Government Climate Change Scientists*. Hearing before the Committee on Oversight and Government Reform, House of Representatives, 110th Congress, First Session, January 30; https://babel.hathitrust.org/cgi/pt?id=pst.000061491786&view=1up&seq=1

United States House of Representatives (2007b). *Disappearing Polar Bears and Permafrost: Is a Global Warming Tipping Point Embedded in the Ice?* Hearing before the Subcommittee on Investigations and Oversight, Committee on Science and Technology, House of

Representatives, 110th Congress, First Session, October 17; https://babel.hathitrust.org/cgi/pt?id=pst.000063523263&view=1up&seq=1

United States House of Representatives (2011). *Climate Science and EPA's Greenhouse Gas Regulations*. Hearing before the Subcommittee on Energy and Power of the Committee on Energy and Commerce, House of Representatives, 112th Congress, First Session, March 8; https://babel.hathitrust.org/cgi/pt?id=umn.31951d03457933v&view=1up&seq=3

United States House of Representatives (2017). *Climate Science: Assumptions, Policy Implications, and the Scientific Method*. Hearing before the Committee on Science, Space, and Technology, House of Representatives, 115th Congress, First Session, March 29; https://www.govinfo.gov/content/pkg/CHRG-115hhrg25098/pdf/CHRG-115hhrg25098.pdf

United States Senate (2005). *The Role of Science in Environmental Policy Making*. Hearing before the Committee on Environment and Public Works, United States Senate, 109th Congress, First Session, September 28; https://babel.hathitrust.org/cgi/pt?id=pst.000063509977;view=1up;seq=1

United States Senate (2013). *Climate Change: It's Happening Now*. Hearing before the Committee on Environment and Public Works, United States Senate, 113th Congress, First Session, July 18; https://www.govinfo.gov/content/pkg/CHRG-113shrg95976/pdf/CHRG-113shrg95976.pdf

United States Supreme Court (2007). *127 S. Ct. 1438 (2007) 549 U.S. 497 Massachusetts et al., Petitioners, v. Environmental Protection Agency et al. No. 05-1120*. Supreme Court of United States. April 2; https://scholar.google.ca/scholar_case?case=18363956969502505811&q=Massachusetts+v.+EPA,+549+U.S.+497&hl=en&as_sdt=2006&as_vis=1

University of St. Thomas (undated): *Einsteinism: Its Fallacies and Frauds*. AR1-28_001, Arvid Reuterdahl Papers; https://cdm16120.contentdm.oclc.org/digital/collection/reuterdahl/id/265/rec/26

Vaillant, George E. (1992). *Ego Mechanisms of Defense: A Guide for Clinicians and Researchers*. Washington, DC: American Psychiatric Press; https://books.google.ca/books?id=ZABas1ynLJkC&vq=denial&source=gbs_navlinks_s

Van der Linden, Sander (2019). "Countering Science Denial." *Nature Human Behaviour* 3, p. 889, June 24; https://www.nature.com/articles/s41562-019-0631-5

Van Dongen, Jeroen (2010). "On Einstein's Opponents and other Crackpots." *Studies in History and Philosophy of Modern Physics* 41 (1), p. 78; https://www.sciencedirect.com/science/article/abs/pii/S1355219809000598?via%3Dihub

Vittachi, Anuradha (1990). "The Denial Syndrome." *New Internationalist*, April 5; https://newint.org/features/1990/04/05/denial

Vosoughi, Soroush, Roy, Deb, and Aral, Sinan (2018). "The Spread of True and False News Online." *Science* 359 (6380), p. 1146, March 9; https://www.science.org/doi/10.1126/science.aap9559

Waldman, Scott (2022). "Climate Misinformation Spreads on Musk's Twitter." *E&E News*, ClimateWire, December 23; https://www.eenews.net/articles/climate-misinformation-spreads-on-musks-twitter/

Walker, Joseph (1998). *1998 American Petroleum Institute Global Climate Science Communications Team Action Plan*. Climatefiles, April 3; https://www.climatefiles.com/trade-group/american-petroleum-institute/1998-global-climate-science-communications-team-action-plan/

Washington, Haydn, and Cook, John (2011). *Climate Change Denial*. London: Routledge.

Wazeck, Milena (Koby, Geoffrey S., trans.) (2014). *Einstein's Opponents*. Cambridge: Cambridge University Press; https://books.google.ca/books?id=w-RRAgAAQBAJ&dq=The+Einstein+Dossiers:+Science+and+Politics+-+Einstein%27s+Berlin+Period.+Berlin&source=gbs_navlinks_s

Weart, Spencer (2011). "Global Warming: How Skepticism Became Denial." *Bulletin of the Atomic Scientists* 67 (1), p. 41; https://journals.sagepub.com/doi/10.1177/0096340210392966

Westman, Robert S. (2011). *The Copernican Question*. Berkeley: University of California Press; https://books.google.ca/books?id=iEueQqLQyiIC&vq=Tolosani&source=gbs_navlinks_s

Whitbourne, Susan Krauss (2011). "The Essential Guide to Defense Mechanisms." *Psychology Today*, October 22; https://www.psychologytoday.com/ca/blog/fulfillment-any-age/201110/the-essential-guide-defense-mechanisms

Wihbey, John (2012). *'Denier,' 'Alarmist,' 'Warmist,' 'Contrarian,' 'Confusionist,' 'Believer.'* Yale Climate Connections, August 16; https://yaleclimateconnections.org/2012/08/denier-alarmist-warmist-contrarian-confusionist-believer/

Wikipedia (2020). *Climate Change Denial*. https://en.wikipedia.org/wiki/Climate_change_denial

Wilhelm, Trevor (2019). "Deniers and Propaganda: Amherstburg Debate Tackles 'Climate Crisis.'" *Windsor Star*, November 13; https://windsorstar.com/news/local-news/deniers-and-propaganda-amherstburg-debate-tackles-climate-crisis/

Williams, Hywel T. P., McMurray, James R., Kurz, Tim, and Lambert, F. Hugo (2015). "Network Analysis Reveals Open Forums and Echo Chambers in Social Media Discussions of Climate Change." *Global Environmental Change* 32, p. 126; https://www.sciencedirect.com/science/article/pii/S0959378015000369?via%3Dihub#bib0015

Wirth, Tim (1990). *Global Warming*. Congressional Record – Senate, p. 7474, April 19; https://www.govinfo.gov/content/pkg/GPO-CRECB-1990-pt5/pdf/GPO-CRECB-1990-pt5-7-2.pdf

WMO (2020). *State of Climate Services 2020 Report: Move from Early Warnings to Early Action*. World Meteorological Organization, October 13; https://public.wmo.int/en/media/press-release/state-of-climate-services-2020-report-move-from-early-warnings-early-action

Wong-Parodi, Gabrielle, and Feygina, Irina (2020). "Understanding and Countering the Motivated Roots of Climate Change Denial." *Current Opinions in Environmental Sustainability* 42, p. 60; https://www.sciencedirect.com/science/article/abs/pii/S1877343519301009

Yu, Chao, Margolin, Drew B., and Allred, Shorna B. (2021). "Tweeting about Climate: Which Politicians Speak Up and What Do They Speak Up About?" *Social Media + Society*, July 19; https://journals.sagepub.com/doi/full/10.1177/20563051211033815

2 Climate Science in Washington

The evolution of climate change in American politics can be divided into two periods: climate science and climate denialism. This chapter deals with the former, the arrival of the science of climate change in Washington. During the second half of the 20th century, an honour guard of climate experts paraded into Washington to teach federal politicians about the latest climate science. Upon their arrival in Congress in the 1950s, scientists warned about the pending dangers of climate change, but more study was necessary. A few decades later, with mounting evidence in support of climate change, scientists were taken aback by unexpected opposition from a small number of their colleagues, the contrarians.

The discussion is organized by administration, first the White House, followed by Congress. This chapter ends with the administration of Ronald Reagan who set the stage for the climate denialism movement and the contrarian scientists. The fossil fuel industry, power producers, and related industries, collectively known as the energy-industrial complex, had decided to fight against the scientific evidence, but their campaign began in earnest with the presidency of George H. W. Bush and grew with each subsequent Republican president, which is discussed in the next chapter.

Before beginning this section on the modern era of climate science in American politics, a brief tribute must be made to the third president of the United States, Thomas Jefferson (1743–1826); in his *Notes on the State of Virginia* was a chapter on "Climate," where he commented[1]: "A change in our climate, however, is taking place very sensibly." Shortly before he died, he recommended further studies[2]: "to shew the effect of clearing and culture towards changes of climate." Jefferson's weather recordings are a milestone in the political history of climate change, as he had been the first leader of the nation to be interested in "changes of climate." Not until the 36th president, Lyndon Johnson, would there be another.

A notable mention must also be made about the pioneering meteorologist Cleveland Abbe (1838–1916), who was the chief meteorologist for the United States Weather Bureau (from its inception during the administration of Ulysses S. Grant through to Woodrow Wilson). Mr. Abbe wrote an article in 1889 titled "Is Our Climate Changing," where he stated[3]: "No important climatic change has yet been demonstrated since human history began … the question of recent changes in climate must be relegated to the future." During these early days of scientific

DOI: 10.4324/9781003455417-3

meteorology, the challenge was to define in scientific terms the climate of the day. This top government official on meteorology also wrote a passionate note about how scientists, especially meteorologists, can face censorship and attacks from businesses protecting their interests; Mr. Abbe concluded[4]:

> It is wrong to mutilate or suppress the record of an observation of a phenomenon of nature, but it is also wrong to make a bad use of the record … Misrepresentations are to be avoided as harmful. The 'Independent Press' … repressing all cheats and hoaxes, defending the truth and the best interests of the whole nation as against the self-interest of a few.

These remarkable statements from over a century ago echo present-day problems facing climate scientists. Climate deniers "mutilate or suppress the record," creating misinformation, and public media must hold firm against "all cheats and hoaxes" and "the self-interest of a few" – or, as we know it, the energy-industrial complex.

President Franklin Roosevelt: 1933–45

There was no political interest in climate change during the Roosevelt administration, but a series of events would soon change things.

The big science story of the day was the atomic bomb which, surprisingly, would be an impetus to opening the modern era on climate research. Before the U.S. had entered World War II, a two-page letter by Albert Einstein (1879–1955) to President Roosevelt forever changed the world, as the famous scientist urged the president to build the atomic bomb.[5] President Roosevelt listened, and the Manhattan Project was born.

A disturbing conversation was taking place among leading scientists involved in the Manhattan Project before the first atomic bomb was about to be tested; the physicist Enrico Fermi (1901–54) was said to have joked: "Now, let's make a bet whether the atmosphere will be set on fire by this test." The dark humour had been sparked by his colleague Edward Teller (1908–2003) who had wondered[6]:

> Well, what would happen to the air if an atomic bomb were exploded in the air? … There's nitrogen in the air, and you can have a nuclear reaction in which two nitrogen nuclei collide and become oxygen plus carbon, and in this process you set free a lot of energy.

Dr. Teller had been mostly correct; indeed, "a nuclear reaction in which two nitrogen nuclei collide and become oxygen plus carbon, and in this process you set free a lot of energy" does actually happen. The good news (as the science predicted) was that this took place without transforming the atmosphere into an end-of-days fireball. What happened, in fact, was the nuclear carbon burned into carbon dioxide. This new type of anthropogenic carbon dioxide was whimsically called "bomb carbon,"[7] which came with its own separate set of problems as it was radioactive.

A nuclear explosion literally creates carbon dioxide from nitrogen, whereas the burning of fossil fuels creates carbon dioxide from the sequestered carbon of fossil fuels. We have, as a result, added two types of anthropogenic carbon dioxide into our atmosphere. The "bomb carbon" inadvertently became a stimulus for scientific investigation into the fate of all carbon dioxide in the atmosphere.

Albert Einstein had also written to President Roosevelt about the limited access scientists had to members of his administration[8]: "[Physicist Leo Szilard (1898–1964)] is now greatly concerned about the lack of adequate contact between scientists who are doing this work and those members of your Cabinet who are responsible for formulating policy." Likewise, along with urging for better communication between scientists and government officials, a group of scientists and other scholars, including Albert Einstein, in 1955, had pleaded with world leaders to adopt a humanistic approach to the nuclear existential threat and commented[9]:

> We have not yet found that the views of experts on this question depend in any degree upon their politics or prejudices. They depend only, so far as our researches have revealed, upon the extent of the particular expert's knowledge. We have found that the men who know most are the most gloomy.

Out of fear of a nuclear Armageddon, the Bulletin of the Atomic Scientists originated the Doomsday Clock and, since 2007, climate change has joined the menacing criteria used to set the time to midnight[10]: "Humanity continues to face two simultaneous existential dangers – nuclear war and climate change." For the moment, only climate change is happening and again scientists are calling for better communication with policymakers and a humanistic approach.

During the Roosevelt administration, global warming was not yet on the government agenda. In fact, quite the opposite, as the U.S. Department of Agriculture, in a report from 1941 titled "Climate and Man," had rejected the theory that carbon dioxide could have any effect on the planet's temperature.[11] The government report had a specific section on "Climatic Change Through the Ages," which included this commentary[12]: "there is no proof that short-time climatic changes, at least, are anything more than matters of chance." It should be pointed out that the report does not reflect science denialism, but science skepticism; only in the following decade did science demonstrate that global warming was taking place.

Guy Stewart Callendar

During this period, there was not much activity on the science of climate change, but one lone major researcher, Guy Stewart Callendar (1898–1964), described the impact of the greenhouse effect in his first paper on the subject in 1938, "The Artificial Production of Carbon Dioxide and Its Influence on Temperature" (today we would have called it "Anthropogenic Global Warming"). He found that the average annual carbon dioxide emissions had been 4,000 million tons per year for the past 20 years, and the average world temperature had increased by 0.07°C compared to the mean temperature of the 19th century. He went on to calculate that the global temperature would be higher by 0.16°C in the 20th century and by 0.39°C in the 21st century, compared to the 19th century. More importantly, instead of just talking about theoretical temperature projections, he found evidence of global warming in his day[13]: "The temperature observations at 200 meteorological stations are used to show that world temperatures have actually increased at an average rate 0.005°C. per year during the past half century." So, the British-Canadian scientist was the first to detect actual global warming from meteorological records.

As his Swedish mentor Svante Arrhenius (1859–1927) before him, Guy Callendar believed that global warming would benefit civilization[14]: "In any case the return of the deadly glaciers should be delayed indefinitely." The melting of the glaciers was never presumed to be a problem by the early researchers. The following year, Guy Callendar wrote about the "grand experiment" that is an "agent of global change" (the dumping of carbon dioxide into the atmosphere) and, famously, presented a graph plotting global temperatures from 1858 to 1939, with a sharp rise beginning in 1910.[15]

Climate change science was still not a hot topic in those days. Overall, the greater scientific community was skeptical. A 1937 report by the meteorologist Joseph Burton Kinser (1874–1954), for example, slammed the concept[16]:

> There is much loose talk these days about changes in climate due to human activity, and various suggestions that man should do this, that, or the other thing to prevent droughts. Such talk is utter nonsense … [two pages] suggestions of man-made changes in climate are relegated to the kindergarten of science.

The early carbon dioxide theory on climatic change had never gained much traction within the scientific community as it was believed that water vapour was the dominant factor, masking any impact of carbon dioxide, which, for example, the above U.S. Department of Agriculture had reported. The science on global warming, consequently, just gathered dust among the piles of obscure studies.

President Harry Truman: 1945–53

There was still only minor interest in the climate sciences during the Truman administration, although the weather was receiving a lot of political attention. In 1947, the government first became involved in weather[17] modification experiments – Project Cirrus involved the seeding of clouds with dry ice (solid carbon dioxide).[18] Irving Langmuir (1881–1957), who had championed the technology, had been an advisor for Project Cirrus; on December 11, 1950, he reported that[19]:

> the government should seize on the phenomenon of weather control as it did on atomic energy … In the amount of energy liberated, the effect of 30 milligrams of silver iodide under optimum conditions equals that of one atomic bomb.

Later, commenting on Project Cirrus, General Donald N. Yates (1909–93), president of the American Meteorological Society, stated[20]:

> The potentialities of artificial weather control and modification have excited the imagination of all of us ... As a result, Federal, State, and municipal authorities have been besieged with requests to make rain, to stop rain, to put out fires, to destroy hurricanes ... even to create major climatic changes.

General Yates had been skeptical about Project Cirrus and had requested a group of experts to review the work; they concluded that no weather modification had been demonstrated by the project. Overall, Project Cirrus was considered a failure and was shut down in 1951.[21] Even though Project Cirrus had flopped, the research on weather modification was only just getting started.

Office of Naval Research

During the first half of the 20th century, there had been little scientific interest in the greenhouse effect, and its impact on the climate continued to be of no interest during the Truman administration. Yet, in the following Eisenhower administration and seemingly out of nowhere, a revelation in the political history of climate change took place on March 8, 1956, when a scientist testified at a congressional hearing that the burning of fossil fuels could warm the planet and "cause a remarkable change in climate," which he famously described as "the greatest geophysical experiment in history."[22] This was a scientist who had not yet published a paper on the greenhouse effect, and his specialty was not the atmosphere or the climate. But, putting a few pieces together from the Truman period may explain this otherwise unexpected visit to Congress during the subsequent Eisenhower administration.

The testimony had been from the noted oceanographer Roger Revelle (1909–91), and his scientific interest had arisen from the problem of a warming planet being a "combination of meteorology and oceanography"[23] – important questions were how much of the carbon dioxide had been taken up by the oceans and how much remained in the atmosphere.[24] Dr. Revelle comprehended the potential dangers of climate change better than most and became a relentless crusader to warn politicians on what lay ahead. A long-time colleague of his, Walter Munk (1917–2019), later stated about Dr. Revelle[25]: "Typical of this was the greenhouse effect, which he really invented, which he was the first to sense was happening, to consider the implications."

While Dr. Revelle was a major contributor to the science of climate change, his influence on American politics was also important; he had been the first to visit Congress to discuss the science of the greenhouse effect during the Eisenhower administration and became a frequent witness in congressional hearings in the following decades. His interest in the climate sciences may have been associated with a group – the Office of Naval Research (ONR) – formed under the Truman presidency.

The ONR was the government organization that financed much of the revival of the climate sciences and was created by President Truman on August 1, 1946, after the idea for the organization had been introduced to him in a report named "Science, the Endless Frontier."[26] This organization opened opportunities for basic research, including in the climate sciences, which did not exist before. The ONR had taken the lead among government agencies on the climate sciences and had sponsored the weather modification experiments of Project Cirrus,[27] the pioneering work of Gilbert Plass (1920–2004),[28] the noted study of Roger Revelle and Hans Suess (1909–93),[29] and the early atmospheric CO_2 measurement program of David Keeling (1928–2005).[30]

Roger Revelle had a relationship with the ONR, working directly for them as head of the ONR's Geophysics Branch just after the Second World War and, later, as a consultant.[31] It was no accident that the close connections between the ONR and Dr. Revelle coincided with government interest in climate science and the funding that came along with it, laying the groundwork for a new generation of climate scientists and the modern era of the science of climate change.[32]

Congress

There were no early hearings in Congress on the greenhouse effect, though there were two references to natural "climatic change."[33] Similarly, at a later hearing in 1951, N. B. Bennett Jr. of the Department of the Interior raised the subject of "Changed Climate – Local, National, Global,"[34] but, again, the context was unrelated to the greenhouse effect and emissions of carbon dioxide from the burning of fossil fuels. The focus of the 1951 hearing was on weather modification, which included the following declaration[35]:

> Research and experimentation in the field of weather modification and control have attained the stage at which the application of scientific advances in this field appears to be practical. The effect of the use of measures for the control of weather phenomena upon the social, economic, and political structures of today, and upon national security, cannot now be determined … The Congress therefore recognizes that experimentation and application of such measures are matters of national concern.

At this Senate hearing, weather modification to cause "major climatic changes" had also been discussed.[36]

During the term of President Truman, the groundwork had been laid for a resurgence in the science of climate change through the initiatives of Roger Revelle and the Office of Naval Research. Presidents and members of Congress were not yet aware of global warming, but that was about to change.

President Dwight Eisenhower: 1953–61

A 1978 study by the Congressional Research Service summarized what initiated government interest in climate research[37]:

> The possibility of climatic alterations by human activity was alluded to in the scientific literature at the beginning of this century [Arrhenius], and again in the late 1930's [Callendar], but it received little serious attention until the 1950's. The first period of thermonuclear testing, 1954 to 1958, generated a great deal of concern about drastic and widespread effects on weather. It was felt that anything which liberated such great energies must somehow influence the atmosphere …
>
> Some evidence that manmade carbon dioxide was accumulating in the atmosphere appeared as early as 1938. This, together with some early systematic data from Scandinavia, led to the inclusion of a carbon dioxide (CO_2) measurement program during the International Geophysical Year (IGY), 1957–1958.

With the geopolitics of the Cold War as the backdrop, and the nuclear arms race underway, government funds flowed into scientific studies on the fate of radiation and "bomb carbon" in the atmosphere, and "weapons" to control the weather for military purposes – weather weapons were all the rage during the Cold War.

President's Advisory Committee on Weather Control

After World War II, legislation on weather modification was proposed in every session of Congress for the next three decades, and the Senate discussed bills for the creation of an Advisory Committee on Weather Control at the end of Truman's term[38]; the first law was passed in the early days of the Eisenhower administration which created this committee on August 13, 1953,[39] headed by Captain Howard Thomas Orville (1901–60), who explained in a popular press interview[40]: "They are quiet men, so little known to the public that the magnitude of their job, when you first hear of it, staggers the imagination. Their object is to control the weather and change the face of the world."

On February 8, 1958, Captain Orville gave a presentation called "The Impact of Weather Control on the Cold War,"[41] where he unexpectedly mentioned a special type of "weather control:"[42]

> The carbon dioxide that is released to the atmosphere by industry in the burning of coal and oil and their derivatives (most of it during the last generation) may have changed the composition of the atmosphere sufficiently to bring about a general warming of the world's temperature by about 1°F. This warming represents a 2-percent increase in the carbon dioxide content of the air. Studies show that when the amount has increased up to 10 percent the icecaps will begin to melt with the resultant rise in the sea levels. Coastal cities and areas such as New York and Holland would be inundated.

These words from over six decades ago are the first warning from a member of an advisory committee to the president of the United States on the serious consequences of climate change.

Breaking News: Global Warming

The modern age of the science of climate change began during the Eisenhower administration, starting with a pivotal study by a Canadian-born researcher, Gilbert Norman Plass (1920–2004), who demonstrated that carbon dioxide was indeed a major factor in global warming.[43] The subject of climate change, at this time, expanded outside of scientific circles into the public domain. Dr. Plass had laid out the connections between the combustion of fossil fuels and global warming at a meeting of the American Geophysical Union in May 1953, and the story was picked up by mainstream media, including *Time*[44]:

> modern man burns nearly 2 billion tons of coal and oil each year … his furnaces belch some 6 billion tons of unseen carbon dioxide into the already tainted air. By conservative estimate, the earth's atmosphere, in the next 127 years, will contain 50% more CO_2.

In a later paper, Gilbert Plass warned of the dangers of climate change[45]: "In fact the temperature rise from this cause may be so large in several centuries that it will present a serious problem to future generations."

In 1956, *Time* also reported on the scientist Roger Revelle[46] (discussed above) who expressed similar ominous warnings[47]:

> In 50 years or so this process, says Director Roger Revelle … may have a violent effect on the earth's climate … they may be able to predict whether man's factory chimneys and auto exhausts will eventually cause salt water to flow in the streets of New York and London.

These magazine articles are early examples of print media framing the threat of the climate change issue for the public. Global warming even made its debut on television on February 12, 1958, with the documentary from Bell Labs, *The Unchained Goddess*, produced by Frank Capra (1897–1991); the hour-long, educational film warned[48]:

> every year of more than six billion tons of carbon dioxide which help air absorb heat from the sun … a few degrees rise in the Earth's temperature would melt the polar ice caps … and if this happened, an inland sea would fill a good portion of the Mississippi valley.

There was not yet a sense of urgency about climate change, but scientists had begun to raise the alarm about the future, and the oil industry had taken careful notice of this threat to its business (discussed in Chapter 5).[49]

Worth noting, in his farewell address, President Eisenhower famously cautioned the American public of the menacing influence of both the "military-industrial complex" and the "scientific-technological elite; regarding the latter he stated:"[50] "YET, in holding scientific research and discovery in respect, as we should, we must also be alert to the equal and opposite danger that public policy could itself become the captive of a scientific-technological elite." Indeed, public policy in America has "become the captive," not of a "scientific-technological elite" as he had feared, but of the energy-industrial complex which has undue influence over policymaking on climate change.

Congress

Anthropogenic climate change suddenly burst onto the political scene in 1956, as mentioned above, when Dr. Revelle testified at a hearing of the House Committee on Appropriations on the International Geophysical Year (IGY)[51] on atmospheric and related research. At this session, he described the already advanced stage of the science[52]:

> This burning of these fuels which were accumulated in the earth over hundreds of millions of years, and which we are burning up in a few generations, is producing tremendous quantities of carbon dioxide in the air ... Now, nobody knows what this will do. Lots of people have supposed that it might actually cause a warming up of the atmospheric temperature and it may, in fact, cause a remarkable change in climate ...
>
> Here we are making perhaps the greatest geophysical experiment in history ...
>
> The increase in the number of hurricanes on the east coast, however, is certainly tied in one way or the other with the general northward movement of the warm air.

The following year, Dr. Revelle gave expert testimony in front of the same committee again, where he explained to politicians the grave consequences of climate change[53]:

> This [carbon dioxide in the atmosphere] might, in fact, make a considerable change in the climate. It would mean that the lines of equal temperature on the earth would move north and the lines of equal rainfall would move north and that southern California and a good part of Texas, instead of being just barely livable as they are now, would become real deserts ...
>
> It raises the temperature. It is like a greenhouse ...

Roger Revelle continued to visit the Capitol frequently in later administrations, teaching the latest crop of political newbies, as well as re-educating veteran politicians, on the science of climate change.

President John Kennedy: 1961–3

The Cold War had made the Americans anxious about weather weapons. Senator John Kennedy had expressed concerns about Russian scientists controlling the weather[54]:

> The terrors of the nuclear age and the current advantages of Soviet science pose threats far more terrifying to contemplate than the perils braved by John Carroll during the Revolution and the War of 1812 … But today John Carroll would read of the possibilities of hydrogen bombs and atomic missiles raining upon Baltimore, Washington and other cities – of Soviet space vehicles conquering outer space – of Russian submarines terrorizing the high seas while their aircraft patrolled the skies – of Communist scientists controlling even our weather and tides.

The danger of weather weapons had also been voiced by Senator Clinton Anderson (1895–1975) (Democrat from New Mexico) to the president in a letter on February 14, 1961[55]:

> You will see under the heading of 'Controlled climate,' he [John von Neumann] points out that weather control and climate control are really much broader than rain making. He realized that a new Ice Age might be manufactured in order to annoy others; that changes could affect level of the seas and hence the habitability of the continental coastal shelves; that hurricanes might be corrected or depressed, and the like.

The bizarre idea that a "new Ice Age may be manufactured in order to annoy others" sounded more like science fiction than science, but Senator Anderson was speaking about a stunning report by the late mathematician John von Neumann (1903–57) called "Can We Survive Technology?" The mathematician had written about global warming and warned of "climatic warfare" and "global climate control," which he concluded would be worse than nuclear warfare.[56]

The White House was starting to take some notice of climate change. We will never know how far President Kennedy would have pursued the matter since his term had been cut tragically short.

Congress

An article in the press had appeared about the Senate Interior Committee and its "far-reaching study of weather control," chaired by Senator Anderson on February 11, 1961, where he stated that he was interested in weather modification because[57]:

> Dr. John Von Neumann, former AEC Commissioner, warned that this — not ICBM — was the ultimately dangerous threat. Other scientists have agreed. But so far, weather control research has dealt only with limited possibility of seeding clouds for rain or snow, not large-scale climate changes.

The article had appeared just before Senator Anderson had written his letter about weather control to President Kennedy.

While weather control remained a popular topic, Congress first reacted to the climate sciences in 1963, when Senator Edmund Muskie (1914–96) (Democrat from Maine) commissioned a staff report called *Study of Air Pollution*, which included warnings of the perils of carbon dioxide in the atmosphere from the burning of fossil fuels.[58]

Just after John Kennedy had been elected, Dr. Revelle was back at yet another congressional hearing. This time he did not specifically discuss the greenhouse effect or carbon dioxide emissions but presented a more general environmental warning[59]:

> I think there is a growing realization among all the peoples of the world that we have to use our whole planet better than we are using it now. It is quite ironic that this is happening at the same time that we are learning how to leave the planet altogether to go out into space. But our human populations are increasing so fast, and we are using up our easily won resources so fast that man's future will be pretty bleak and dim if he can't achieve some kind of harmony with his own planet.

Soon, Dr. Revelle would be joined by other scientists bringing messages about climate change to Washington.

President Lyndon Johnson: 1963–9

There was little evidence that the White House had given much attention to climate change until the arrival of Lyndon Johnson. He had long been outspoken on the menacing potential of weather weapons. In the late 1950s, while a senator, he commented[60]:

> The masters of infinity could have the power to control the earth's weather, to cause drought and flood, to change the tides and raise the level of the sea, to divert the Gulf Stream, and change temperate climates to frigid.

And later, as Vice President for John Kennedy, he made another stark statement[61]: "He who controls the weather will control the world."

President Lyndon Johnson was the first president to directly acknowledge the climate change problem; in a moving speech to Congress, on February 8, 1965, he warned[62]:

> The modern technology, which has added much to our lives can also have a darker side. Its uncontrolled waste products are menacing the world we live in, our enjoyment and our health …
>
> Air pollution is no longer confined to isolated places. This generation has altered the composition of the atmosphere on a global scale through radioactive materials and a steady increase in carbon dioxide from the burning of fossil fuels.

A few months later, the president gave another speech to Congress on this subject[63]:

> The development of methods for altering weather and climate to the benefit of mankind is a subject of quickening interest in the Congress and the

> Executive Branch of the Government of the United States – as, indeed, it is to all of the human race … Furthermore, it is essential for us to investigate the manner in which man may at present be inadvertently changing weather and climate.

President Johnson had a remarkable grasp of the threat of climate change over half a century ago that would surpass many of his successors.

The Oval Office would authorize one or two reviews on the climate sciences beginning with this president in 1965. President Johnson had asked the President's Science Advisory Committee for an environmental report, for which he had written in the forward[64]:

> Ours is a nation of affluence. But the technology that has permitted our affluence spews out vast quantities of wastes and spent products that pollute our air …I am asking the appropriate Departments and Agencies to consider the recommendations and report to me on the ways in which we can move to cope with the problems cited in the Report.

Tucked away in an appendix of the report, "Carbon Dioxide from Fossil Fuels – The Invisible Pollutant," was a warning about carbon dioxide emissions[65]:

> By the year 2000 the increase in atmospheric CO_2 will be close to 25%. This may be sufficient to produce measurable and perhaps marked changes in climate … The climatic changes that may be produced by the increased CO_2 could be deleterious from the point of view of human beings.

Although relegated to the appendix, the implications of the information on climate change had not gone unnoticed by the petroleum industry which wanted to keep the discussion on a "rational track,"[66] at least rational in their eyes (this is covered in a later chapter).

The year 1965 was a milestone in the political history of climate change. Lyndon Johnson was the first president to acknowledge the problem with his two speeches, and the first government report at the behest of the president, outlining the serious consequences of climate change, was issued. President Johnson's appreciation of global warming exceeded that of other world leaders at the time. From here on, American presidents would all have climate change on their agenda – some would welcome it; some would have it forced upon them; and some would just deny it.

Congress

Political activity on climate change also increased significantly in Congress during Johnson's presidency. In the latter half of the 1960s, testimony on "climatic change" from expert witnesses had become more frequent at congressional hearings, including these interesting examples:

- Senate hearing – 1966 – geophysicist Gordon MacDonald (1929–2002)[67]:

As civilization grows more complex, society places greater burdens on the atmosphere. We are just beginning to appreciate that the atmosphere is not a dump of unlimited capacity, but we do not yet know what the critical capacity of the atmosphere is or by what effects it should be measured. For example, we can consider carbon dioxide introduced into the atmosphere as a result of the Industrial Revolution.

- House hearing – 1966 – Thomas Malone (VP of Research, The Travelers Insurance Co.)[68]:

If the earth is warmed, the ice melts and the sea level would be raised so high that, were it to happen, we would probably have to swim home from this building this morning … The degree of danger which exists from the warming of the earth is something we must resolve in a matter of decades. The situation could become serious by the end of the century.

- Senate hearing – 1966 – atmospheric scientist Louis Battan (1923–86)[69]:

The need for investigation of the causes for climatic change and the feasibility of exerting influences on the climate are of great importance. In this regard, the requirement for further international cooperation becomes increasingly obvious. The climate on one part of the globe cannot be changed without causing a change on other areas.

- Senate hearing – 1967 – Robert M. White (Administrator, U.S. Department of Commerce)[70]:

This study of the possible global deterioration of clean air must include analysis of the effects of these changes on climate and weather. There has been much speculation on the long-term effects of increasing atmospheric carbon dioxide on the world's climate. Recent evaluations raise some question about the significance of this effect …

- House hearing – 1968 – Emilio Daddario (1918–2010)[71]:

As an example, I would interject here that there have been alarming stories in the past few years that the accumulation of carbon dioxide in the atmosphere from fossil fuel combustion might cause a 'greenhouse effect' and raise the temperature of the earth. Such a climate change would upset the world's weather, melt icecaps, and so forth.

The last quote is especially significant, for Mr. Daddario (Democrat from Connecticut) was not giving expert testimony but was a politician and the chair of the House session. However, a complication had arisen with global warming: The world's temperature had been slightly cooling since World War II; Emilio Daddario, in continuing with his opening remarks, complained about these conflicting reports[72]:

> Last summer, HEW [Health, Education and Welfare] scientists reported that the mean annual world temperature actually was falling. The reason given … was that particles [aerosols] in the atmosphere were increasing, thereby reflecting more sunlight and decreasing solar radiation reaching the ground. This may be a lesson for us that early impressions of environmental effects are often incomplete and that public statements should be tempered until adequate studies are completed.

For the next decade, congressional hearings on climate change were distracted by the debate of global warming versus global cooling, but Washington was taking an interest in the science of climate change.

President Richard Nixon: 1969–74

Even when a Republican won the next election, the political momentum continued on the environment and climate change. President Nixon established the EPA in late 1970, and many within his administration had been vocal on the threat of climate change. The Chairman, Task Force for the Environment for the president-elect, Russell Train (1920–2012), for example, had stated[73]: "The carbon dioxide content of the atmosphere is increasing, a portent of possible disaster to the world." Russell Train became the first chair of the Council of Environmental Quality, and a report from the Council appeared on August 3, 1970; herein, President Nixon, himself, penned a passionate message to Congress[74]:

> The recent upsurge of public concern over environmental questions reflects a belated recognition that man has been too cavalier in his relations with nature. Unless we arrest the depredations that have been inflicted so carelessly on our natural systems – which exist in an intricate set of balances – we face the prospect of ecological disaster …
>
> However, the report will, I think, be of great value to the Congress (and also to the Executive Branch) by assembling in one comprehensive document a wealth of facts, analyses and recommendations concerning a wide range of our most pressing environmental challenges … [several pages]
>
> The accompanying report by the Council describes the principal problems we face now and can expect to face in the future, and it provides us with perceptive guidelines for meeting them. These deserve the most careful consideration. They point the directions in which we must move as rapidly as circumstances permit …

> We should strive for an environment that not only sustains life but enriches life, harmonizing the works of man and nature for the greater good of all.

Among the "principal problems" discussed within the report of the Council was a chapter called "Man's Inadvertent Modification of Weather and Climate," and a specific section, "Carbon Dioxide—An Earth Warmer?" which concluded[75]:

> If one-half of that carbon dioxide were added to the atmosphere and there were no compensating effects, then the earth's average temperature would increase about by 2° to 3°F. Such a rise, if not counteracted by other effects, could in a period of a few decades, lead to the start of substantial melting of ice caps and flooding of coastal regions.

The threat of climate change had been clearly laid out by this report endorsed by the president.

Earlier, on September 17, 1969, Nixon's advisor Patrick Moynihan (1927–2003) had written a foreboding message to John Ehrlichman (1925–99), the White House Counsel[76]:

> As with so many of the more interesting environmental questions, we really don't have very satisfactory measurements of the carbon dioxide problem. On the other hand, this very clearly is a problem, and, perhaps most particularly, is one that can seize the imagination of persons normally indifferent to projects of apocalyptic change …
>
> It is now pretty clearly agreed that the CO_2 content will rise 25% by 2000. This could increase the average temperature near the earth's surface by 7 degrees Fahrenheit. This in turn could raise the level of the sea by 10 feet. Goodbye New York. Goodbye Washington, for that matter.

At a Senate hearing, "man-made climate change" had been directly mentioned[77]:

> Large-scale engineering projects which are feasible for major world powers may, inadvertently or by design, produce changes in world climate damaging to our security … The ARPA [Advanced Research Projects Agency] program, closely focused on military security aspects of man-made climate change, employs existing models wherever possible, modifying them as needed to make them practical tools for climate prediction.

However, "man-made climate change," in this case, was not in reference to global warming but weather modification which was part of the investigations of Nile Blue and had been funded by the Department of Defence.[78] Nile Blue boosted the development of computer modelling (general circulation models, GCMs), and a major cost of the project was a state-of-the-art supercomputer for climate modelling to simulate deliberate modifications (i.e., geoengineering); for example, if a country attempted "to melt large portions of the Arctic ice," what would happen?

On August 1, 1974, the Department of Commerce was requested to head up the Subcommittee on Climate Change[79]:

> Studies are being conducted to assess the large-scale impact of man's activities on climate (inadvertent global climate modification) … A global network of monitoring stations is being established to determine baseline values of small, variable trace constituents of the atmosphere and to determine the effect of long-term trends in these trace constituents on the global weather and climate. These data will provide input for diagnostic and prognostic numerical global models being developed to predict climate changes resulting from man's activities.

This report of the Subcommittee on Climate Change initiated under President Nixon – "The Federal Role in Weather Modification" – had been sent to President Ford and Congress.

President Nixon was the last Republican president to demonstrate concern over climate change. The Watergate scandal soon overshadowed his environmental achievements.

Global Cooling Schism

By the late 1960s, the scientific community generally accepted that the planet's temperatures would rise from the burning of fossil fuels, but there was a problem: Measurements of the Earth's temperature had shown a slight cooling trend since 1940. The observed cooling, rather than necessarily disproving the theory of global warming, pointed to something else happening as well. The question was whether global warming (greenhouse gases) or global cooling (aerosols) was going to win out in the long run; the science at the time was not sure. This caused the global cooling schism of the 1970s.

On December 3, 1972, professors George Kukla (1930–2014) and Robley K. Matthews raised the profile of global cooling in stunning fashion by sending a letter to President Nixon about the cooling threat.[80] The White House took the note seriously, and the Department of State reacted in early 1973 with the formation of the ad hoc Panel on the Present Interglacial, under the Interdepartmental Committee for Atmospheric Sciences, to study the threat of global cooling. The report of the Panel was issued in August 1974 (the month that President Nixon resigned) which concluded[81]:

> It is a remarkable coincidence that the CO2 warming effect and the maximum particles cooling effect turn out to be very nearly equal and opposite. Keeping in mind that the particle cooling effect is to be regarded as an upper limit, however, it follows that the CO2 warming effect will according to this analysis be likely to dominate in the years ahead, resulting in a modest net warming of climate.

The Panel, then, did not support the fears of global cooling. And the next year, a colleague of Dr. Kukla, Wallace Broecker (1931–2019) projected that global cooling was about to give way to global warming.[82]

As this was going on, another zealous proponent, Reid Bryson (1920–2008), was also selling global cooling in Washington to anyone who would listen in the Nixon administration; during the summer of 1973, his "Wisconsin Plan for Climatic Research" had been presented to the National Oceanic and Atmospheric Administration (NOAA), the National Science Foundation, and the National Security Council.[83] George Kukla and Reid Bryson stood out as exceptions to the rule, as most climate scientists who came to Washington had been invited to give testimony in congressional hearings, but these two scientists lobbied directly, and government departments paid attention to them. The level of interest from government was a bit strange, since there was no scientific consensus on global cooling being a threat.

Reid Bryson was a source of information[84] for a CIA report, called "A Study of Climatological Research as it Pertains to Intelligence Problems." This report, also issued in August 1974, was arguably the worst climate study ever produced by a government agency or department. While the report acknowledged the threats of a changing climate, anthropogenic climate change was completely ignored, and the report promoted Bryson's pet theory on global cooling.[85] While the science of the day would not have excluded the possibility of global cooling, the notable omission of global warming would have been controversial even at that time.

The study had the markings of climate denial written all over it. How did such a biased and improper report come to be issued by the CIA? Nothing within the report itself provided any clues. However, a suspect may be found – James Schlesinger (discussed in a later chapter), who subsequently proved himself to be an ardent climate denier, was briefly the Director of the CIA in mid-1973. Although speculative, I would not be surprised if he personally had ordered the review to be carried out by Dr. Bryson.

Considering the expert advice Drs. Kukla and Bryson offered on global cooling, the question arises whether they were among the first of the contrarians (those scientists who dismissed the scientific evidence for global warming). During the early 1970s, any scientific skepticism about global warming had dissipated, but there were still some questions whether global warming or global cooling would dominate. The narrow views of George Kukla and Reid Bryson, which suggested bias in favour of global cooling, were more indicative of climate denialism than climate skepticism – an assessment which is further reinforced by their opinions later morphing into full-blown climate denial.[86]

The global cooling schism should have faded away but has continued to haunt the discussions of climate change. Climate deniers fabricated a scientific consensus on global cooling to "prove" scientific consensus on global warming was shaky at best; they used media reports instead of peer-reviewed scientific literature to "support" their claims. In the House of Representatives, Ted Poe (1948–) (Republican from Texas), for example, was still flogging the tired global cooling hypothesis as a reason to distrust the science of climate change in 2009.[87] Worse still, a decade later, Clay Higgins (1961–) (Republican from Louisiana) remarked[88]:

> I'm old enough to recall when the science of the time stated that very soon, global cooling would overtake the planet. Then the language changed to climate change. And now in today's hearings, we have heard more about global warming again.

Such glib comments on global cooling appear daily on my Twitter feed. They are just distractions from the real issues of climate change. A thorough review of the literature[89] has debunked the phoney consensus on global cooling that is endlessly promoted by climate deniers in Twitter and Congress.

Congress

The global cooling schism was brought up at a Senate hearing in 1973 by Reid Bryson[90]:

> Much of the discussion of the last 5 years with regard to climatic change has been centered on the question of whether the Earth is going to get warmer because we are adding carbon dioxide to the atmosphere or is the Earth going to get colder because we are adding particulate pollution to the atmosphere?

At the hearing, his testimony demonstrated his climate denial leanings as he also commented[91]:

> Even if we were to say 'let us stop using fossil fuels so that we do not add carbon dioxide to the atmosphere, because that impacts the world's climate,' how on earth could you stop using fossil fuels? … it is in the use of fossil fuels that one gains affluence.

His opinion clearly reveals a certain reverence for fossil fuels and totally discounts the consequences of climate change, which is all the more shocking because Dr. Bryson was a climate scientist who was well aware of the evidence. Senator Hubert Humphrey (1911–78) (Democrat from Minnesota), who knew of the controversial views of Dr. Bryson, asked him if his opinion agreed with that of his peers. Dr. Bryson skirted the issue and never directly answered the question.[92] Senator Humphrey was probably the first in Congress to be in a climate brawl, albeit a very mild one, by challenging the climate denial point of view of a contrarian.

Others came to Congress testifying about climate change. Earlier, at a House hearing, the Associate Administrator of NASA, John Naugle (1923–2013), warned of the pollution of carbon dioxide[93]:

> We know that man, through industrial revolution and an increasing population, is not only polluting the atmosphere, but may be changing its very nature. We have testified in previous years that the amount of carbon dioxide in the atmosphere has changed by a significant amount in the past 50 years … [several pages]

> As we look at our planet, as we look at the population that is increasing, we know that man is not only polluting, but possibly beginning to change the very fundamental nature of our atmosphere on the earth.

At another hearing, Harry Perry (Research Advisor to the Assistant Secretary of Mineral Resources) expressed similar concerns[94]:

> The carbon dioxide which results from the combustion of fossil fuels, which is the major source of carbon dioxide in the air, is considered a potential pollutant. The best data available indicate that the carbon dioxide content of the atmosphere has been increasing so that it appears that the balance between carbon dioxide in the air, sea water, and biosphere has been disturbed.

Dr. Lee DuBridge (1901–94), the Science Advisor to President Nixon, was also a witness at the same hearing and talked about the greenhouse effect[95]:

> If there is a very substantial increase of carbon dioxide in the atmosphere, then the greenhouse effect would become important and the temperature balance of the earth might change and this might have very important consequences … One projection based on a faster increase than I just mentioned estimates that the CO_2 concentration will be 25 percent higher by the year 2000.

These excerpts on global warming serve to highlight how the threat of climate change was being presented to Congress despite the distraction and background noise caused by the global cooling schism which hampered political discussions on climate change for several more years.

President Gerald Ford: 1974–7

During the Ford administration, the National Academy of Sciences (NAS) examined the state of the science of "climatic change" but could not settle the global cooling schism, stressing limited information; there was a small section of the report that looked directly at the subject of global warming and global cooling, called "Carbon Dioxide and Aerosols." The NAS report, nevertheless, did express concern about human influence on the climate's "heat balance."[96]

Before this report, on August 22, 1974, Congressman Morris Udall (1922–98) (Democrat from Arizona) had explained to the new president[97]:

> We all know (sometimes to our sorrow when they are misused) of the enormous power and potential of modern science and technology. Within just the past year, the longterm problems of worldwide food and protein shortages, non-renewable natural resource management, climate change, and energy conservation and resource development, have taken on frightening new dimensions …

For which President Ford thanked him. While President Ford, in general, did not have much to say about climate change, activity was picking up in Congress.

Congress

So far, reading the minutes of the earlier congressional hearings on climate change, one gets the impression that the political reception was polite with perhaps mild curiosity. Overall, there was a lack of interest, and only a few members of a subcommittee, usually Democrats, even bothered to show up for these hearings. Climate change had not yet reached a level of importance within political circles and was usually only included in hearings dealing with broader environmental issues – that was about to change during Ford's presidency.

Another milestone in the political history of climate change took place in May of 1976, when the first hearing specifically on the climate took place, and it was a big one; the lengthy session covered six days of testimony by leading climate experts. The chair George Brown (1920–99) (Democrat from California) introduced the purpose of the hearing[98]:

> While some scientists are warning that we might be starting off on a long-term cooling trend, making the plunge into another Little Ice Age, others seem to be concerned that the polar ice caps might permanently melt, drowning our coastal cities in the rising seas. Some scientists are concerned that mankind may be inadvertently modifying the climate by polluting the environment with carbon dioxide, particulates, and heat, but there seems to be no agreement as to whether the net result will be to heat us up or cool us off. On the other hand, other scientists seem especially concerned about climatic changes arising from natural causes, but don't seem to have reached a consensus over what these causes are.

Mr. Brown's comment of "other scientists seem especially concerned about climatic changes arising from natural causes" suggests that there were early contrarians making their presence felt in Washington.

At the historic session, the climate scientist Stephen Schneider (1945–2010) testified about "climatic crises:"[99]

> In conclusion, while it cannot be stated with certainty that increased effort and resources in the field of climatic studies will provide immediate payoffs in terms of forecast capability, it must be said that the costs of our ignorance of the knowledge and workings of the climate system may already be unacceptably high, and that while assurance of success cannot be guaranteed, the value of obtaining better understanding of the climatic system makes this scientist comfortable with the feeling that support for climatic research is a worthwhile hedge against potential climatic crises.

A special guest at this hearing was the Swedish scientist Bert Bolin (1925–2007), who would later head the IPCC. His testimony ended with the warning[100]: "We must accept that there are limits to the exploitation of the Earth's resources. One such limit is set by the principle that man's natural environment must not be changed drastically and irreversibly." And during the question period, he acknowledged[101]:

> But it is a typical situation in the environmental area that we may occasionally have to make decisions at a stage when we really are not quite sure. We must play the game in a flexible manner, supporting scientific development to obtain the kind of information we need, and constantly trying to keep the activities of society as flexible as possible to be able to adjust.
>
> But in the field of energy, which is so fundamental to the whole of our society, this is, of course, much more difficult than anywhere else.

There was also a discussion on aerosols and global cooling, but Dr. Bolin was more concerned about global warming; he stressed that if we stopped emitting aerosols, the problem would disappear within weeks, but carbon dioxide in the atmosphere would stay there for centuries or longer. Twenty years after Dr. Revelle first testified, climate science finally had its first big day in Congress, and more would follow.

President Jimmy Carter: 1977–81

The first time that climate change became an active talking point in the Oval Office was during the Carter administration.[102] The chairman of President Carter's Council on Environmental Quality, James Gustave Speth (1942–), summarized the political milieu surrounding climate change in the early Carter administration[103]:

> In my expert opinion, in the period shortly after President Carter took office in 1977, there was a growing sense of concern and indeed urgency within the federal government that fossil fuel burning was heating the planet and causing the climate to change in many ways that could be catastrophic.

The future president had been prepped about the climate change issue by a primer when he was running for office, written by Stephen Schneider who advised[104]:

> The costs to the next generation of an irreversiably [sic] damaged climate could be measured from billions to trillions of dollars. Unfortunately, in present economic accounting systems these uncertain costs are "externalized", therefore, not included in the cost/benefit analysis over issues such as pollution controls or growth rate limitation regulations. If the potential costs to the future were internalized in present accounting, perhaps through the use of "next-generation environmental impact statements", then we would sustain less risk of doing irreparable harm to the world of our children and grandchildren.

After the election, another report chaired by Roger Revelle was issued by the National Academy of Sciences (NAS) called "Energy and Climate," which began with strong warnings about the burning of fossil fuels[105]: "the human consequences of a marked climatic change that might be brought about by the addition of large quantities of carbon dioxide to the atmosphere, it may be concluded that world society could probably adjust itself, given sufficient time and a sufficient degree of international cooperation. But over shorter times, the effects might be adverse, perhaps even catastrophic." The NAS report clearly laid out that a crisis from climate change was on the horizon and was definitely coming.

The NAS informed Frank Press (1924–2020), Director of the Office of Science and Technology Policy, about the study, who sent a note to President Carter with the ominous subject line[106]: "Release of Fossil CO_2 and the Possibility of Catastrophic Climate Change." A note with such a scary title, however, failed to get the president's serious attention as no action was taken, and almost half a century later, an article referring to this note was called[107]: "The 1977 White House Climate Memo that Should Have Changed the World."

The Department of Energy

A geopolitical event, the Cold War, had first spurred political interest in climate change, and this time another geopolitical event – the energy crisis – did the same. Yet, what should have been an excellent opportunity for renewable energy, instead turned into one for coal. The energy crisis had led to the formation of the Department of Energy by President Carter, and its first secretary was James Schlesinger (discussed further in Chapter 4).

In July 1979, a report for the Council on Environmental Quality, "The Carbon Dioxide Problem: Implications for Policy in the Management of Energy and Other Resources," was written by an all-star cast of climate experts – George Woodwell (1930–), Gordon MacDonald, Roger Revelle and David Keeling, and the report was presented to President Carter by James Speth. After two decades of warnings about the future threat of climate change, this report to the president was the first to communicate a sense of urgency as well[108]: "Man is setting into motion a series of events that seem certain to cause a significant warming of world climates … Enlightened policies in the management of fossil fuels and forests can delay or avoid these changes, but the time for implementing the policies is fast passing." Some within the government did not see it that way. Just as with the earlier memo from Frank Press ("Release of Fossil CO_2 and the Possibility of Catastrophic Climate Change"), the Department of Energy again downplayed the need for any action.[109]

President Carter sided not with the climate scientists but with the Department of Energy. Just days later, he announced sweeping reforms in support of liquid fuels from coal by the creation of the Energy Security Corporation (aka Synthetic Fuels Corporation[110]).[111] Gordon MacDonald, one of the authors of "The Carbon Dioxide Problem," was frustrated by the decision of the president[112]: "I would have to say that there has been no substantive reaction [from the White House] … I briefed that report to many people in high places in this administration. Yet, as you heard in Sunday's speech and yesterday, the President wishes to continue with his proposals for synthetic fuels. I think this is a mistake." But the political situation was messy and these massive investments appealed to a panicky public.

Meanwhile, Frank Press had requested an assessment of the scientific basis of climate change from the National Academy of Sciences. Their findings appeared in a landmark report called "Carbon Dioxide and Climate: A Scientific Assessment" (often called "the Charney Report"), which concluded that during the first half of the 21st century, there would be a doubling of carbon dioxide levels in the atmosphere that would increase global temperatures by about 3°C.[113] In the forward to the report, Verner Suomi (1915–95) laid out in bold language the situation[114]:

> We now have incontrovertible evidence that the atmosphere is indeed changing and that we ourselves contribute to that change. Atmospheric concentrations of carbon dioxide are steadily increasing, and these changes are linked with man's use of fossil fuels …
>
> The conclusions of this brief but intense investigation may be comforting to scientists but disturbing to policymakers. If carbon dioxide continues to increase, the study group finds no reason to doubt that climate changes will result … A wait-and-see policy may mean waiting until it is too late.

Only six months later, Frank Press rushed for a second opinion from the National Academy of Sciences for the president, this time to assess the social and economic consequences of an increasing concentration of atmospheric carbon dioxide. In response, the NAS formed the Ad Hoc Study Panel on Economic and Social Aspects of Carbon Dioxide Increase and responded with a letter which started by addressing some recent controversial studies of "estimates of extremely small effects" – two papers, one by Reginald Newell and Thomas Dopplick, and the other by Sherwood Idso (who was an early contrarian on the science of climate change[115]) – that were dismissed by the NAS.[116] The letter from the NAS also warned of "political manipulation, controversy and division" that would soon come into play. Then the NAS concluded[117]:

> To sum up, carbon dioxide will pose exceedingly difficult and divisive policy questions for all the world's nations individually and collectively. We do not know enough to address most of these questions right now. We believe we can learn faster than the problem can develop.

Members of the Panel included Roger Revelle, George Woodwell, and William Nierenberg.

The NAS letter grasped the pending crisis of climate change only too well but probably killed any chances of immediate action with these five words: "We do not know enough." Frank Press sent a memo (including the NAS letter) to President Carter, on May 5, 1980[118]:

> The NAS [National Academy of Sciences] believes that better understanding of the climatic consequences will emerge in 5-10 years. The Nation's principal efforts in the near term, therefore, must be focused on research to acquire this improved knowledge, but with as low a political profile as possible until we reduce the uncertainty. Meanwhile, the Nation would be well advised to keep open all relevant policy and technological options.

In a hand-written response, President Carter called it a "good memo."

During his public speeches on climate change, President Carter had delivered mixed messages. On August 2, 1979, in his Environmental Message, he mentioned the need to examine increasing levels of carbon dioxide in the atmosphere, and, on February 29, 1980, he stressed this global environmental threat.[119] But in his State of the Union Address on January 23, 1980, the focus was on coal and other fossil fuels (and solar power was mentioned once).[120] Then, a year later, the carbon dioxide build-up in the atmosphere was mentioned for the first time in a State of the Union Address, and he pushed his solar energy agenda.[121]

President Carter had famously installed a solar heating system in the West Wing of the White House at the end of April 1979, where the system of 32 solar-thermal panels supplied hot water to the kitchen; in celebration of the event, he predicted[122]:

> In the year 2000, the solar water heater we are dedicating today will still be here, supplying cheap, efficient energy. A generation from now, this solar heater can either be a curiosity, a museum piece, an example of a road not taken, or it can be just a small part of one of the greatest adventures in our nation's proud history – harnessing the power of the sun to enrich our lives as we move away from our crippling dependence on foreign oil. I am determined that America will move toward the solar age.

The solar water heating system did not make it to the year 2000, as Ronald Reagan removed it in 1986.[123] The panels ended up in a government warehouse, but in 1991, half of the panels were acquired by Unity College in Maine (some were installed to heat water in their cafeteria). Three of the panels ended up in museums,[124] just as President Carter had thought might happen. The solar panels were largely a symbolic gesture which President Carter justified, not to help stop climate change, but to help end the energy crisis. While he championed solar energy, his decision to pursue coal over renewable energy speaks for itself. President Carter was, at best, lukewarm about acting on climate change.[125]

Congress

In a report by the Congressional Research Service, an alarm was sounded about the dangers of global warming[126]:

> What would happen if society elected to ignore the problem of carbon dioxide until it manifested itself (perhaps in another 20 years) in the form of a clear signal that a global warming trend had begun that was unmistakably attributable to the further accumulation of carbon dioxide in the atmosphere? Delaying until then a mandated action to phase over the principal energy sources from fossil fuels to other alternative kinds of fuels and taking into account another several decades for the transition to be completed would put us half-way into the next century before the problem could be

> shut off at its source. But perhaps the most disturbing aspect of the carbon dioxide problem is that the effects of carbon dioxide would endure for hundreds of years, even after the abandonment of the fossil fuel economy, because of the long recovery time associated with the processes that would rid the atmosphere of excess carbon dioxide and establish an equilibrium condition.

The report to Congress emphasized the urgency in dealing with climate change.

As the issue of the day was the energy crisis, and Carter's administration had opted for the greater use of fossil fuels ("synthetic fuels"), a few politicians pushed back citing the dangers of climate change. For example, Paul Tsongas (1941–97) (Democrat from Massachusetts) stated that consuming more fossil fuels could cause climate change with serious consequences[127]:

> Finally, there is the fundamental question as to whether we can burn all our oil, natural gas, and coal without causing abiding changes in the world's climate. These changes could severely limit our ability to grow food and could produce flooding of our coastal cities.

Another example of this push back was during a Senate hearing on the "Synthetic Fuels Production Act of 1979," where the chair Senator Abraham Ribicoff (1910–98) (Democrat from Connecticut) asked[128]: "Have you … given any attention to the doomsday forecast of carbon dioxide out of the synthetic fuels program?" At the same hearing, the outspoken scientist Gordon MacDonald made the general comment[129]:

> Man, through the burning of carbon-based fuels, is setting in motion a series of events that seem certain to cause a significant warming of world climates over the next decades. The use of synthetic fuels will accelerate and intensify these climatic changes … [page]
>
> The CO_2 problem is one of the most important contemporary environmental problems, is a direct product of industrialization, threatens the stability of climates worldwide and, therefore, the stability of all nations, and the problem can be controlled, not technically but by policy.

Dr. MacDonald had argued that action was required now, and he recommended carbon taxes or limits be set on carbon dioxide emissions – such direct advocacy from a scientist rarely had been heard in Congress on climate change. In response, Senator Ribicoff stated[130]: "So, we have a very grim dilemma that we and the entire world face."

Senator Ribicoff listened to the expert testimony and followed up. Just two weeks later, on July 30, 1979, he organized a symposium for a small group of senators on "Carbon Dioxide Accumulation in the Atmosphere, Synthetic Fuels and Energy." The session involved some of the leading climate scientists, including

Wallace Broecker, Stephen Schneider, Roger Revelle, and George Woodwell; the latter started off the session with[131]:

> There is no question but that the carbon dioxide problem is one of the most important world environmental problems. It is sufficiently important to enter into all of the decisions that we make in the exploitation of fossil fuels, oil, and coal, for energy.

At this symposium, Roger Revelle raised a special political obstacle when confronting climate change[132]:

> We tend to best think of the immediate future and as we go further and further into the future, things become less interesting and we begin to take them much less into account. One of the reasons, of course, is that it is so hard to predict the future. But the CO_2 problem is very much like the inexorable advance of a glacier. From year to year, you cannot notice much difference, but in 50 years, you may be overwhelmed by the glacier.

After being involved with politicians for a quarter of a century, Dr. Revelle understood the political challenges of dealing with climate change better than most scientists.

A year later, climate change was headlining another congressional hearing named "Effects of Carbon Dioxide Buildup in the Atmosphere." Presiding over the session, on April 3, 1980, was again Senator Tsongas, who stressed[133]:

> Of particular interest to this committee is the extent to which the burning of fossil fuels to meet our country's energy needs will result in higher atmospheric carbon dioxide concentrations and thereby result in serious adverse effects, both direct and indirect, on our climate, environment, and indeed society …
>
> Possible? Probable? We really don't know, but if it happens, it means goodby Miami, goodby Corpus Christi, goodby Sacramento, goodby Boston, which obviously is of much more concern, goodby New Orleans, goodby Charleston, Savannah, and Norfolk. On the positive side, it means we can enjoy boating at the foot of the Capitol and fishing on the South Lawn …
>
> I think this may well turn out to be the single most disastrous example of short-term planning by I think a very indulgent society …

Witnesses at the hearing included the climate experts James Speth, George Woodwell, Gordon MacDonald, and Wallace Broecker. Another witness, MIT professor David Rose (1922–85), put the problem in a fuller perspective[134]:

> I have seen problems as large as this totally ignored until the civil sages are washed over, and here we are talking about the whole world. An environmental, conceivable environmental calamity would make every other one

> look like nothing—back to Noah and the flood, and furthermore, we have been warned this time.

The turnout at the session had been low; of the eighteen Committee members, only three attended: Paul Tsongas, Bill Bradley (1943–) (Democrat from New Jersey), and Dale Bumpers (1926–2016) (Democrat from Arkansas) who explained the poor turnout[135]: "It is very difficult to get politicians to get excited about some ultimate cataclysm happening 20 to 30 years down the pike. They operate in a crisis atmosphere." And this climate "crisis atmosphere" is still absent in Congress – over forty years later.

Apathy on climate change existed among the public as well,[136] but this was beginning to change at the start of the 1980s. A report of this sparsely attended congressional session was televised on the *CBS Evening News*, hosted by the well-respected broadcast journalist Walter Cronkite (1916–2009), who opened the story with[137]: "The Senate Energy and Natural Resources Committee heard warning today that a coal burning society may be making things hot for itself;" a reporter, Nelson Benton (1924–88), then described how the greenhouse effect worked and the "possible disruptive changes in the Earth's climate 50 to 70 years from now … scientists and a few politicians are beginning to worry that global energy planning does not take the greenhouse effect seriously enough."

While climate change was being discussed much more in the White House and the Capitol, there was still no political will to act, and Americans remained largely apathetic about the issue. Perhaps, if re-elected, Jimmy Carter would have moved forward with climate legislation at last, but that one slim hope was quickly dashed when Ronald Reagan won.

President Ronald Reagan: 1981–9

On November 13, 1979, Governor Reagan announced his candidacy and laid out his plans to solve the energy crisis[138]:

> The answer obvious to anyone except those in the administration, it seems, is more domestic production of oil and gas. We must also have wider use of nuclear power within strict safety rules, of course. There must be more spending by the energy industries on research and development of substitutes for fossil fuels.
>
> In years to come solar energy may provide much of the answer but for the next two or three decades we must do such things as master the chemistry of coal.

Then, in a campaign address, he supported fossil fuels and renewables[139]: "We should streamline and reform regulations which block the use of coal and safe nuclear power, and we should encourage the development of alternative fuels, such as solar and biomass." However, his interest in renewables was short lived.

Ronald Reagan mocked what he called "solar socialism" and was reported to have said "go away with this shit of renewables."[140] His reasons for rejecting solar energy may be related to it being a pet project of the previous administration, but his principal objection to it was his general disdain for big government.[141] Ronald Reagan believed that corporations, in this case, the energy-industrial complex, could and should manage itself without government regulation.

Early in his presidency, two major studies were issued on climate change, one from the National Academy of Sciences (NAS) and the other from the EPA. Both reports had struck a very different tone from those previously seen in Washington. The Office of Science and Technology Policy of the Executive Office of the President had requested the NAS for a study on the growing carbon dioxide levels in the atmosphere. And near the end of the summary, the conclusion was made that the scientific evidence would not[142]: "support steps to change current fuel-use patterns away from fossil fuels." The chair of the committee, William Nierenberg (1919–2000) – soon to be a founder of the right-wing think tank, the George C. Marshall Institute – wrote in the introduction[143]:

> There is a broad class of problems that have no "solution" in the sense of an agreed course of action that would be expected to make the problem go away. These problems can also be so important that they should not be avoided or ignored until the fog lifts … The payoff is that we will have had the chance to consider alternative courses of action with some degree of calm before we may be forced to choose among them in urgency or have them forced on us when other – perhaps better – options have been lost. Increasing atmospheric CO_2 and its climatic consequences constitute such a problem …
>
> Our stance is conservative: we believe there is reason for caution, not panic.

Comparable comments appeared in the EPA report – "Can We Delay a Greenhouse Warming:"[144]

> Responses to the threat of a greenhouse warming are polarized. Many have dismissed it as too speculative or too distant to be of concern. Some assume that technological options will emerge to prevent a warming or, at worst, to ameliorate harmful consequences. Others argue that only an immediate and radical change in the rate of CO_2 emissions can avert a worldwide catastrophe. The risks are high in pursuing a "wait and see" attitude on one hand, or in acting impulsively on the other.

The EPA's advice against "acting impulsively" was similar to the "not panic" comment of the NAS study. But the EPA did acknowledge that there was a risk[145]:

> Although much uncertainty remains concerning the magnitude, timing, and possible effects of rising levels of CO_2, this issue is considered to be one of the most important facing the scientific community, and one that raises significant questions for policymakers … [many pages]
>
> Though a ban on coal … is the only policy that could significantly delay a 2°C warming, it appears to be economically and politically infeasible … [page]

> A 2°C increase in temperature by (or perhaps well before) the middle of the next century leaves us only a few decades to plan for and cope with a change in habitability in many geographic regions. Changes by the end of the 21st century could be catastrophic taken in the context of today's world. A soberness and sense of urgency should underlie our response to a greenhouse warming.

The summaries of both the NAS and the EPA on the serious consequences of climate change are subdued compared to their earlier reports, and any mention of dangerous events was tempered by frequent reminders of the uncertainty of the science. A contrarian influence seemed present in both studies, as neither one pushed for action against fossil fuels, and both just promoted more research. President Reagan's response to the studies, if any, is not known.

In 1986, Senator Al Gore (1948–) (Democrat from Tennessee) sent a letter to President Reagan with the following plea[146]:

> One of the most serious long-term environmental problems facing the United States and the world is the greenhouse effect. This global warming, which is the result of the accumulation of carbon dioxide and trace gases in the atmosphere, could have disastrous consequences ranging from alteration of weather patterns to a rise in sea levels …

Our children and grandchildren will benefit from our actions – or suffer from our procrastination.

The letter did not stir the president.

Ronald Reagan had remained silent about climate change until his last year in office. In early 1988, he had been compelled to report on the government's research on the greenhouse effect to Congress, and he was brief[147]:

> In accordance with Section 9 of Public Law 99 – 383 (100 Stat. 816), I transmit herewith a report on current government activities in the area of research on the so-called "Greenhouse Effect."
>
> While you will note that extensive investigations of the phenomenon are in progress,[148] we do not plan to establish an International Year of the Greenhouse Effect as suggested in the language of Public Law 99 – 383.

However, by mid-1988, the president began talking a lot about climate change. At the Moscow Summit between Ronald Regan and Mikhail Gorbachev (1941–) on June 1, 1988, the joint statement, surprisingly, included a passing comment about global warming[149]:

> a possible global warming trend. They emphasized their desire to make more active use of the unique opportunities afforded by the space programs of the two countries to conduct global monitoring of the environment and the ecology of the Earth's land, oceans and atmosphere.

Then, on August 20th, President Reagan, uncharacteristically, went out of his way to describe a particular benefit of nuclear energy[150]: "To replace this energy [100 GW of nuclear energy] with electricity produced by oil would require two million barrels of oil per day, pump 350 million tons of carbon dioxide into the atmosphere each year." Ronald Reagan was obliquely acknowledging the problem of climate change from the burning of fossil fuels in his promotion of nuclear energy.

What had nudged President Reagan to talk about global warming? One can speculate from the timing that it was because it was an election year, and he wanted to help the campaign of George H. W. Bush, his VP. Climate change was suddenly on the mind of the electorate, especially after June 23rd, when NASA scientist Jim Hansen announced at a Senate hearing that he had measured global warming,[151] which was broadcasted widely by the press. So, we may surmise, Ronald Regan was willing to give the impression that he supported the trendy voting issue to help keep Republicans in power.

Later in August, George H. W. Bush made his famous "White House effect" statement[152]:

> Those who think we are powerless to do anything about the 'greenhouse effect' are forgetting about the 'White House effect.' In my first year in office, I will convene a global conference on the environment at the White House. It will include the Soviets, the Chinese …The agenda will be clear. We will talk about global warming.

His climate change message was reinforced by his running mate, Dan Quayle (1947–), during the vice-presidential debate on October 5th[153]:

> Now the greenhouse effect is an important environmental issue. It is important for us to get the data in, to see what alternatives we might have to the fossil fuels, and make sure we know what we're doing.
>
> And there are some explorations and things that we can consider in this area. The drought highlighted the problem that we have, and therefore, we need to get on with it, and in a George Bush Administration, you can bet that we will.

Even after the successful election campaign of George H. W. Bush, President Reagan continued to recognize the threat of climate change in his final budget address to Congress[154]:

> Because changes in the earth's natural systems can have tremendous economic and social effects, global climate change is becoming a critical concern. Our ability to understand and predict these changes is currently limited, and a better understanding is essential for developing policies.

Reagan's call for more research became the standard climate "action plan" for all future Republican presidents.

President Reagan was generally indifferent about the climate change issue, but future Republican presidents would be openly hostile to it. The strategy of the Republican Party to deal with this pesky issue was climate denialism. As seen in the next chapter, no democracy has been more corrupted by science denialism, in this case climate denialism, than the United States of America.

The IPCC – The Greatest Scrutiny of Science in History

An historic moment in the science of climate change was the formation of the Intergovernmental Panel on Climate Change (IPCC) which can be partially attributed to the Reagan administration. Mostafa Tolba (1922–2016), the Executive Director of UNEP (United Nations Environment Programme), directly contacted George Shultz (1920–2021), the Secretary of State during the Reagan administration, to push for policies to combat climate change. In response, the U.S. recommended an expert assessment of the science on climate change as a strategy to delay taking any material action.[155] The suggestion was adopted, and the task of organizing the global review fell to a new organization – the IPCC, officially formed by UNEP and the World Meteorological Organization (WMO), which held its initial meeting in November 1988. The first report of the IPCC, issued in 1990, concluded that industrial emissions were causing GHG levels to rise in the atmosphere and were contributing to climate change.

A popular misconception is that the IPCC is a scientific organization; not really, as the official members are political appointees, often bureaucrats. However, the reports are expert assessments, as scientists do the essential work of preparing the reviews on climate change for politicians. While the writers are experts in their scientific fields, the final editors of the reports don't have to be. In the critical summary document, called the "Synthesis Report," every sentence from the scientists is carefully dissected by political appointees and the final language is carefully worded. These meetings turn into wordsmith marathons that are arduous for the participants. Because of such scrutiny, any final documented statements tend to be more moderate and watered down than those coming from the scientists. The formation of the IPCC was unprecedented, as it is the largest international review of science in history, which is continuing to this day, more than 30 years since its formation.

What had started as a political ploy by the Reagan administration to mollify the U.N. and to delay action on climate change backfired and turned into the worst nightmare for Republicans. Evidence mounted in every series of IPCC reports that the burning of fossil fuels was warming the planet. Like the monster created by Frankenstein that turned against its creator, Reagan's monster had turned against the Republicans which ultimately forced them to discredit the IPCC and its scientific contributors by accusing them of being corrupt, alarmist, or globalist. But the IPCC shrugged off such attacks and chugged along, summarizing the growing scientific evidence in the peer-reviewed literature that climate change was a threat to humanity. The first U.S. president to have to deal with the reports of the IPCC was George H. W. Bush, and his response was to embrace climate denial, as discussed in the next chapter.

Congress

While the White House during Reagan's administration was relatively mute about climate change, Congress was not. With the national and global movement on climate change growing, congressional hearings became livelier and more heated. The scientific warnings from expert witnesses had been joined by a chorus of Democratic political leaders, especially Al Gore, who pushed the climate change agenda forward, despite indifference from the White House.

Al Gore and James Scheuer (1920–2005) (Democrat from New York) co-chaired three important hearings on climate change in the House of Representatives during the first half of the 1980s. At the first hearing on July 31, 1981, Mr. Gore described his initial reluctance to accept the science[156]:

> Quite frankly, my first reaction to it [global warming] several years ago was one of disbelief. Since then I have been waiting patiently for it to go away, but it has not gone away. The evidence continues to indicate that this Nation may be on the way to a natural disaster of unprecedented proportions.

At the second hearing on March 25, 1982, he commented on the growing scientific consensus[157]:

> The debate over the greenhouse effect has undergone a very important transition over the past decade. Ten years ago, it was viewed as a pet theory of a few scientists, perhaps on the fringes of science, but it has slowly and inexorably moved into the mainstream of scientific thought. There is now a broad consensus in the scientific community that the greenhouse effect is a reality.

And at the last of the three hearings on February 28, 1984, Mr. Gore stated[158]:

> Indeed, some still find it difficult to discuss this issue [the greenhouse effect] in a totally rational way, in view of its almost unthinkable potential. Can we really imagine a New York City with the climate of Palm Beach, a Kansas resembling Central Mexico, or 40 percent of Florida under water? It's difficult.

A witness at this last hearing was the popular astronomer Carl Sagan (1934–96), who testified[159]: "Nations are concerned about themselves, not about the planet, and this kind of problem will keep coming at us as we discover more and more such potential catastrophic consequences."

Later, Al Gore appeared as a witness at a Senate hearing, where he laid out the challenges in dealing with the climate crisis[160]:

> As has been indicated already, this is a difficult issue to approach as a policymaker. We in the House and Senate, respond to shorter-term deadlines, generally speaking, such as the next budget cycle, and an issue that forces us

> to look into the next century and contemplate consequences of a magnitude far beyond those we are accustomed to dealing with is very difficult to approach …
>
> For those unfamiliar with the greenhouse effect, it may sound more like a plot for a bad science fiction novel than a serious environmental issue, deserving public policy review of the highest order. Given its serious and potentially drastic impacts, Federal research and study efforts must place the greatest priorities on solving the mystery surrounding the greenhouse effect. Otherwise, future generations may experience a science fiction coming true.

This "science fiction" has since come true and, just like in the movies, the government is still ignoring the scientists.

Next up was Carl Sagan who was back again testifying[161]:

> Because the effects occupy more than a human generation, there is a tendency to say that they are not our problem. Of course, then they are nobody's problem. Not on my tour of duty, not in my term of office. It's something for the next century. Let the next century worry about it.
>
> But the problem is that there are effects, and the greenhouse effect is one of them, which have long-term consequences. If you don't worry about it now, it's too late later on. So in this issue, as in so many other issues, we are passing on extremely grave problems for our children when the time to solve the problems, if they can be solved at all, is now ….[a few pages]
>
> The solution to these problems requires a perspective that embraces the planet and the future, because we are all in this greenhouse together.

Dr. Sagan eloquently nailed the problem, but not enough politicians listened.

A significant turning point in the politics of climate change took place in a Senate hearing near the end of Reagan's term called "Greenhouse Effect and Global Climate Change," when Jim Hansen (1941–), Director, NASA Goddard Institute for Space Studies, proclaimed that he had measured global warming. The first session took place on November 9 and 10, 1987, and according to Dr. Hansen[162]: "The scientific evidence for the greenhouse effect is overwhelming. The greenhouse effect is real, it is coming soon, and it will have major effects on all peoples."

Then, on June 23, 1988, the second session opened with a passionate statement by J. Bennett Johnston (1932–) (Democrat from Louisiana)[163]:

> We have only one planet. If we screw it up, we have no place else to go. The possibility, indeed, the fact of our mistreating this planet by burning too much fossil fuels and putting too much CO2 in the atmosphere and thereby causing this greenhouse effect is now a major concern of Members of Congress and of people everywhere in this country.

A similar message was given by Timothy Wirth (1939–) (Democrat from Colorado)[164]:

> The scientific community has done an outstanding job of compiling and analyzing mountains of evidence about global climate change. As I read it, the scientific evidence is compelling. The global climate is changing as the earth's atmosphere gets warmer. Now the Congress must begin to consider how we are going to slow or halt the warming trend, and how we're going to cope with the changes that may already be inevitable.
>
> In essence, this is an issue that has moved from the world of science to the policy arena in the United States and throughout the world.

History was made at this hearing[165] when Jim Hansen summarized his own research[166]:

> I would like to draw three main conclusions. Number one, the earth is warmer in 1988 than at any time in the history of instrumental measurements. Number two, the global warming is now large enough that we can ascribe with a high degree of confidence a cause and effect relationship to the greenhouse effect. And number three, our computer simulations indicate that the greenhouse effect is already large enough to begin to effect the probability of extreme events such as summer heat waves…

In my opinion, that the greenhouse effect has been detected, and it is changing our climate now.
Hansen's speech sent shock waves around the world, and the greenhouse effect became headline news.

Overall, the Reagan-era hearings – notably, the sessions involving Mr. Gore, Dr. Sagan, and Dr. Hansen – demonstrated the pushback against Reagan's apathy on climate change. In September 1988, alone, there had been four hearings and several bills introduced on climate change.[167] Democrats, finally, were getting on board. But a paradigm shift in the politics of climate change took place in the late 1980s when the energy-industrial complex launched a massive propaganda campaign against the science of climate change, unleashing a tsunami of disinformation and a parade of contrarian scientists at congressional hearings. As discussed in the next chapter, climate denialism spread quickly through the ranks of the Republican Party.

Notes

1 Jefferson 1954, p. 80.
2 Jefferson 1824.
3 Abbe 1889, pp. 687–668.
4 Abbe 1907.
5 Einstein 1939.
6 Horgan 2015.

7 Early studies were by Wallace Broecker (Broecker, Schulert and Olson 1959; Broecker and Walton 1959).
8 Einstein 1945.
9 Born et al. 1955; the letter was known as the Russel-Einstein Manifesto. Einstein died the following month.
10 Bulletin of the Atomic Scientists 2020.
11 United States Department of Agriculture 1941, p. 94.
12 United States Department of Agriculture 1941, p. 9; also see pp. 7–8, 67.
13 Callendar 1938, p. 223; for a review of the work of Callendar see Fleming 1998, pp. 113–118; Dee 2022.
14 Callendar 1938, p. 236.
15 See Fleming 1998, pp. 115–116.
16 Kincer 1937, pp. 29, 31; also see p. 34.
17 President Truman had been interested in the weather (Harry S. Truman Library 1961–3).
18 United States Senate 1951, p. 40.
19 See Novak 2011.
20 United States Senate 1951, p. 43.
21 For details on Project Cirrus, see United States Senate 1951, pp. 40–44, 51, 56–69, 187–189; United States Senate 1957, various pages; United States House of Representatives 1976b, p. 161; United States Senate 1978, pp. 37–41. The Defense Department also sponsored the Artificial Cloud Nucleation Project from 1951 to 1953; see, for example, United States Senate 1957, p. 124.
22 United States House of Representatives 1956, p. 473.
23 United States House of Representatives 1956, p. 472.
24 The science historian Spencer Weart provided a summary of Dr. Revelle's research during this period (Weart 2022, "Roger Revelle's Discovery"); the science historian Fleming, though, is less generous as to the influence of Revelle (Fleming 1998, p. 123).
25 Wallace and Smollar 1991.
26 Old 1961, p. 35; this article details the story behind the creation of the ONR.
27 United States Senate 1951, p. 40.
28 Plass 1956; Plass had reported preliminary results back in 1953.
29 Revelle and Suess 1957.
30 Keeling 1960.
31 United States House of Representatives 1957, p. 6; also see United States House of Representatives 1956, p. 466.
32 Weart 2022, "Government: The View from Washington, DC."
33 "Climatic Change" was briefly mentioned at a House of Representatives session in 1909 (United States House of Representatives 1909, p. 67) and at a Senate hearing in 1935 (United States Senate 1935, p. 21). These hearings are early examples where "climatic change" had been raised in Congress, but the term was not in reference to the greenhouse effect. "Climatic change" had been used by John Tyndall to describe changes in the climate by greenhouse gases.
34 United States Senate 1951, p. 110.
35 United States Senate 1951, p. 3.
36 United States Senate 1951, p. 43.
37 United States Senate 1978, p. 149.
38 Case 1951, p. 12738, 1952, pp. 7777–7778.
39 United States Senate 1957, pp. 28–59; United States Senate 1978, pp. 195–196; Novak 2011, and references therein. Their first report to President Eisenhower was on February 8, 1956 (United States Senate 1957, p. 28), and the final report of the committee was sent to President Eisenhower on December 31, 1957. The members of the committee are named in United States Senate 1957, pp. 34–35. The weather-control program led by Orville ended on June 30, 1958. For an early article, see Orville 1954.

40 Orville 1958, p. 9479; also see Novak 2011. Programs on weather control continued, usually under the Department of Defense (see, for example, Gerald R. Ford Library 1976; United States Senate 1978).
41 See United States Senate 1958, pp. 280–285.
42 See United States Senate 1958, p. 282.
43 Plass 1956 and references therein.
44 Anon. 1953a; also see Anon. 1953b. His work and the impact of the burning of fossil fuels on the climate were discussed by Putnam 1953, pp. 170, 454–459. For a review of Plass see Fleming 1998, pp. 121–122.
45 Plass 1956. The New York Times also had an article on the paper.
46 For a review of Revelle see Fleming 1998, pp. 122–128; Weart 2022, "Roger Revelle's Discovery."
47 Anon. 1956.
48 Anon. 1958.
49 See Matthews 1959; Nevins 1960.
50 Eisenhower 1961, p. 18.
51 For information on the formation of the IGY see Korsmo 2007. Note that all the hearings that Dr. Revelle attended during the '50s were connected to the IGY.
52 United States House of Representatives 1956, pp. 472–474.
53 United States House of Representatives 1957, pp. 106–108; for other testimony by Dr. Revelle, see United States House of Representatives 1958, pp. 71, 75 (here, he specifically mentioned the "greenhouse effect"); United States House of Representatives 1959, pp. 73–74, 92.
54 John F. Kennedy Library 1958; once president, he encouraged international cooperation on "weather prediction" (John F. Kennedy Library 1961a,b).
55 See Hulac 2018 and references therein.
56 Von Neumann 1955, pp. 666, 669, 673. A related article – "How to Wreck the Environment" – appeared in 1968, where Gordon MacDonald discussed "geophysical warfare," including climate modification (MacDonald 1968); at the time, MacDonald was a member of the Science Advisory Committee for President Johnson.
57 United States Senate 1964, p. 15. The earlier actions of Congress on weather modification were summarized in this session (pp. 2–5).
58 United States Senate 1963, pp. 2, 22.
59 United States Senate 1961, p. 89.
60 United States Senate 1958, p. 283; from a speech on January 21, 1958.
61 WMH 2017; this is an excellent source of information on the history of weather modification; also see United States Senate 1978; Fleming 2010.
62 Johnson 1965a.
63 Johnson 1965b.
64 White House 1965, p. iii. In response to the call by the president, the Department of Commerce reported that it had a program on "studying the CO2 content of the atmosphere … and they are measuring stratospheric temperature (Office of Science and Technology 1967, p. 6). Members of committee included Roger Revelle, Gordon MacDonald, and Frederick Seitz who later became a noted contrarian.
65 White House 1965, pp. 126–127; see Appendix Y4, pp. 112–127. Note that during the Kennedy administration, Roger Revelle had been the Science Advisor to Secretary of the Interior, Stewart Udall (1920–2010).
66 Ikard 1965.
67 United States Senate 1966a, p. 148; for the comments of others see pp. 172–173.
68 United States House of Representatives 1966, p. 289; also see United States Senate 1966a, p. 135, 1966b, p. 99.
69 United States Senate 1966b, p. 233.
70 United States Senate 1967, p. 813.

71 United States House of Representatives 1968, p. 2; note the reference to "climate change," an early use of the term in a government document. Mr. Daddario had also been the chair the House hearing with Mr. Malone, two years earlier.
72 United States House of Representatives 1968, p. 2.
73 Oreskes 2019, p. 2. Mr. Train became the second administrator of the EPA.
74 White House 1970, pp. v–vi, xiv, xv; contributors to the report were Russell Train, Robert Cahn, and Gordon MacDonald.
75 White House 1970, p. 97. The document predicted that the CO2 levels would reach 400 ppm in about 2010 (p. 96); this threshold was reached in 2016. The following section was "Particle Pollution – An Earth Cooler?"
76 Moynihan 1969.
77 United States Senate 1972, p. 823.
78 Leitenberg 1984, "Case Study 2 – Weather Modification," pp. 12–15; for more on Nile Blue see United States Senate 1972, pp. 823–824; United States House of Representatives 1976b, pp. 429–431, 441–443; Weinberger 2018. The Nile Blue program was later renamed "Climate Dynamics."
79 United States House of Representatives 1976b, p. 136; also see pp. 132, 168; Reeves 2015.
80 See Reeves 2015; also see, Kukla and Matthews 1972. In 1982, Dr. Kukla later gave testimony in Congress, where he supported the theory of global warming (United States House of Representatives 1982, pp. 69–70, 88), but he later was again promoting global cooling (see DeSmog, "George Kukla").
81 Federal Council for Science and Technology 1974, p. 14.
82 Broecker 1975, p. 460.
83 CIA 1974a, p. 28; also see CIA 1974b.
84 CIA 1974a. Bryson is not directly identified in the CIA studies as their author, but he was connected to it at a later Congressional hearing (United States House of Representatives 1976a, p. 211).
85 For comments by Stephen Schneider on the CIA study see Schneider 2009, pp. 73–74.
86 See DeSmog, "George Kukla; DeSmog, "Reid Bryson."
87 Poe 2009a,b.
88 United States House of Representatives 2019, p. 12; also see p. 15. For a more recent mention see United States of Representatives 2022, p. 57.
89 Peterson, Connolley, and Fleck 2008, p. 1326.
90 United States Senate 1973, p. 120.
91 United States Senate 1973, p. 123.
92 United States Senate 1973, p. 125.
93 United States House of Representatives 1969, pp. 28, 90.
94 Congress of the United States 1969, p. 323.
95 Congress of the United States 1969, pp. 26–27; for more on DuBridge, see Oreskes 2019, pp. 2–3.
96 United States Committee for the Global Atmospheric Research Program 1975, p. 43. George Kukla and Richard Lindzen (see below) contributed to the study.
97 Udall 1974.
98 United States House of Representatives 1976a, pp. 1–2.
99 United States House of Representatives 1976a, p. 62.
100 United States House of Representatives 1976a, p. 31; others giving testimony included: William Nierenberg, and Reid Bryson; for a related hearing see United States Senate 1977.
101 United States House of Representatives 1976a, p. 32.
102 For an insider look at the later years of the Carter administration see Rich 2018, Part One, 1979–1982.
103 Speth 2021, p. 11; he reviews the Carter administration in Chapter 2 of his book.

104 Jimmy Carter Library 1976; the primer also included Dr. Schneider's testimony to Congress, discussed above (United States House of Representatives 1976a, p. 37).
105 National Academy of Sciences 1977, p. 11. In early 1977, Revelle, Schneider, Woodwell, Broecker, among others, attended a workshop associated with the future Department of Energy (Elliott and Machta 1979; also see Dahlman et al. 1980).
106 Jimmy Carter Library 1977.
107 Pattee 2022.
108 Woodwell et al. 1979, pp. 7–8.
109 Woodwell et al. 1979, pp. 1–2. During the Carter administration, the CEQ also issued various reports to Congress (Speth 2021, pp. 20–22).
110 The Synthetic Fuels Corporation was established on June 30, 1980 (abolished by Ronald Reagan in April 1986) and had a budget of $22 billion per year from 1980 to 1984.
111 Jimmy Carter Library 1979b.
112 United States Senate 1979a, p. 66; for more on Gordon MacDonald, see Rich 2018.
113 National Academy of Sciences 1979, p. 1; for more on the "Charney Report," see Rich 2018.
114 National Academy of Sciences 1979, pp. vii–viii.
115 DeSmog, "Sherwood B. Idso."
116 Jimmy Carter Library 1980, April 18, p. 2.
117 Jimmy Carter Library 1980, April 18, pp. 9, 11.
118 Jimmy Carter Library 1980, May 5.
119 United States Senate 1980, p. 174.
120 Carter 1980.
121 Carter 1981.
122 Jimmy Carter Library 1979a, pp. 6–7.
123 Murse 2021.
124 Biello 2010.
125 For a more positive view on President Carter's stance see Alter 2020.
126 United States Senate 1978, p. 186. Another report from the Comptroller General had been sent to Congress, which briefly mentioned "global warming" (Staats 1977, pp. 6.19, 9.9).
127 United States House of Representatives 1977, p. 2.
128 United States Senate 1979a, p. 34.
129 United States Senate 1979a, pp. 52, 54.
130 United States Senate 1979a, p. 66.
131 United States Senate 1979b, p. 3.
132 United States Senate 1979b, p. 18.
133 United States Senate 1980, pp. 1–4.
134 United States Senate 1980, p. 160.
135 United States Senate 1980, p. 111.
136 United States Senate 1980, p. 197; Rich 2018.
137 Cronkite 1980.
138 Reagan 1979.
139 Reagan 1980a.
140 Davidson 2012.
141 Reagan 1980b.
142 National Academy of Sciences 1983, p. 4.
143 National Academy of Sciences 1983, p. xiii; also see p. 61; for further comments on the study see Oreskes, Conway, and Shindell 2008, p. 109; Rich 2018.
144 Seidel and Keyes 1983, p. i.
145 Seidel and Keyes 1983, pp. 1–1, 7–5, 7–7.
146 Gore 1986.
147 Reagan 1988a.

148 See Committee on Earth Sciences 1988; also see Committee on Earth Sciences 1989 (the reports detailed the research and budgets of various government agencies); Congress had requested the reports through the National Science Foundation Authorization Act for Fiscal Year 1987. Studies were also being conducted by the EPA (which would not be completed until the Bush administration); see Smith and Tirpak 1989; Lashof and Tirpak 1990.
149 Ronald Reagan Library 1988.
150 Reagan 1988b.
151 United States Senate 1988, p. 39; for an earlier session with Dr. Hansen, see United States Senate 1986 (this hearing was one of the first with "Climate Change" in its title).
152 See Hudson 2018; also see Bodansky in Mintzer and Leonard 1994, p. 49; Hecht and Tirpak 1995, p. 383; Rich 2018.
153 Commission on Presidential Debates 1988, pp. 89–90.
154 Reagan 1989; the announcement by the president referenced the report of the Committee on Earth Sciences (Committee on Earth Sciences 1988; also see Committee on Earth Sciences 1989).
155 Hecht and Tirpak 1995, pp. 380–381; some within the administration viewed the initiative as an excuse to stall the push for a Framework Convention, see p. 400. George Shultz later supported action on climate change (Shultz 2015).
156 United States House of Representatives 1981, p. 2; also see Schneider 2009, pp. 87–96.
157 United States House of Representatives 1982, p. 2; Jim Hansen was a witness, as was George Kukla.
158 United States House of Representatives 1984, p. 1.
159 United States House of Representatives 1984, p. 19; also see Schneider 2009, pp. 95–108.
160 United States Senate 1985, p. 4.
161 United States Senate 1985, pp. 7, 10.
162 United States Senate 1987, p. 53.
163 United States Senate 1988, p. 1; also see Rich 2018.
164 United States Senate1988, p. 5.
165 See, for example, Fleming 1998, p. 134; Bolin 2008, p. 49.
166 United States Senate 1988, pp. 39–40.
167 United States House of Representatives 1988, p. 1.

References

Abbe, Cleveland (1889). "Is Our Climate Changing?" *The Forum*, Vol. VI, p. 678. New York: The Forum Publishing Co.; https://books.google.ca/books?id=3aLPAAAAMAAJ&dq=%22is+our+climate+changing%22,+abbe&source=gbs_navlinks_s

Abbe, Cleveland (1907). "Is Not Honesty the Wisest Policy?" *Monthly Weather Review* 35 (1), p. 7; https://journals.ametsoc.org/view/journals/mwre/35/1/1520-0493-35_1_7b.xml

Alter, Jonathan (2020). "Climate Change was on the Ballot with Jimmy Carter in 1980—Though No One Knew It at the Time." *Time*, September 29; https://time.com/5894179/jimmy-carter-climate-change/

Anon. (1953a). "Science: Invisible Blanket." *Time*, May 25; http://content.time.com/time/magazine/article/0,9171,890597,00.html; also see Anon. 1953. "Industrial Gases Warming Up Earth, Physicist Notes Here." *Washington Post*, May 5; Anon. 1953. "How Industry May Change Climate." *New York Times*, May 24.

Anon. (1953b). "Growing Blanket of Carbon Dioxide Raises Earth's Temperature." *Popular Mechanics*,August,p.119;https://books.google.ca/books?id=8NsDAAAAMBAJ&pg=PA119&lpg=PA119&dq=%E2%80%9CGrowing+Blanket+of+Carbon+Dioxide+Raises+Earth%E2%80%99s+Temperature.%E2%80%9D+Popular+Mechanics&source=bl&ots=

wPwryUmv1Q&sig=ACfU3U1aXd2E9d8l9EtuqL_ukAXlFolPkg&hl=en&sa=X&ved=2ahUKEwiSycCOgqv4AhUthYkEHb_CChoQ6AF6BAgCEAM#v=onepage&q&f=false

Anon. (1956). "Science: One Big Greenhouse." *Time*, May 28; http://content.time.com/time/magazine/article/0,9171,937403,00.html

Anon. (1958). *The Unchained Goddess*. YouTube; https://www.youtube.com/watch?v=T6YyvdYPrhY

Biello, David (2010). "Where did the Carter White House's Solar Panels Go?" *Scientific American*, August 6; https://www.scientificamerican.com/article/carter-white-house-solar-panel-array/

Bolin, Bert (2008). *A History of the Science and Politics of Climate Change*. Cambridge: Cambridge University Press.

Born, Max, et al. (1955). *Statement: The Russell-Einstein Manifesto*. Pugwash Conferences on Science and World Affairs, July 9; https://pugwash.org/1955/07/09/statement-manifesto/

Broecker, Wallace S. (1975). "Climatic Change: Are We on the Brink of a Pronounced Global Warming?" *Science* 189 (4201), p. 460; https://blogs.ei.columbia.edu//files/2009/10/broeckerglobalwarming75.pdf

Broecker, Wallace S., Schulert, Arthur, and Olson, Edwin A. (1959). "Bomb Carbon-14 in Human Beings." *Science* 130 (3371), p. 331; http://science.sciencemag.org/content/130/3371/331.2

Broecker, Wallace S., and Walton, Alan (1959). "Radiocarbon from Nuclear Tests." *Science* 130 (3371), p. 309; http://science.sciencemag.org/content/130/3371/309

Bulletin of the Atomic Scientists (2020). *Closer Than Ever: It Is 100 Seconds to Midnight*. 2020 Doomsday Clock Announcement, January 23; https://thebulletin.org/doomsday-clock/2020-doomsday-clock-statement/

Callendar, Guy Stewart. (1938). "The Artificial Production of Carbon Dioxide and Its Influence on Temperature." *Quarterly Journal Royal Meteorological Society* 64 (275), p. 223, April; http://www.met.reading.ac.uk/~ed/callendar_1938.pdf

Carter, Jimmy (1980). *Jimmy Carter State of the Union Address 1980*. Jimmy Carter Presidential Library and Museum, January 23; https://www.jimmycarterlibrary.gov/assets/documents/speeches/su80jec.phtml

Carter, Jimmy (1981). *Jimmy Carter State of the Union Address 1981*. Jimmy Carter Presidential Library and Museum, January 16; https://www.jimmycarterlibrary.gov/assets/documents/speeches/su81jec.phtml

Case, Francis (1951). *Committee to Study and Evaluate Public and Private Experiments in Weather Modification*. Congressional Record – Senate, p. 12738, October 8; https://www.govinfo.gov/content/pkg/GPO-CRECB-1951-pt10/pdf/GPO-CRECB-1951-pt10-4-1.pdf

Case, Francis (1952). *Research and Experimentation in Weather Modification*. Congressional Record – Senate, p. 7776, June 21; https://www.govinfo.gov/content/pkg/GPO-CRECB-1952-pt6/pdf/GPO-CRECB-1952-pt6-9.pdf

CIA (1974a). *A Study of Climatological Research as it Pertains to Intelligence Problems*. CIA, August; https://babel.hathitrust.org/cgi/pt?id=mdp.39015022217494&view=1up&seq=3

CIA (1974b). *Potential Implications of Trends in World Population, Food Production, and Climate*. CIA, August; https://babel.hathitrust.org/cgi/pt?id=uiug.30112018865672&view=1up&seq=3

Commission on Presidential Debates (1988). *The 1988 Vice Presidential Debate*. Gerald R. Ford Presidential Library & Museum, Digital Collections; https://www.fordlibrarymuseum.gov/library/document/0375/1682723.pdf

Committee on Earth Sciences (1988). *Our Changing Planet, A U.S. Strategy for Global Change Research*. Office of Science and Technology Policy; https://www.nap.edu/catalog/18703/our-changing-planet-a-us-strategy-for-global-change-research

Committee on Earth Sciences (1989). *Our Changing Planet, the FY 1990 Research Plan*. Office of Science and Technology Policy, July; https://www.carboncyclescience.us/sites/default/files/documents/2013/ocp1990.pdf

Congress of the United States (1969). *Environmental Effects of Producing Electric Power*. Hearings before the Joint Committee on Atomic Energy, Congress of the United States, 91st Congress, First Session, October–November; https://babel.hathitrust.org/cgi/pt?id=uc1.$b642258&view=1up&seq=7

Cronkite, Walter (1980). *1980: Walter Cronkite on Climate Change*. YouTube, April 3; https://www.youtube.com/watch?v=MU9s0XyEctI&feature=youtu.be

Dahlman, Roger C., Gross, Thomas, Machta, Lester, and MacCracken, MIchael (1980). *Carbon Dioxide Effects Research and Assessment Program, Carbon Dioxide Research Progress Report, Fiscal Year 1979*. May, United States Department of Energy; https://books.google.ca/books?id=Jzz0AAAAMAAJ&dq=%22Carbon+dioxide+effects+research+and+Assessment+Program:+workshop%22&source=gbs_navlinks_s

Davidson, Osha Gray (2012). "Can the U.S. Create Its Own Energy Revolution." *Inside Climate News*, November 20; http://insideclimatenews.org/news/20121120/germany-energiewende-clean-energy-economy-solar-wind-electric-grid-utilities?page=show

Dee, Sylvia G. (2022). *A Mild-mannered Biker Triggered a Huge Debate over Humans' Role in Climate Change – in the Early 20th Century*. The Conversation, February 21; https://theconversation.com/a-mild-mannered-biker-triggered-a-huge-debate-over-humans-role-in-climate-change-in-the-early-20th-century-170954

DeSmog (undated). "George Kukla"; https://www.desmogblog.com/george-kukla

DeSmog (undated). "Reid Bryson"; https://www.desmogblog.com/reid-bryson

DeSmog (undated). "Sherwood B. Idso"; https://www.desmog.com/sherwood-b-idso/

Einstein, Albert (1939). *Einstein Letter to FDR Regarding Atomic Bomb*. Franklin D. Roosevelt Presidential Library and Museum, August 2; https://www.fdrlibrary.org/documents/356632/390886/document007.pdf/3483329d-7b68-442d-953d-eb91e0c5c9b1

Einstein, Albert (1945). *Albert Einstein to Franklin D. Roosevelt, March 25, 1945*. Harry S. Truman Library & Museum, March 25; https://www.trumanlibrary.gov/library/research-files/albert-einstein-franklin-d-roosevelt

Eisenhower, Dwight D. (1961). *The Farewell Address*. Dwight D. Eisenhower Presidential Library, Museum & Boyhood Home, January 17; https://www.eisenhowerlibrary.gov/sites/default/files/research/online-documents/farewell-address/reading-copy.pdf

Elliott, William P., and Machta, Lester (eds.) (1979). *Carbon Dioxide Effects: Research and Assessment Program, Workshop on the Global Effects of Carbon Dioxide from Fossil Fuels, Miami Beach, Florida, March 7-11*. May, United States Department of Energy; https://www.osti.gov/servlets/purl/6385084

Federal Council for Science and Technology (1974). *Report of the Ad Hoc Panel on the Present Interglacial*. Interdepartmental Committee for Atmospheric Sciences, August; https://babel.hathitrust.org/cgi/pt?id=uc1.31822000471953&view=1up&seq=9&skin=2021&size=125

Fleming, James R. (1998). *Historic Perspectives on Climate Change*. New York: Oxford University Press.

Fleming, James R. (2010). *Fixing the Sky: The Checkered History of Weather and Climate Control*. New York: Columbia University Press; https://books.google.ca/books?id=yRhHAAAAQBAJ&source=gbs_navlinks_s

Gerald R. Ford Library (1976). *Box 63, Folder "10/13/76 S3383 National Weather Modification Policy Act of 1976" of the White House Records Office*. Gerald R. Ford Presidential Library and Museum; https://www.fordlibrarymuseum.gov/library/document/0055/1669610.pdf

Gore, Al (1986). Letter to *Honorable Ronald W. Reagan*. May 21, 1986; https://static1.squarespace.com/static/571d109b04426270152febe0/t/60ff55269076e220ee40e5a0/1627346215563/Gore.Letter+to+Reagan.1986.pdf

Harry S. Truman Library (1961–3). "Motion Picture MP2002-432." Harry S. Truman Library & Museum; https://www.trumanlibrary.gov/movingimage-records/mp2002-432-screen-gems-collection-outtakes-television-series-decision-conflicts

Hecht, Alan D., and Tirpak, Dennis (1995). "Framework Agreement on Climate Change: A Scientific and Policy History." *Climatic Change* 29, p. 371; https://link.springer.com/article/10.1007/BF01092424

Horgan, John (2015). "Bethe, Teller, Trinity and the End of Earth." *Scientific American*, August 4; https://blogs.scientificamerican.com/cross-check/bethe-teller-trinity-and-the-end-of-earth/

Hudson, Marc (2018). *George Bush Sr Could Have Got in on the Ground Floor of Climate Action – History Would Have Thanked Him*. The Conversation, December 5; https://theconversation.com/george-bush-sr-could-have-got-in-on-the-ground-floor-of-climate-action-history-would-have-thanked-him-108050

Hulac, Benjamin (2018). "Every President since JFK Was Warned about Climate Change." *E&E News*, November 6; https://www.eenews.net/stories/1060105233

Ikard, Frank N. (1965). "Meeting the Challenges of 1966." *American Petroleum Institute, Proceedings* 45 (1), p. 12; http://www.climatefiles.com/trade-group/american-petroleum-institute/1965-api-president-meeting-the-challenges-of-1966/

Jefferson, Thomas (1824). *From Thomas Jefferson to Lewis E. Beck, 16 July 1824*. National Archives, Founders Online; https://founders.archives.gov/documents/Jefferson/98-01-02-4410

Jefferson, Thomas (1954). *Notes on the State of Virginia*. Chapel Hill: University of North Carolina Press; https://books.google.ca/books?id=2bvqCQAAQBAJ&vq=climate&source=gbs_navlinks_s

Jimmy Carter Library (1976). *Science and Technology Task Force Papers. Collection: Records of the 1976 Campaign Committee to Elect Jimmy Carter; Series: Noel Sterrett Subject File; Folder: Science and Technology Task Force Papers [2]; Container 93*. Jimmy Carter Presidential Library and Museum; https://www.jimmycarterlibrary.gov/digital_library/campaign/564806/93/76C_564806_93_02.pdf

Jimmy Carter Library (1977). *Collection: Office of Staff Secretary; Series: Presidential Files; Folder: 7/18/77 [1]; Container 31*. Jimmy Carter Presidential Library and Museum; https://www.jimmycarterlibrary.gov/digital_library/sso/148878/31/SSO_148878_031_07.pdf

Jimmy Carter Library (1979a). *Collection: Office of Staff Secretary; Series: Presidential Files; Folder: 6/20/79; Container 121*. Jimmy Carter Presidential Library and Museum; https://www.jimmycarterlibrary.gov/digital_library/sso/148878/121/SSO_148878_121_12.pdf

Jimmy Carter Library (1979b). *Collection: Office of Staff Secretary; Series: Presidential Files; Folder: 9/12/79 [2]; Container 130, Energy Security Corporation*. Jimmy Carter Presidential Library and Museum; https://www.jimmycarterlibrary.gov/digital_library/sso/148878/130/SSO_148878_130_04.pdf

Jimmy Carter Library (1980). *Collection: Office of Staff Secretary; Series: Presidential Files; Folder: 5/5/80 [1]; Container 160*. Jimmy Carter Presidential Library and Museum; https://www.jimmycarterlibrary.gov/digital_library/sso/148878/160/SSO_148878_160_11.pdf

John F. Kennedy Library (1958). *Remarks of Senator John F. Kennedy at the John Carroll Society Dinner, Washington, D.C., February 13, 1958*. John F. Kennedy Presidential Library and Museum, February 13; https://jfklibrary.org/Research/Research-Aids/JFK-Speeches/John-Carroll-Society-Washington-DC_19580213.aspx

John F. Kennedy Library (1961a). *State of the Union Address*. John F. Kennedy Presidential Library and Museum, 30 January; https://www.jfklibrary.org/Asset-Viewer/Archives/JFKWHA-006.aspx

John F. Kennedy Library (1961b). *Address at United Nations General Assembly*. John F. Kennedy Presidential Library and Museum, September 25; https://www.jfklibrary.org/Asset-Viewer/DOPIN64xJUGRKgdHJ9NfgQ.aspx

Johnson, Lyndon (1965a). *Special Message to the Congress on Conservation and Restoration of Natural Beauty*. The American Presidency Project, February 8; https://www.presidency.ucsb.edu/documents/special-message-the-congress-conservation-and-restoration-natural-beauty

Johnson, Lyndon (1965b). *Message to the Congress Transmitting Sixth Annual Report on Weather Modification*. The American Presidency Project, May 24; https://www.presidency.ucsb.edu/documents/message-the-congress-transmitting-sixth-annual-report-weather-modification

Keeling, Charles D. (1960). "The Concentration and Isotopic Abundances of Carbon Dioxide in the Atmosphere." *Tellus* 12 (2), p. 200; https://scrippsco2.ucsd.edu/assets/publications/keeling_tellus_1960.pdf

Kincer, J. B. (1937). *Speaking of the Weather: Is Our Climate Changing*. Springfield: Illinois Farmers' Institute; https://babel.hathitrust.org/cgi/pt?id=mdp.39015066907596&view=1up&seq=3

Korsmo, Fae L. (2007). "The Genesis of the International Geophysical Year." *Physics Today* 60 (7), p. 38; https://physicstoday.scitation.org/doi/10.1063/1.2761801

Kukla, G. J., and Matthews, R. K. (1972). "When Will the Present Interglacial End?" *Science* 178 (4057), p. 190; https://www.science.org/doi/10.1126/science.178.4057.190

Lashof Daniel, A., and Tirpak, Dennis A. (ed.) (1990). *Policy Options for Stabilizing Global Climate, Report to Congress, Main Report*. United States Environmental Protection Agency, Office of Policy, Planning and Evaluation. December; https://nepis.epa.gov/Exe/ZyPDF.cgi/91014BJ0.PDF?Dockey=91014BJ0.PDF

Leitenberg, Milton (1984). *Studies of Military R&D and Weapons Development, Case Study 2 – Weather Modification: The Weather of an R&D Program into a Military Operation*. Federation of American Scientists; https://fas.org/man/eprint/leitenberg/

MacDonald, Gordon J. F. (1968). "How to Wreck the Environment." In: Nigel Calder, ed., *Unless Peace Comes*, New York: Viking Press; https://coto2.files.wordpress.com/2013/11/1968-macdonald-how-to-wreck-the-planet.pdf

Matthews, M. A. (1959). "The Earth's Carbon Cycle." *New Scientist* 6 (151), p. 644, October 8; https://books.google.ca/books?id=qYkP2bThsFEC&dq=%22The+Earth%27s+Carbon+Cycle%22,+teller&source=gbs_navlinks_s

Mintzer, Irving M., and Leonard, J. Amber (ed.) (1994). *Negotiating Climate Change*. Cambridge: Cambridge University Press.

Moynihan, Daniel P. (1969). *Memorandum, For John Ehrlichman*. Richard M. Nixon Presidential Library and Museum, September 17; https://www.nixonlibrary.gov/sites/default/files/virtuallibrary/documents/jul10/56.pdf

Murse, Tom (2021). *A Brief History of White House Solar Panels*. ThoughtCo, August 3; http://usgovinfo.about.com/od/thepresidentandcabinet/tp/History-of-White-House-Solar-Panels.htm

National Academy of Sciences (1977). *Energy and Climate: Studies in Geophysics*. National Research Council, Washington, DC: National Academy of Sciences; https://www.nap.edu/download/12024

National Academy of Sciences (1979). *Carbon Dioxide and Climate: A Scientific Assessment*. National Research Council, Wood Hole, MA, July 23–27; https://nap.nationalacademies.org/catalog/12181/carbon-dioxide-and-climate-a-scientific-assessment

National Academy of Sciences (1983). *Changing Climate*. National Research Council. Washington, DC: National Academy Press; https://www.nap.edu/catalog/18714/changing-climate-report-of-the-carbon-dioxide-assessment-committee

Nevins, Allan (1960). *Energy and Man: A Symposium*. New York: Appleton-Century-Crofts; https://books.google.ca/books?id=gl2LDwAAQBAJ&source=gbs_navlinks_s

Novak, Matt (2011). "Weather Control as a Cold War Weapon." *Smithsonian Magazine*, December 5; https://www.smithsonianmag.com/history/weather-control-as-a-cold-war-weapon-1777409/

Office of Science and Technology (1967). *Responses of the Federal Departments and Agencies to the President's Science Advisory Committee Report, "Restoring the Quality of Our Environment."* May; https://static1.squarespace.com/static/571d109b04426270152febe0/t/60ff1ea5e66d19425f113804/1627332291983/OST.Responses+of+the+Fed+Departments+and+Agencies.1967.pdf

Old, Bruce S. (1961). "The Evolution of the Office of Naval Research." *Physics Today* 14 (8), p. 30; https://physicstoday.scitation.org/doi/pdf/10.1063/1.3057690

Oreskes, Naomi (2019). *Testimony before the House Committee on Oversight and Reform*. House of Representatives, Committee on Oversight and Reform, Subcommittee on Civil Rights and Civil Liberties, Hearing on "Examining the Oil Industry's Efforts to Suppress the Truth about Climate Change," October 23; https://oversight.house.gov/legislation/hearings/examining-the-oil-industry-s-efforts-to-suppress-the-truth-about-climate-change

Oreskes, Naomi, Conway, Erik M., and Shindell, Matthew (2008). "From Chicken Little to Dr. Pangloss: William Nierenberg, Global Warming, and the Social Deconstruction of Scientific Knowledge." *Historical Studies in Natural Sciences* 38 (1), p. 109; https://online.ucpress.edu/hsns/article-abstract/38/1/109/105584/From-Chicken-Little-to-Dr-Pangloss-William?redirectedFrom=fulltext

Orville, Howard T. (1954). "Weather Made to Order?" *Collier's Weekly*, May 28, p. 25; http://www.unz.com/print/Colliers-1954may28-00025

Orville, Howard T. (1958). *Who Will Win the Weather War?* Congressional Record – Senate, p. 9479, May 26; https://www.govinfo.gov/content/pkg/GPO-CRECB-1958-pt7/pdf/GPO-CRECB-1958-pt7-10-1.pdf

Pattee, Emma (2022). "The 1977 White House Climate Memo That Should Have Changed the World." *Guardian*, June 14; https://www.theguardian.com/environment/2022/jun/14/1977-us-presidential-memo-predicted-climate-change

Peterson, Thomas C., Connolley, William M., and Fleck, John (2008). "The Myth of the 1970s Global Cooling Scientific Consensus." *Bulletin of the American Meteorological Society* 89 (9), 1325, September; https://journals.ametsoc.org/doi/pdf/10.1175/2008BAMS2370.1

Plass, Gilbert N. (1956). "Carbon Dioxide and the Planet." *American Scientist* 44, 302; https://www.americanscientist.org/article/carbon-dioxide-and-the-climate; also see Plass, Gilbert N. (1956). "The Carbon Dioxide Theory of Climatic Change." *Tellus* 8 (2), p. 140, May; https://onlinelibrary.wiley.com/doi/abs/10.1111/j.2153-3490.1956.tb01206.x

Poe, Ted (2009a). *The Fear of Global Warming*. Congressional Record – House, p. H3631, March 18; https://www.govinfo.gov/content/pkg/CREC-2009-03-18/pdf/CREC-2009-03-18-pt1-PgH3631-2.pdf

Poe, Ted (2009b). *The Cooling World*. Congressional Record–House, p. H14948, December 15; https://www.govinfo.gov/content/pkg/CREC-2009-12-15/pdf/CREC-2009-12-15-pt1-PgH14948.pdf

Putnam, Palmer Cosslett (1953). *Energy in the Future*. Princeton, NJ: D. Van Nostrand Company, Inc.; https://babel.hathitrust.org/cgi/pt?id=mdp.39015006902194&view=1up&seq=9&q1=climate

Reagan, Ronald (1979). *Ronald Reagans Announcement for Presidential Candidacy, 1979*. Ronald Reagan Presidential Library & Museum, January 13; https://www.reaganlibrary.gov/archives/speech/ronald-reagans-announcement-presidential-candidacy-1979

Reagan, Ronald (1980a). *Televised Campaign Address 'A Vital Economy: Jobs, Growth, and Progress for Americans*.' Ronald Reagan Presidential Library & Museum, October 24; https://www.reaganlibrary.gov/archives/speech/televised-campaign-address-vital-economy-jobs-growth-and-progress-americans

Reagan, Ronald (1980b). *1980 Ronald Reagan and Jimmy Carter Presidential Debate*. Ronald Reagan Presidential Library & Museum, October 28; https://www.reaganlibrary.gov/archives/speech/1980-ronald-reagan-and-jimmy-carter-presidential-debate

Reagan, Ronald (1988a). *Message to Congress Transmitting a Report on Federal Greenhouse Effect Research*. Ronald Reagan Presidential Library & Museum, January 26; https://www.reaganlibrary.gov/012688c

Reagan, Ronald (1988b). *Statement on Signing the Price-Anderson Amendments Act of 1988*. Ronald Reagan Presidential Library & Museum, August 20; https://www.reaganlibrary.gov/research/speeches/082088c

Reagan, Ronald (1989). *Letter to the Speaker of the House of Representatives and the President of the Senate Transmitting the Fiscal Year 1990 Budget*. Ronald Reagan Presidential Library & Museum, January 9; https://www.reaganlibrary.gov/research/speeches/010989d

Reeves, Robert W. (2015). *Scientific, Environmental, and Political Context Leading to Concept of a Climate Diagnostic Workshop*. October; https://www.cpc.ncep.noaa.gov/products/outreach/CDPW40/CD&PW_reeves_denver.pdf

Revelle, Roger R., and Suess, Hans E. (1957). "Carbon Dioxide Exchange between Atmosphere and Ocean and the Question of an Increase of Atmospheric CO_2 during the Past Decades." *Tellus* 9 (1), p. 18; https://pdfs.semanticscholar.org/d014/06a57bff758203390e36247bd96e0c9f8102.pdf

Rich, Nathaniel (2018). "Losing Earth: The Decade We Almost Stopped Climate Change." *New York Times*, August 1; https://www.nytimes.com/interactive/2018/08/01/magazine/climate-change-losing-earth.html

Ronald Reagan Library (1988). *Joint Statement Following the Soviet-United States Summit Meeting in Moscow*. Ronald Reagan Presidential Library & Museum, June 1; https://www.reaganlibrary.gov/research/speeches/060188b

Schneider, Stephen H. (2009). *Science as a Contact Sport*. Washington, DC: National Geographic Society; https://books.google.ca/books?id=gC2xlxwYfYkC&dq=stephen+schneider,+contact+sport&source=gbs_navlinks_s

Seidel, Stephen, and Keyes, Dale (1983). *Can We Delay a Greenhouse Warming?* EPA, September; https://nepis.epa.gov/Exe/ZyPDF.cgi/9101HEAX.PDF?Dockey=9101HEAX.PDF

Shultz, George P. (2015). "A Reagan Approach to Climate Change." *Washington Post*, March 13; https://www.washingtonpost.com/opinions/a-reagan-model-on-climate-change/2015/03/13/4f4182e2-c6a8-11e4-b2a1-bed1aaea2816_story.html

Smith, Joel B., and Tirpak, Dennis (ed.) (1989). *The Potential Effects of Global Climate Change on the United States, Report to Congress*. United States Environmental Protection

Agency, Office of Policy, Planning and Evaluation, Office of Research and Development, December; https://www.nrc.gov/docs/ML1434/ML14345A597.pdf

Speth, James Gustav (2021). *They Knew, The US Federal Government's Fifty-Year Role in Causing the Climate Crisis*. Cambridge: MIT Press.

Staats, Elmer B. (1977). *Report to the Congress, By the Comptroller General of the United States, U.S. Coal Development – Promises, Uncer*tainties. United States General Accounting Office, September 22; https://www.gao.gov/assets/emd-77-43.pdf

Udall, Morris K. (1974). *Letter to the President, August 22, Box 7, Folder "Science and Technology Adviser: October 8, 1974-February 5, 1975" of the White House Special Files Unit Files at the Gerald R. Ford Presidential Library*. Gerald R. Ford Presidential Library & Museum; https://www.fordlibrarymuseum.gov/Library/document/0010/1554458.pdf

United States Committee for the Global Atmospheric Research Program (1975). *Understanding Climatic Change, a Program for Action*. National Research Council. Washington, DC: National Academy of Science; https://ia803200.us.archive.org/18/items/understandingcli00unit/understandingcli00unit.pdf

United States Department of Agriculture (1941). *Climate and Man (Part One)*. U.S. Department of Agriculture, Reprint 2004, Honolulu: University Press of the Pacific; https://books.google.ca/books?id=vPHHk2t1Tq4C&dq=%22the+theory+received+a+fatal+blow+when+it+was+realized+that+carbon+dioxide%22,+department+of+agriculture,+1941&source=gbs_navlinks_s

United States House of Representatives (1909). *Agricultural Appropriations Bill*. House of Representatives, United States, Hearings before the Committee on Agriculture, 61st Congress, Second Session; December 13; https://babel.hathitrust.org/cgi/pt?id=chi.096068800&view=1up&seq=9

United States House of Representatives (1956). *National Science Foundation, International Geophysical Year*. Hearings before the Subcommittees of the Committee on Appropriations, House of Representatives, 84th Congress, Second Session, March 8, p. 426; https://babel.hathitrust.org/cgi/pt?id=umn.31951d00742016t;view=1up;seq=476

United States House of Representatives (1957). *National Science Foundation, Report on the International Geophysical Year*. Hearings before the Committee on Appropriations, 85th Congress, First Session, May 1; https://babel.hathitrust.org/cgi/pt?id=mdp.39015036820101&view=1up&seq=11&q1=greenhouse

United States House of Representatives (1958). *National Science Foundation*. Hearings before the Subcommittee of the Committee on Appropriations, House of Representatives, 85th Congress, Second Session, June 2; https://babel.hathitrust.org/cgi/pt?id=uc1.$b654289&view=1up&seq=7

United States House of Representatives (1959). *National Science Foundation, National Academy of Sciences*. Hearings before the Subcommittee of the Committee on Appropriations, House of Representatives, 86th Congress, First Session, February; https://books.google.ca/books?id=7FcPCEi7aYsC&dq=%22amount+of+carbon+dioxide+in+the+air+controls,+at+least+to+some+extent,+the+average+air+temperature+and+the+loss+of+heat+from+the+earth%22&source=gbs_navlinks_s

United States House of Representatives (1966). *The Adequacy of Technology for Pollution Abatement*. Hearings before the Subcommittee on Science, Research, and Development of the Committee on Science and Astronautics, House of Representatives, 89th Congress, Second Session, August 4, p. 257; https://babel.hathitrust.org/cgi/pt?id=mdp.35112104242914&view=1up&seq=635&skin=2021

United States House of Representatives (1968). *Environmental Quality*. Hearings before the Subcommittee on Science, Research, and Development of the Committee on Science and

Astronautics, U.S. House of Representatives, 90th Congress, Second Session, January 17; https://babel.hathitrust.org/cgi/pt?id=uc1.$b654593&view=1up&seq=7

United States House of Representatives (1969). *1970 NASA Authorization*. Hearings before the Subcommittee on Space Science and Applications of the Committee on Science and Astronautics, U.S. House of Representatives, 91st Congress, First Session, March–April; https://babel.hathitrust.org/cgi/pt?id=mdp.39015035474496&view=1up&seq=5

United States House of Representatives (1976a). *The National Climate Program Act*. Hearings before the Subcommittee on the Environment and the Atmosphere of the Committee on Science and Technology, U.S. House of Representatives, 94th Congress, Second Session, May 18–27; https://babel.hathitrust.org/cgi/pt?id=mdp.39015068355620;view=1up;seq=3

United States House of Representatives (1976b). *Weather Modification*. Hearings before the Subcommittee on the Environment and the Atmosphere of the Committee on Science and Technology, U.S. House of Representatives, 94th Congress, Second Session, June 15-8; https://babel.hathitrust.org/cgi/pt?id=mdp.39015022400660&view=1up&seq=3&size=125

United States House of Representatives (1977). *Constraints on Coal Development*. Oversight Hearing before the Subcommittee on Energy and the Environment of the Committee on Interior and Insular Affairs, House of Representatives, 95th Congress, First Session, June 9; https://babel.hathitrust.org/cgi/pt?id=uiug.30112104079055&view=1up&seq=1

United States House of Representatives (1981). *Carbon Dioxide and Climate: The Greenhouse Effect*. Hearing before the Subcommittee on Natural Resources, Agricultural Research and Environment and the Subcommittee on Investigations and Oversight of the Committee on Science and Technology, U.S. House of Representatives, 97th Congress, First Session, July 31; https://babel.hathitrust.org/cgi/pt?id=mdp.39015082344238;view=1up;seq=1

United States House of Representatives (1982). *Carbon Dioxide and Climate: The Greenhouse Effect*. Hearing before the Subcommittee on Natural Resources, Agriculture Research and Environment and the Subcommittee on Investigations and Oversight of the Committee on Science and Technology, U.S. House of Representatives, 97th Congress, Second Session, March 25; https://babel.hathitrust.org/cgi/pt?id=mdp.39015011554253;view=1up;seq=1

United States House of Representatives (1984). *Carbon Dioxide and the Greenhouse Effect*. Hearing before the Subcommittee on Investigations and Oversight and the Subcommittee on Natural Resources, Agricultural Research and the Environment of the Committee on Science and Technology, U.S. House of Representatives, 98th Congress, Second Session, February 28; https://babel.hathitrust.org/cgi/pt?id=uc1.31210024955054;view=1up;seq=1

United States House of Representatives (1988). *Implications of Global Warming for Natural Resources*. Oversight Hearings before the Subcommittee on Water and Power Resources of the Committee on Interior and Insular Affairs, House of Representatives, 100th Congress, Second Session, September 27, October 17; https://babel.hathitrust.org/cgi/pt?id=umn.31951d00283282e;view=1up;seq=3

United States House of Representatives (2019). *Climate Change, Part I: The History of a Consensus and the Causes of Inaction*. Hearing before the Subcommittee on Environment of the Committee on Oversight and Reform, House of Representatives, 116th Congress, First Session, April 9; https://www.govinfo.gov/content/pkg/CHRG-116hhrg36637/pdf/CHRG-116hhrg36637.pdf

United States House of Representatives (2022). *What More Gulf of Mexico Oil and Gas Leasing Means for Achieving U.S. Climate Targets*. Oversight Hearing before the Subcommittee on Energy and Mineral Resources of the Committee on Natural Resources,

U.S. House of Representatives, 117th Congress, Second Session, January 20; https://www.govinfo.gov/content/pkg/CHRG-117hhrg46588/pdf/CHRG-117hhrg46588.pdf

United States Senate (1935). *Protection of Land Resources Against Soil Erosion*. Hearings before a Subcommittee of the Committee on Agriculture and Forestry, United States Senate, 74th Congress, First Session, April 2–3; https://babel.hathitrust.org/cgi/pt?id=mdp.39015063988383&view=1up&seq=5

United States Senate (1951). *Weather Control and Augmented Water Potable Water Supply*. Joint Hearings before Subcommittees of the Committees on Interior and Insular Affairs Interstate and Foreign Commerce and Agriculture and Forestry, United States Senate 82nd Congress, First Session, March–April; https://babel.hathitrust.org/cgi/pt?id=uiug.30112119746383&view=1up&seq=5&size=125

United States Senate (1957). *Experimental Research Program in Cloud Modification*. Hearings before a Subcommittee of the Committee on Interstate and Foreign Commerce, United States Senate, 85th Congress, First Session, March 26, May 13; https://babel.hathitrust.org/cgi/pt?id=uiug.30112121381104&view=1up&seq=5&q1=%22climatic%20change%22

United States Senate (1958). *Compilation of Materials on Space and Astronautics No. 2*. Special Committee on Space and Astronautics, United States Senate, 85th Congress, Second Session, April 14th; https://books.google.ca/books?id=0nohAAAAMAAJ&dq=%22weather+control+means+that+our+knowledge%22&source=gbs_navlinks_s

United States Senate (1961). *Marine Science*. Hearings before the Committee on Interstate and Foreign Commerce, United States Senate, 87th Congress, First Session, March–May; https://babel.hathitrust.org/cgi/pt?id=uc1.31822010116788&view=1up&seq=3

United States Senate (1963). *A Study of Pollution – Air*. A Staff Report to the Committee on Public Works, September; https://babel.hathitrust.org/cgi/pt?id=mdp.39015007374369&view=1up&seq=10&q1=%22Carbon%20dioxide%20is%20also%20produced%20whenever%20we%20burn%20carbonaceous%20fuels%22

United States Senate (1964). *Weather Modification*. Hearing before the Subcommittee on Irrigation and Reclamation of the Committee on Interior and Insular Affairs, United States Senate, 88th Congress, Second Session, May 21; https://babel.hathitrust.org/cgi/pt?id=umn.31951d02087009a&view=1up&seq=3&skin=2021

United States Senate (1966a). *Weather Modification*. Hearings before the Committee on Commerce, United States Senate, 89th Congress, First Session, February 24, p. 133; https://babel.hathitrust.org/cgi/pt?id=uc1.31822006846109&view=1up&seq=147&size=125

United States Senate (1966b). *Weather Modification*. Hearings before the Subcommittee on Water and Power Resources of the Committee on Interior and Insular Affairs, United States Senate, 89th Congress, Second Session, March–April; https://babel.hathitrust.org/cgi/pt?id=uc1.31822019235076&view=1up&seq=1&q1=%22climatic%20change%22

United States Senate (1967). *Air Pollution – 1967 (Air Quality Act)*. Hearings before the Subcommittee on Air and Water Pollution of the Committee on Public Works, United States Senate, 90th Congress, First Session, April 4, p. 949; https://babel.hathitrust.org/cgi/pt?id=uiug.30112106886309&view=1up&seq=3

United States Senate (1972). *Department of Defense Appropriations for Fiscal Year 1973*. Hearings before a Subcommittee of the Committee on Appropriations, United States Senate, 92nd Congress, Second Session; https://babel.hathitrust.org/cgi/pt?id=umn.31951p012130496&view=1up&seq=7&size=125

United States Senate (1973). *U.S. and World Food Situation*. Hearings before the Subcommittee on Agricultural Production, Marketing, and Stabilization of Prices, and

Subcommittee on Foreign Agricultural Policy of the Committee on Agriculture and Forestry, United States Senate, 93rd Congress, First Session, October 17–18; https://babel.hathitrust.org/cgi/pt?id=uiug.30112019044384;view=1up;seq=3

United States Senate (1977). *National Climate Program Act*. Hearings before the Subcommittee on Science, Technology, and Space of the Committee on Commerce, Science, and Transportation, United States Senate, 95th Congress, First Session, June–July; https://babel.hathitrust.org/cgi/pt?id=mdp.39015068355463&view=1up&seq=1&size=125

United States Senate (1978). *Weather Modification: Programs, Problems, Policy and Potential*. Prepared at the Request of Howard W. Cannon, Chairman, Committee on Commerce, Science, and Transportation, United States Senate, 95th Congress, Second Session, May; https://babel.hathitrust.org/cgi/pt?id=mdp.39015076085656&view=1up&seq=3

United States Senate (1979a). *Synthetic Fuels*. Hearings before the Committee on Governmental Affairs, U.S. Senate, 96th Congress, First Session, July 17–20; https://babel.hathitrust.org/cgi/pt?id=mdp.39015083100290;view=1up;seq=3

United States Senate (1979b). *Carbon Dioxide Accumulation in the Atmosphere, Synthetic Fuels and Energy Policy, a Symposium*. Committee on Government Affairs, U.S. Senate, 96th Congress, First Session, July 30; https://babel.hathitrust.org/cgi/pt?id=uc1.31822005677661;view=1up;seq=5

United States Senate (1980). *Effects of Carbon Dioxide Buildup in the Atmosphere*. Hearing before the Committee on Energy and Natural Resources, United States Senate, 96th Congress, Second Session, April 3; https://babel.hathitrust.org/cgi/pt?id=mdp.39015081188222;view=1up;seq=1

United States Senate (1985). *Global Warming*. Hearing before the Subcommittee on Toxic Substances and Environmental Oversight of the Committee on Environment and Public Works, United States Senate, 99th Congress, First Session, December 10; https://babel.hathitrust.org/cgi/pt?id=umn.31951d00275504i;view=1up;seq=1

United States Senate (1986). *Ozone Depletion, the Greenhouse Effect, and Climate Change*. Hearings before the Subcommittee on Environmental Pollution of the Committee on Environment and Public Works, United States Senate, 99th Congress, Second Session, June 10–11; https://babel.hathitrust.org/cgi/pt?id=mdp.39015035756843&view=1up&seq=1

United States Senate (1987). *Greenhouse Effect and Global Climate Change*. Hearings before the Committee on Energy and Natural Resources, United States Senate, 100th Congress, First Session, November 9–10; https://babel.hathitrust.org/cgi/pt?id=uc1.b5127806;view=1up;seq=3

United States Senate (1988). *Greenhouse Effect and Global Climate Change*. Hearing before the Committee on Energy and Natural Resources, United States Senate, 100th Congress, First Session, June 23, Part 2; https://babel.hathitrust.org/cgi/pt?id=uc1.b5127807;view=1up;seq=3

Von Neumann, John (1955). "Can We Survive Technology?" *Fortune*, June; for a copy of the article see http://geosci.uchicago.edu/~kite/doc/von_Neumann_1955.pdf

Wallace, Amy, and Smollar, David, (1991). "Roger Revelle, Founder of UCSD, Is Dead at 82." *Los Angeles Times*, July 16; https://www.latimes.com/archives/la-xpm-1991-07-16-mn-2362-story.html

Weart, Spencer (2022). *The Discovery of Global Warming*. American Institute of Physics; https://history.aip.org/climate/index.htm#contents

Weinberger, Sharon (2018). *Chain Reaction: How a Soviet A-Bomb Test Led the U.S. into Climate Science*. UNDARK, April 20; https://undark.org/2018/04/20/wilo-imagineers-of-war/

White House (1965). *Restoring the Quality of Our Environment*. Washington, DC: The White House, November; https://ozonedepletiontheory.info/Papers/Revelle1965AtmosphericCarbonDioxide.pdf

White House (1970). *Environmental Quality, the First Annual Report of the Council on Environmental Quality*. Washington, DC: The White House, August; https://www.slideshare.net/whitehouse/august-1970-environmental-quality-the-first-annual-report-of

WMH (2017). *President Lyndon Johnson Approves Weather Warfare*. Weather Modification History, November 26; https://weathermodificationhistory.com/president-lyndon-johnson-weather-wafare/

Woodwell, George M., MacDonald, Gordon J., Revelle, Roger, and Keeling, C. David (1979). *The Carbon Dioxide Problem: Implications for Policy in the Management of Energy and Other Resources*. Council on Environmental Quality, July; http://graphics8.nytimes.com/packages/pdf/science/woodwellreport.pdf

3 Climate Denialism in Washington

We can never say that science did not warn us. While the scientific evidence on climate change mounted every year, and leading climate scientists continued to appear as expert witnesses at congressional hearings (and still do), Congress did not heed their warnings. Why? Propaganda from climate denial was muting the messaging of the climate sciences and corrupting American politics. Shockingly, a handful of contrarians triumphed over thousands of scientists, not because they were better scientists or communicators, but because Republicans were cheering for them. Instead of appearing anti-science, Republicans could claim that they were merely supporting the other side of the science as presented by the contrarians. Consequently, the development of a bipartisan policy to prevent the climate crisis was impossible after Reagan's term, as discussed in this chapter. Organized climate denial and its propaganda appeared in political circles for the first time in Washington when President George H. W. Bush was in office and was an immediate success for the energy-industrial complex.

President George H. W. Bush: 1989–93

Two conferences featured prominently during President George H. W. Bush's time in office which defined his climate change legacy – the White House effect conference and the Rio Conference. A conference about the "White House effect" on the "greenhouse effect" that President Bush had proudly promised during his campaign was planned for mid-April 1990. This would be followed by the international treaty negotiations at Rio two years later.

The White House Effect Conference

At the end of 1989, President Bush had even promoted the upcoming White House conference on climate change in a meeting with Mikhail Gorbachev.[1] It is astonishing that the issue of climate change was even raised at this historic meeting – the Malta Summit – where the two leaders discussed the lifting of the iron curtain and declared that the Cold War was over.

DOI: 10.4324/9781003455417-4

Shortly before the presidential climate conference took place, President Bush had welcomed the IPCC to Washington on February 5, 1990. Here, he instructed the IPCC on the best approach to take[2]:

> As experts, you understand that economic growth and environmental integrity need not be contradictory priorities. One reinforces and complements the other; each, a partner. Both are crucial …
>
> If we hope to promote environmental protection and economic growth around the world, it will be important not to work in conflict but with our industrial sectors. That will mean moving beyond the practice of command, control, and compliance toward a new kind of environmental cooperation and toward an emphasis on pollution prevention rather than mere mitigation and litigation.

At this meeting with the IPCC, President Bush once more announced his special upcoming conference, again raising hope that America was prepared to lead in the fight against climate change.

When the highly anticipated big day of the "Science and Economics Research Related to Global Change" conference finally arrived, President Bush revealed at last where he stood, leaving much of the world dumbfounded. The White House effect was a classic "AmBush,"[3] because he was going to teach climate experts a thing or two about science, instead of them teaching him. In welcoming delegates to his conference on April 17th, President Bush went out of his way to stress that there were views on the science of climate change in disagreement with the IPCC[4]:

> Some of you may have seen two scientists just on one of our talk shows on Sunday -- respected men debating global change. One scientist argued that if we keep burning fossil fuels at today's rate, and I quote, "By the end of the next century, Earth could be nine degrees Fahrenheit warmer than today." And the other scientist saw no evidence of rapid change and warned against a drastic reordering of our economy that could cause us, in his words, "to end up the impoverished nation awaiting a warming that never comes." Two scientists, two diametrically opposed points of view -- now, where does that leave us?

Bush's comments were a direct slap in the face of the IPCC (and their chairman Bert Bolin who was sitting in the audience). The IPCC had just finished reviewing the science of climate change and had not found such "opposed points of view" in the peer-reviewed literature. According to President Bush, the IPCC had missed important scientific information that he had found simply by watching TV.

But there was much more to come. The president had issued directives to control the narrative of the conference. The briefing notes from his administration for American officials at the conference directed them to avoid certain issues[5]: "not beneficial to discuss whether there is or is not warming … In the eyes of the public we will lose this debate." Following a text-book strategy on propaganda, the notes went on to suggest[6]: "A better approach is to raise the many uncertainties that need to be better understood on this issue." When the secret directives were uncovered, Senator Al Gore (1948–) objected to these antics in an op-ed in the *New York Times*, under the title "To Skeptics on Global Warming …"[7]

What, then, was the "White House effect" on the greenhouse effect? The answer was clear – to deny the science. Bush's "Science and Economics Research Related to Global Change" conference was, in fact, the first major climate-denial conference in the world (the climate denial of George H. W. Bush is discussed further in the next chapter).

Although President Bush wanted to present himself as more knowledgeable than climate scientists, other observers disagreed; for example, he was found to be[8]:

> detached, uninterested, and as his brief remarks in the April meeting showed, responsive only to the politics of a complex issue. He never sat for a full-dress scientific briefing on it or exercised control over administration policy, even after infighting among administration officials became public, or leaders of other industrialised nations pledged action.

William A. Nitze, the Deputy Assistant Secretary of State for Environment, offered a similar assessment[9]:

> President Bush's failure to come to grips with the climate issue is in large part the result of personal ignorance … Instead he chose to rely on the knowledge, opinions and judgment of a few key advisors, particularly his former Chief of Staff, Mr. Sununu.

While President Bush had avoided scientific briefings, he was nevertheless more engaged with climate change than any previous president had been, as he had little choice because of the upcoming Rio Conference.

The Rio Conference

Political action on climate change would be an important topic for the high-profile Earth Summit (United Nations Conference on Environment and Development) in Rio for mid-June 1992.[10] Candidate Bush had strong words on climate change action, as discussed in the previous chapter, and, once in office, his Secretary of State James Baker (1930–) likewise expressed support for the cause[11]: "We can probably not afford to wait until all of the uncertainties have been resolved before we do act. Time will not make the problem go away." A few months later, another high-ranking official, William K. Reilly (1940–), the Administrator of the EPA, announced that the president had agreed to participate in the Framework Convention process for Rio.[12]

On August 27, 1989, encouraging words came from the president[13]: "We need to work more cooperatively to develop constructive solutions to global warming, including measures to promote energy efficiency and conservation and greater protection of forest resources." President George H. W. Bush had been saying all the right things, but Al Gore, then the Senator from Tennessee, had his misgivings[14]:

> Once again, the President has been dragged slowly and reluctantly toward the correct position when the White House should have been providing leadership. I welcome their decision finally to begin recognizing the importance of global climate change as an international issue demanding attention, but it's too early for rave reviews of a policy developed in an atmosphere of political damage control in response to Congressional pressure.

Gore's doubts were confirmed, as the president soon adopted a hard line against climate initiatives under pressure from "forceful interests at work in Washington."[15]

Negotiations on the upcoming Rio Conference were not going well, mainly thanks to the U.S.; the message from Washington was clear: no firm, legal commitments.[16] At the Noordwijk Ministerial Conference in November 1989, and again at the Second World Climate Conference in Geneva a year later, the U.S. rejected any attempts to define the level of emission reductions or timeline.[17] The Swedish minister candidly voiced his frustration[18]: "Your government [American] is fucking this thing up!" The Bush administration maintained this obstructionist posture despite warnings from members of the National Academy of Sciences presented by the Nobel Laureate Henry Way Kendall (1926–99).[19]

By early 1990, international pressure was building for the U.S. to do more on climate change at the upcoming Rio Conference, and President Bush let other world leaders know that he was unhappy about it. Just before his notorious White House effect conference, on April 13, 1990, he met with Prime Minister Thatcher (1925–2013) of the UK and told her[20]: "I'm worried about extreme environmentalists throwing people out of work. I want to get cost and science in there. We need study, expertise. We can't deal just with pure emotion. I'm trying to find the proper balance." And four days later, President Bush complained to Prime Minister Ruud Lubbers (1938–2018) of The Netherlands[21]: "Your environment guy is pushing pretty hard … I don't like to complain, but he is giving us a little grief" (Prime Minister Lubbers agreed to try to control him). Later in the summer, George H. W. Bush also had a similar discussion with Helmut Kohl (1930–2017) of Germany.[22]

World leaders, including President Bush,[23] had yet to commit to even attending the Rio conference. On June 18, 1991, the host of the conference, President Fernando Collor de Mello (1949–) of Brazil, expressed his worries.[24] The following month at the London Economic Summit, the lack of commitment was raised again, this time by John Major (1943–), the Prime Minister of the United Kingdom. At the Summit, President Bush seemed to surprise some leaders when he asked[25]: "Is [Amazon deforestation] what the [Rio] conference will focus on?" Prime Minister Brian Mulroney (1939–) of Canada reminded him that: "The Brazilian Conference [Rio] won't deal just with forests: other issues include climate change," and the prime ministers of The Netherlands and Italy agreed. Also, a surprising remark was made by John Major[26]: "There is not real political input on the work of the conference. It may need political brokering. We can't just leave it to experts without input. The implications of our decisions are profound." With Rio only a year away, apparently few world leaders had given it much thought. George Bush was not the only one dragging his feet.

As Rio approached, President Bush had to deal with more pressing matters, as the conference fell amid the election campaign. In a meeting with Helmut Kohl on March 21, 1992, George H. W. Bush explained his predicament[27]:

> Now, I have to say that we have a major problem with this Rio Conference. Given my schedule and the campaign I can't commit at this time and I certainly I am not going to commit to things that will halt our economy. Maybe our experts can resolve these problems. But we don't want a big bill at the end of the day. We will continue to work in the run-up at this UN meeting. The U.S. and Germany should work very closely together on this. This conference is at a bad time for me politically …
>
> But I can't go to Rio and get myself embarrassed because we can't satisfy the demands of the LDCs [Least Developed Countries] and the environmentalist advocates. We have a particular problem with global warming commitments. All things being equal, I would like to go, but I need to see more results from the working group first. If I am not there, you can always blame things on me!

Overall, there was a friendly tone to the meeting, with the two leaders trying to work together.

While many world leaders empathized with the president, the bureaucrats (especially within the EU) tasked with putting the Rio agreement together, were openly contemptuous of him. The American delegation had alienated itself from the rest,[28] and the Brazilian delegate José Goldemberg (1928–), for example, described the "hard-line positions of the US."[29] President Bush provided clear messages on what the U.S. could accept at Rio. There was no room for any further negotiation. In the end, the bureaucrats relented and drafted a proposal with no legally binding targets.[30]

With the new draft for Rio in hand, President Bush quickly contacted other world leaders. With Canadian Prime Minister Brian Mulroney, on May 1, 1992, the president let him know that the new deal could work[31]:

> I know the completion of that treaty is important to make the Rio summit a success. I'm getting all kinds of pressure on whether to go to Rio. Given the size of the U.S. and all, I know it would disappoint lot of friends (including Kohl) if I didn't go … I haven't actually looked at it yet, but going by what I'm told – that we can live with the language. I'm going to catch hell from the right wing, but I'm willing to take it.

His admission that he would "catch hell from the right wing" for going too far was a reminder of the powerful forces at work within his own party. On the other hand, Brian Mulroney (who was also a conservative politician) expected trouble with the left-wing environmentalists in Canada for not going far enough. President Bush also had a cordial conversation with Chancellor Kohl a few days later.[32]

The watered-down proposal had done the trick. George H. W. Bush attended the Rio Conference, where he defended his position[33]:

> Let's face it, there has been some criticism of the United States. But I must tell you, we come to Rio proud of what we have accomplished and committed to extending the record on American leadership on the environment …
>
> It's been said that we don't inherit the Earth from our ancestors, we borrow it from our children. When our children look back on this time and this place, they will be grateful that we met at Rio, and they will certainly be pleased with the intentions stated and the commitments made. But they will judge us by the actions we take from this day forward. Let us not disappoint them.

As the children from the time of Rio are now adults, it is hard to imagine how they can be anything but disappointed with the actions that were taken, or rather not taken, by America. Of course, one of his own children was not disappointed by the lack of action against climate change and must have made his father proud when he also became president and took no action.

As the election was approaching, on October 15, 1992, the U.S. became the first industrialized nation to ratify the United Nations Framework Convention on Climate Change (UNFCCC), a part of the Rio Convention. Just a few weeks later, George Bush lost the election to Bill Clinton.

Congress

In 1988, Congress, itself, had requested the National Academy of Sciences (NAS) to prepare an analysis on climate change: [34] "This study should establish the scientific consensus on the rate and magnitude of climate change, estimate the projected impacts, and evaluate policy options for mitigating and responding to such changes;" completed in 1991, the report "Policy Implications of Greenhouse Warming" recommended[35]:

> Despite the great uncertainties, greenhouse warming is a potential threat sufficient to justify action now. Some current actions could reduce the speed and magnitude of greenhouse warming; others could prepare people and natural systems of plants and animals for future adjustments to the conditions likely to accompany greenhouse warming.

The NAS study was different from any previous one done for the government, in that Congress this time had specifically requested[36]: "Tell us what to do." At a Senate hearing on the report, Senator Gore attacked the Bush administration[37]:

> The report from the Academy is important because it helps build consensus for taking action. Here in the Congress, many of us have been trying to move the United States to action and international leadership.

> Unfortunately, we have been opposed at every turn by President Bush, by his administration, and by his Chief of Staff [John Sununu], who keeps saying we do not need to worry about global warming. Well, this study says we do.

At the hearing, Frank Press, now the president of NAS, made the remark[38]: "Public discussions of greenhouse warming, and its implications have often been inconclusive. It has proven difficult to distinguish scientific fact from conjecture." His passing comment was actually a bombshell, in my opinion. What was going on? Why couldn't the NAS "distinguish scientific fact from conjecture?" How could some of the most prestigious scientists in the nation, members of the NAS, not be able to do this? While Dr. Press did not explain the source of "conjecture," a new type of scientist had arrived in Washington around the same time as the NAS study was started – the "contrarian" (also called "skeptic," especially in these early days). Contrarians were scientists who disagreed with mainstream scientific views on climate change. Dr. Richard Lindzen had presented to the NAS study[39] and would later be known as a leading contrarian scientist.

Dr. Lindzen and other well-known contrarian scientists who frequently testified at congressional hearings are discussed below; their testimonies took place across different administrations, but each contrarian's testimonies have been collected and presented only under the administration where their services were often called upon by Republicans[40]:

- Patrick J. Michaels (1950–2022) – under *President George H. W. Bush*
- Richard S. Lindzen (1940–) – under *President George H. W. Bush*
- Bjorn Lomborg (1965–) – under *President George W. Bush*
- Judith Curry (1953–) – under *President Barack Obama*
- John R. Christy (1951–) – under *President Donald Trump*

The contrarians were a fringe group among the scientific community with limited impact on the science of climate change but a disproportionate influence on the political will to do anything about it. Experts had been teaching politicians about climate change at congressional hearings for the past few decades, and just when political action seemed inevitable, the contrarians showed up and disrupted the momentum. The contrarians often had academic credentials but seldom discussed their own peer-reviewed studies, and only offered their opinions on alleged flaws of the published literature, thereby casting doubt on the science of climate change. Their purpose was to give Republicans a "legitimate" excuse to do nothing about the global threat.

Repeatedly testifying at hearings, contrarians signaled to politicians that there was debate within the scientific community (which there was not, apart from the few members of this fringe group). Their dissenting views produced a false controversy and gave an opportunity for Republicans to insist that the argument over the science needed to be settled by Congress. The proper role of Congress was to decide on what actions were required, if any, based on the science, not to judge the science per se. Yet, Congress resurrected the inquisitions from Galileo's days and passed political judgement over the science of climate change.

The trenches of climate politics in the United States were the congressional hearings and, with the infiltration of contrarians, the debates became messy. The contrarian strategy was coy and non-confrontational as it merely sought to create doubt among politicians. After decades of mounting evidence in support of the science and extreme weather events taking place, contrarians are still defiantly peddling this "doubt."

Even before Frank Press of the National Academy of Sciences dropped the bombshell regarding "conjecture" at the 1991 hearing, an earlier example of the influence of contrarians on Congress could possibly be seen in 1985, when Senator Dave Durenberger (1934–) (Republican from Minnesota) introduced a session[41]:

> My conclusion is, that grappling with this problem is going to be just about as easy as nailing jello to the wall … Depending upon whom you ask, the consequences could be a disaster of biblical proportions or just maybe nothing. Unfortunately, more and more of the bets are on biblical and less and less on nothing.

The "just maybe nothing" likely reflected the opinion of contrarians.

In a House hearing late in Reagan's term, the noted climate scientist Stephen Schneider (1945–2010) provided an early warning about the new scientific "skepticism:"[42]

> However, the global warming 'experiment' now underway is not only an academic debate, for we and other inhabitants of Earth will have to live with the outcome should too much scientific skepticism cause further delays in implementing actions to prevent or adapt more effectively to global warming.

Senator Gore also criticized the new breed of "skeptics:"[43]

> Around the globe the public and the policy makers must know and understand the threat we face … [page]
>
> The skeptics use uncertainty as an excuse for inaction and when the uncertainty begins to narrow they step in and say wait a minute. Let us keep it as uncertain as possible because if people get the idea that this problem is real, we will have to do some thing, and we do not want to do anything. That is the approach they are taking. That is why they are scared of the truth.

Al Gore and Stephen Schneider used the term "skeptics" (and variants) as the term "climate deniers" did not yet exist, but this is undoubtedly what they meant.

By early 1989, contrarians were testifying at congressional hearings. In February, Philip Sharp (1942–) (Democrat from Indiana) welcomed the new type of scientist at a House hearing[44]:

> Today we ask whether our global environment is in jeopardy as well. We will hear from scientists who differ as to the magnitude of risk we face and

> the wisdom of changing energy policy based upon this risk. I hope they will help us sort through the conflicting accounts that have appeared in the news media ...
>
> I hope the distinguished scientists before us today will help us to understand how much weight we should give to the global warming threat as we formulate our Nation's energy policy.

Sharp's casual introduction to "scientists who differ" revealed that the contrarians were already known to members of Congress. Two months later, contrarians were back as witnesses, this time in the Senate, where they were introduced as "lively writers on some of these issues."[45]

Dr. Patrick Michaels

An historical congressional hearing for climate change took place in the early days of the administration of George H. W. Bush, called "Global Warming." This session, chaired by Philip Sharp, marked the beginning of the climate denial period in Congress, as a contrarian took a seat beside other expert witnesses. The testimony of Patrick Michaels[46] stressed the uncertainty of the scientific projections on climate change,[47] and during the question period, he added[48]:

> I might emphasize that the problem is simply not scientific in what we would consider the traditional scientific realm. There's a problem that eventually devolves into economics and cost, and consequently that means that we are dealing with what may be the greatest systems analysis problem to ever come across the line. It's not simply, will it warm up, but, in fact, what will that do, and what will that cost?
>
> And that's why we have to be very careful about the ranges of estimates that are given on this. There is a tolerable warming. Nobody knows what that is, at which it might, in fact, be counterproductive to be overly interventionist.

Two months later, Patrick Michaels returned to Congress, this time the Senate, where he again mentioned the uncertainty of the science[49]:

> I do realize that concrete policy cannot require "perfect science" because such a thing will certainly never exist. However, with the great uncertainty that surrounds the climatic change problem, I agree with others that policymakers should pursue a portfolio of options – I underline options – that will pay off in different future states.
>
> Under that rubric, one of many options is that no action other than accelerated research could be the optimum policy.

Another contrarian, who was also from the University of Virginia, S. Fred Singer (1924–2020),[50] had testified as well and questioned whether the world was warming at all. At the end of the session, the chair Senator Claiborne Pell (1918–2009)

(Democrat from Rhode Island) politely acknowledged their contrarian views[51]: "It's a little off the main stream of today's world, but time will prove either that you are correct or not correct." One could say that the comments of Senator Pell were too polite, and time has since proven the contrarians "not correct."

At a House hearing in 1991, Dr. Michaels was back as a witness. The chairman of the session, Henry Waxman (1939–) (Democrat from California) listed the many failed promises of President George H. W. Bush on climate change[52]:

> I remembered his promise in 1988 to fight the greenhouse effects with the White House effect. In his promise in 1989 to take decisive action to limit carbon dioxide. I remembered Secretary Baker's pledge not to wait until all the uncertainties have been resolved. I remembered that just last month EPA Administrator Bill Reilly called global warming the greatest environmental threat facing the country. I hoped and expected to see these commitments reflected in the President's global warming policy.

Then, during the question period, Dr. Michaels stated[53]: "What I would like to know as a scientist is why the Earth has been so reluctant to warm up … And if the Earth warmed up just one-half a degree, that is profound." Despite the average global temperature having since risen by more than 1°C – double the level he had personally considered "profound" in 1991 – Dr. Michaels remained a contrarian right up until his death in 2022.

Patrick Michaels continued to provide congressional testimony during the terms of later presidents. During the Clinton administration, he and William Nierenberg were both present. Dr. Nierenberg had emphasized the uncertainty of climate modelling and recommended[54]:

> On the other, those who take a more conservative view may argue that one can now safely wait until the climate changes become clearer and more definitely negative before taking action since the effect of the mitigating actions would show up in reasonable time.
>
> Taking into account this argument and the currently accepted gestation time of one century, I incline to the latter view. That inclination is strengthened by lack of certainty in the inevitability of all climate changes being harmful.

And Dr. Michaels defended the position of contrarians[55]:

> Unfortunately, this issue has evolved in a highly politicized climate. For the last decade, a community of scientists, often referred to as a small minority, has argued that, based upon the data on climate change, the modeled warming was too large, and therefore any intrusive policy would not be based upon reliable models of global warming.
>
> This view has been cast in a very negative political light, which has had a chilling effect on scientific free speech.

By "small minority," he was referring to the contrarians who were being, in effect, censored. His comment on "chilling effect on scientific free speech" was quite telling, for science has peer review, not public review as his use of "free speech" implies, and the work of the contrarians on climate change seldom stood up to the scrutiny of peer review. The point cannot be stressed enough that the contrarians had no contrary scientific results, just contrary opinions on the already peer-reviewed science, but that was enough for most Republicans. This hearing was noteworthy in that the term "skeptics" appeared over two dozen times in the transcript.

During the George W. Bush administration, there was a hearing to review the National Climate Assessment which, once again, was another inquisition with politicians judging the peer-reviewed science; the session was called: "Do the Climate Models Project A Useful Picture of Regional Climate?" Climate models are an essential part of analyzing the threat of climate change and had been under development for decades, so for politicians to question their value represented a serious overreach of their authority and skills.

James Greenwood (1951–) (Republican from Pennsylvania) opened the session by questioning climate models.[56] Among the six witnesses in this hearing, there were two contrarians, Patrick Michaels and Roger A. Pielke Sr. (1946–).[57] Of course, neither thought the climate models were any good; Dr. Pielke summed up his presentation with[58]: "The difficulty of prediction and the impossibility of verification of predictions decades into the future are important factors that allow for competing views of a long term climate future." And Patrick Michaels, who introduced himself as a "Cato scholar," followed[59]: "I believe our change to a less carbon based economy is an historical inevitability. All we have to do is get out of the way." James Greenwood, the chair of the session, then confirmed: "You would command us to command less;" indeed, Mr. Greenwood had correctly understood their denial talk – no government action on climate change was necessary. The entire session – "Do the Climate Models Project a Useful Picture of Regional Climate?" – was a farce. The judgement of science is carried out in the peer-review process and not the partisan halls of Congress.

After Dr. Michaels' death, the noted climate scientist Jim Hansen described his long-time adversary[60]: "Overall, he was selling bullshit, and perhaps he knew that, but he thought that climate scientists were exaggerating, so he continued even as facts piled up against him. If only he had been right." Climate scientists do hope they are wrong about the deadly threat of climate change, but the evidence keeps reaffirming that they are right, unfortunately.

Dr. Richard Lindzen

An article in *Science* in 1992 declared[61]: "The brawl had begun," when the MIT scientist Richard Lindzen[62] shocked many by claiming there was no evidence of greenhouse warming. Earlier, in June 1991, he had been a featured speaker at an early climate denial conference sponsored by the Cato Institute – "Global Environmental Crisis: Science or Politics?" In the brochure for the conference Dr. Lindzen made the outlandish claim that[63]:

> The notion that global warming is a fact and will be catastrophic is drilled into people to the point where it seems surprising that anyone would question it, and yet, underlying it is very little evidence at all … I feel it discredits science.

A year later, Dr. Lindzen was a witness at a congressional hearing. Senator Bennett Johnston set the tone by describing the new reality for sessions on climate change with the arrival of the contrarians[64]:

> The issue of global climate change has become one of the most contentious scientific and political debates of the century. On one side are those who forecast a scenario of rising seas, recurrent drought, and blistering heat. On the other side are those who claim that policies to control emissions of greenhouse gases are premature given both the uncertainty of the science and the economic cost. Most likely the truth lies somewhere in between.

Mr. Johnston acknowledged an IPCC report but also referred to a recent report from a right-wing think tank[65]:

> On the other side of the spectrum is the recent George C. Marshall Institute report which indicates that theoretical estimates of the greenhouse effect seriously exaggerate the threat of global climatic change. The Marshall Institute report observes that although the Earth's temperature has risen somewhat, increasing concentrations of greenhouse gases may not have been the cause.
>
> I hope this hearing will help explain such discrepancies in the science of global climatic change. As members of the scientific community, your task is to provide the best possible assessment of the expected rate and severity of global climate change. It is crucial that we be in a position fully to understand both the scientific theory underlying the greenhouse effect, as well as the uncertainties that remain, in order to make an informed decision regarding domestic and international policy responses.

Dr. Lindzen, who was a member of the Scientific Advisory Board of the Marshall Institute, was the main contrarian at the session. He began his testimony attempting to defend the report from the Marshall Institute[66]:

> I have great regard for Bill Nierenberg, Fred Seitz, who I think were involved, and I think as a document, it is an extremely different document from the IPCC. There is no question that it is the document a few very distinguished scientists approaching information that all of us have to approach and reasonably looking at problems with it. They reach conclusions in some instances that I don't agree with, and in some instances, I feel they have bent over backwards to accept the common scenario. In some instances, I think they have gone the other way. So, I don't feel that there is a profound bias in that. They are pointing to obvious discrepancies that people should be aware of.

Afterwards, he dismissed the impact of the greenhouse effect and claimed that any small rise in temperature would not matter.

The premise of the session was, of course, absurd. The IPCC had assessed thousands of studies from the peer-reviewed scientific literature, whereas the report from the George C. Marshall Institute was an opinion piece from a few senior contrarians, and yet these two sources were being treated as somehow equal by Congress.

A later session during the Clinton administration was extraordinary for bringing three well-known contrarians together as witnesses – Patrick Michaels, John Christy, and Richard Lindzen (among others). Dr. Michaels, as usual, complained about the climate models, while Dr. Christy argued global warming was rising more slowly than most thought, and Dr. Lindzen raised the uncertainty of the science on climate change. During the question period, it was mentioned that contrarians like Dr. Lindzen had been labelled by the Union of Concerned Scientists as "junk scientists."[67]

Dr. Lindzen was also active in congressional hearings early in the term of President George W. Bush[68]:

> I come here usually designated as a skeptic. I am not sure what that means. I think in dealing with this, people are correct in saying that the science is complex, and I think the complexity is not only intrinsic, but has also resulted from the presentation of the issue, which in many ways has forced confusion and irrationality to dominate the discussion. It is presented as a multifaceted problem involving atmospheric composition, heat transfer, weather, temperature, ocean dynamics, hydrology, sea level, glaciology, ecology, and even epidemiology. All of these are subjects filled with uncertainty.
>
> On the other hand, and I do not say any of my colleagues here today have done this, but you know that it is frequently said the science is settled. This is often said without any statement as to exactly what is meant by this, and what relevance it has to the forecast being made …
>
> … I would add to that that man, like the butterfly, has some impact on climate.

At a separate hearing, the following day, Dr. Lindzen attempted to disparage the IPCC[69]:

> The IPCC does deserve some consideration, if not necessarily to criticize it severely, at least understand what the procedures mean. The procedures are in many ways extremely opaque, and certainly the claim of support by thousands of outstanding scientists is more a mantra than a reasonable statement …
>
> I would maintain that we have, to a very large extent, built into our scientific process a predilection for alarmism. There is no easier way to justify science than alarmism. The very fact that meetings such as this do in general endorse more research will convince the scientists that the way to get more

> support for research is to promote alarmism. I think one of the main things we can do is figure out how to support science without causing it to have this bias.

At this hearing, Senator Inhofe (1934–) (Republican from Oklahoma), similarly, had attacked the IPCC. According to him, if the U.S. initiated policies based on the IPCC report, it would be[70]: "a real live horror story." However, Senator Hillary Clinton (1947–) (Democrat from New York) defended the mainstream science[71]:

> Now how are we going to judge the sound science? Obviously, we are going to listen to researchers and scientists … the debate is over. It is just a question of what we are going to do in order to address rising temperatures and their impacts.

Later in the session, Dr. Lindzen added in response to questions[72]:

> We have, in my opinion, at least a century to monitor the system in order to see if actions will be needed to preclude such a possibility. This will leave, in my opinion, adequate time to take suitable measures especially since we can reasonably assume that we will have greater resources at that time to do so. Should I prove wrong, evidence over the next 30 years will show this.

At the start of the millennium, for a scientist to suggest that we wait a hundred years more before acting on climate change was as contrarian as anyone could get. The denial speak of Dr. Lindzen was intended to befuddle the politicians which was not hard to do (as many did not want to hear the scientific truth). As with most contrarians, he never changed his extreme opinions, even after these were found to be demonstrably wrong.

President Bill Clinton: 1993–2001

The beginning of 1993 ushered in a new administration and high expectations that the U.S. would have a more progressive approach to the climate change negotiations under the leadership of Bill Clinton (1946–). The energy-industrial complex must have been worried when Bill Clinton got elected; not only was the new Democratic president looking for aggressive action on climate change, Al Gore, the leading climate change champion in American political circles, was now vice president. Bill Clinton and Al Gore had been the dream team for action on climate change. In an incredible turn of events, what should have led to vigorous climate policies turned into none. What happened?

President Clinton had wasted no time jumping into the fire of climate change politics: In February 1993, he tabled the "Btu tax" on energy.[73] Then a more aggressive policy was announced on April 21st (Earth Day)[74]:

> We must take the lead in addressing the challenge of global warming that could make our planet and its climate less hospitable and more hostile to

> human life. Today, I reaffirm my personal, and announce our nation's commitment to reducing our emissions of greenhouse gases to their 1990 levels by the year 2000.

These proposed policies were the first initiatives by an American president to tackle climate change head on, far exceeding what most other countries were doing, and they were a complete failure. The "Btu tax" was defeated in the Senate,[75] and his Climate Change Action Plan, issued in October 1993, still had a target to reduce emissions to 1990 levels by 2000, but only with voluntary reductions.[76] The Democratic president had been moving too fast, even for his own party. A greater loss for the dream team, however, was the failure in international negotiations for the Kyoto Protocol. The details of this fiasco and the poisoning of the relationship between the White House and the Senate, which culminated around the Byrd-Hagel Resolution, are discussed in the next section.

As the millennium and his term in office were ending, President Clinton discussed climate change one last time in his final State-of-the-Union Address[77]:

> The greatest environmental challenge of the new century is global warming. The scientists tell us the 1990's were the hottest decade of the entire millennium. If we fail to reduce the emission of greenhouse gases, deadly heat waves and droughts will become more frequent, coastal areas will flood, and economies will be disrupted. That is going to happen, unless we act.
>
> Many people in the United States, some people in this Chamber, and lots of folks around the world still believe you cannot cut greenhouse gas emissions without slowing economic growth. In the industrial age, that may well have been true. But in this digital economy, it is not true anymore.

Clinton's warning to the nation went unheeded. But all was not lost, as some hope remained that Al Gore might win the upcoming election.

Congress

The big climate event during Clinton's term was Kyoto. The road to Rio during George H. W. Bush's term had been a rough ride, but at least America made it there, not so for Kyoto. A troubling issue was that the bureaucrats learned nothing from the previous deal; in the coming negotiations for Kyoto, they continued to tell the Americans what to do, ensuring that Kyoto would ultimately fail. Even with a pro-environment government, the political chains holding back the United States were still in place, and the next agreement at Kyoto would never be ratified by the Americans.[78]

Congressional hearings on climate change increased dramatically during the Clinton administration; however, a good number of these were related specifically to the heated negotiations leading up to Kyoto (and later COP meetings), and not the science itself. When Clinton's Kyoto negotiating team showed up to testify, the hearings turned into a cross-examination. The political divide on climate change between Congress and the White House during Clinton's administration had reached new heights, as illustrated by the personal bias shown in the comment of Bob Inglis (1959–) (Republican from South Carolina)[79]: "In my first six years in Congress from 1993-1999, I had said that climate change was hooey. I hadn't looked into the science. All I knew was that Al Gore was for it, and therefore I was against it." However, the most dramatic example of the congressional divide with President Clinton was the unanimous passing of the Byrd-Hagel Resolution.

Byrd-Hagel Resolution

President Clinton had damaged his relationship with Congress, even among his own party, by pushing aggressively for climate change legislation. As time passed, the relationship worsened, which led to a notorious moment in American climate policy development: Only months before the Kyoto meeting was about to begin, on July 25, 1997, the Senate took draconian action with the Byrd-Hagel Resolution (the Senate would not agree to any deal unless developing countries had GHG reduction targets as well, which was not going to happen).[80] The breakdown of the relationship between the White House and Capitol Hill is clearly illustrated by the unanimous passing of the Resolution. Among the Democrats who voted for the Resolution were Joe Biden (Democrat from Delaware), Barbara Boxer (Democrat from California), and John Kerry (Democrat from Massachusetts). This was the only time that Republicans and Democrats acted in a bipartisan manner on climate change.

On November 12, 1998, Al Gore let the world know that the Kyoto Protocol would not be ratified by the U.S. government[81]:

> Signing the Protocol, while an important step forward, imposes no obligations on the United States. The Protocol becomes binding only with the advice and consent of the U.S. Senate. As we have said before, we will not submit the Protocol for ratification without the meaningful participation of key developing countries in efforts to address climate change.

Chuck Hagel (1946–) (Republican from Nebraska), who was a co-sponsor of the Resolution, had continued to espouse the denialist view of many Republicans in Congress[82]:

> There is a reason the organization is called the Intergovernmental Panel on Climate Change. The summaries are political documents drafted by government representatives after intense negotiating sessions …
>
> The working group reports vary widely in their scientific conclusions and predictions for global warming during the next century, but the summaries tend to take very alarmist viewpoints which are then used to justify the draconian measures of the Kyoto Protocol. The IPCC summaries are not science, they are summaries. Furthermore, the predictions made by the IPCC are based on computer models, which have already been shown to be inadequate, and vary widely in their interpretations.

However, Chuck Hagel (along with Bob Inglis) would be one of the few political climate deniers to undergo the conversion into an advocate for climate action, for in a much later congressional hearing in 2019, Mr. Hagel told a different story[83]:

> As scientists reduced uncertainty about climate change over the last two decades, it became clear, very clear, that the U.S. must implement policies to address the challenge, prepare, because climate change is threatening our economy, our environment, and our national security.

And, when Alexandria Ocasio-Cortez (1989–) (Democrat from New York) directly asked Mr. Hagel if "denial [of climate change] or even delaying in that action could cost us American lives," he answered "yes."[84]

In 2022, Chuck Hagel shared what really had happened at the discussions over the Byrd-Hagel Resolution a quarter century before[85]:

> What we now know about some of these large oil companies' positions … they lied. And yes, I was misled. Others were misled when they had evidence in their own institutions that countered what they were saying publicly. I mean they, lied.

While there were missteps made by President Clinton in his handling of the climate agenda with the Senate,[86] Congress had been bamboozled by the energy-industrial complex.

President George W. Bush: 2001–9

Although his father had started the climate denial ball rolling in the White House, it really gained momentum during the presidency of George W. Bush, when the infiltration of climate denial within the Republican Party rapidly progressed; this was not led by George W. Bush, but he willingly supported it.

In the Republican election platform, the President-elect George W. Bush outlined his position on the Kyoto Protocol[87]:

> Complex and contentious issues like global warming call for a far more realistic approach than that of the Kyoto Conference … More research is needed to understand both the cause and the impact of global warming. That is why the Kyoto treaty was repudiated in a lopsided, bipartisan Senate vote. A Republican president will work with businesses and with other nations to reduce harmful emissions through new technologies without compromising America's sovereignty or competitiveness, and without forcing Americans to walk to work.

The stock phrase "more research is needed" is typical of the climate inaction platforms of many conservative parties.

The door to an international agreement on climate change had been slammed shut during Clinton's administration, and now the door was locked, and the key was thrown away. Early in his term, on March 28, 2001, President Bush officially declared that the U.S. would not implement the Kyoto Protocol. Global reaction, surprisingly, was one of shock and dismay.[88] However, all through the negotiations, the U.S. position had been clear and consistent on what was required for it to be part of an international agreement on climate change, and Kyoto did not cut it. If Al Gore had won the election, the U.S. still would not have ratified Kyoto.

The White House, predictably, requested another study from the National Academy of Sciences[89]:

> The Administration is conducting a review of U.S. policy on climate change. We seek the Academy's assistance in identifying the areas in the science of climate change where there are the greatest certainties and uncertainties. We would also like your views on whether there are any substantive differences between the IPCC Reports and the IPCC summaries.

This request was less sincere than usual, as the White House was indirectly accusing the IPCC of chicanery: "any substantive differences between the IPCC Reports and the IPCC summaries." The summaries had been prone to political sanitization, but this was well known to the Americans, who were often doing the sanitizing. The administration was especially interested in the uncertainties of the science, as a justification for not taking climate action.

The NAS study was supportive of the IPCC and concluded[90]:

> The IPCC's conclusion that most of the observed warming of the last 50 years is likely to have been due to the increase in greenhouse gas concentrations accurately reflects the current thinking of the scientific community on this issue … Despite the uncertainties, there is general agreement that the observed warming is real and particularly strong within the past 20 years …
>
> The committee finds that the full IPCC Working Group I (WGI) report is an admirable summary of research activities in climate science, and the full report is adequately summarized in the Technical Summary.

Following the study done by the National Academy of Sciences, President Bush praised the report and recognized the importance of climate change[91]:

> The issue of climate change respects no border. Its effects cannot be reined in by an army nor advanced by any ideology. Climate change, with its potential to impact every corner of the world, is an issue that must be addressed by the world …
>
> My Cabinet-level working group has met regularly for the last 10 weeks to review the most recent, most accurate, and most comprehensive science. They have heard from scientists offering a wide spectrum of views. They have reviewed the facts, and they have listened to many theories and suppositions …

> The policy challenge is to act in a serious and sensible way, given the limits of our knowledge. While scientific uncertainties remain, we can begin now to address the factors that contribute to climate change.

The speech may have sounded reasonable, but there are climate denial talking points sprinkled throughout and no formal actions were proposed which became a standard approach on climate change for right-wing politicians in America and elsewhere.

Still President Bush had started something new for a Republican president: He was going to talk a great deal about climate change; on February 14, 2002, for example, he had more to say[92]:

> In pursuit of this goal, my government has set two priorities: we must clean our air, and we must address the issue of global climate change. We must also act in a serious and responsible way, given the scientific uncertainties …
>
> My administration is committed to cutting our nation's greenhouse gas intensity – how much we emit per unit of economic activity – by 18 percent over the next 10 years. This will set America on a path to slow the growth of our greenhouse gas emissions and, as science justifies, to stop and then reverse the growth of emissions …
>
> Addressing global climate change will require a sustained effort over many generations. My approach recognizes that economic growth is the solution, not the problem. Because a nation that grows its economy is a nation that can afford investments and new technologies.

This was a lengthy sermon full of good spin, stating he would do nothing; of greater concern was Bush's alarming notion that "addressing global climate change will require a sustained effort over many generations;" science already knew that we did not have many generations to address this problem.

There were other examples of President Bush spinning the issue, including during the 2007 State of the Union Address[93]:

> America is on the verge of technological breakthroughs that will enable us to live our lives less dependent on oil. And these technologies will help us be better stewards of the environment, and they will help us to confront the serious challenge of global climate change.

To which Bernie Sanders (1941–) retorted[94]: "I notice that some have said, 'Well, isn't it great; the President of the United States actually uttered the words global climate change.'"

On April 16, 2008, President Bush boasted about his achievements on climate change[95]:

> So we've pursued a series of policies aimed at encouraging the rise of innovative as well as more cost-effective clean energy technologies that can help

> America and developing nations reduce greenhouse gases, reduce our dependence on oil, and keep our economies vibrant and strong for decades to come …
>
> I believe that Congressional debate should be guided by certain core principles and a clear appreciation that there is a wrong way and a right way to approach reducing greenhouse gas emissions. Bad legislation would impose tremendous costs on our economy and on American families without accomplishing the important climate change goals we share …
>
> The growth in emissions will slow over the next decade, stop by 2025, and begin to reverse thereafter, so long as technology continues to advance …

President Bush had cautioned Congress about "bad legislation" on dealing with greenhouse gas emissions. Congress had heeded his warning and never passed any bad legislation; Congress also did not pass any good legislation to deal with greenhouse gas emissions (until 2022).

Late in Bush's second term, there was a milestone in the development of climate policymaking when the U.S. Supreme Court declared that carbon dioxide and other greenhouse gases were an "air pollutant."[96] In response to the court decision, the EPA amended the Clean Air Act.[97] Greenhouse gas emissions finally had to be considered under existing pollution legislation which forced the hand of the executive branch to deal with climate change, and the next president embraced the opportunity.

Congress

In early 2007, a hearing, "Senators' Perspectives on Global Warming," provided a unique opportunity to hear the views on climate change from the major political actors of the day. In this extraordinary session, all senators had been invited to give their opinions on climate change, and a surprisingly good number of them (about one-third) showed up; the all-star cast included: Barbara Boxer, Hillary Clinton, James Inhofe, John Kerry, John McCain, Barrack Obama, Bernie Sanders, and Sheldon Whitehouse. Senator George Voinovich (1936–2016) (Republican from Ohio) jokingly commented on the disparity of views within the committee, where climate change had been called everything from the "greatest hoax" to the "greatest problem."[98] Senator Larry Craig (1945–) (Republican from Idaho) had asked the Congressional Research Service to examine how much had been spoken in the Senate on climate change since the 102nd Congress (1991), and the report was almost 1,000 pages long.[99]

During this session, viewpoints generally followed traditional party lines. One of the few exceptions was John McCain (1936–2018) (Republican from Arizona) who came out in favour of climate action. At the other extreme of the G.O.P. was the prepared statement by Senator Michael Enzi (1944–) (Republican from Wyoming)[100]:

> We do not trust our weathermen to predict the temperature a week in advance, and so it is difficult for me to believe that individuals can predict the

> weather 100 years from now. Particularly given that just a few decades ago, we were told that the world was entering the next ice age, I struggle to see how some can discuss the issue with absolute certainty.
>
> Because the science is not settled on the issue of climate change, I will not support any actions that will put the United States at an economic disadvantage without any guarantees that the problem is real and without any guarantees that these so-called solutions will address the issue.

Even for a Republican, Senator Enzi, by milking the global cooling schism popular in the 1970s, was behind the times.

Another major event in the history of climate-policy making was Al Gore's documentary, *An Inconvenient Truth*, in late spring of 2006; the pushback from the Republicans began immediately with a series of congressional hearings featuring contrarians (of course). Noteworthy attacks came from Joe Barton (1949–) (Republican from Texas) and Ralph Hall (1923–2019) (Republican from Texas); both were members of the denial cabal, discussed further in the next chapter. At one such hearing, Al Gore, no longer a member of Congress, greeted them both, saying[101]: "We were close friends before you went over to what we jokingly refer to as 'the dark side.' And we are friends still." Mr. Barton did not hesitate to criticize his friend by providing science-sounding rhetoric[102] and complaining[103]: "Your suggestion of a carbon tax is something that would harm our competitiveness, raise costs to American families, export jobs and actually do very little to improve our environment," which was a standard response used by the denial cabal. And Mr. Hall followed with a similar argument, referencing a new contrarian Bjorn Lomborg.[104]

At the same session, John Barrow (1955–) (Democrat from Georgia) made an interesting observation on the changes in the style of testimonies over the past 20 years[105]:

> And just in the hearings we have had, I detected something of a pattern on this subject. We opened up with a hearing with the representatives of the scientific community who came before us and said in no uncertain terms there is a problem. We are behind it. And we have to do something about it. Then we had a series of hearings with what I might call the impact community or the solution community, a series of hearings where we are consulting with various sectors of the economy …
>
> And in the course of these hearings, in the course of them, I detected sort of a pattern that has emerged. When the experts on the problem were here, there wasn't much going on. There wasn't much back and forth on the subject of whether there was a problem or not. But when we get into the solution hearings and impact hearings, we get a lot of folks pooh-poohing or putting down the issue and raising questions …

He was describing, basically, the political transition from climate scientists to contrarians testifying at hearings.

Dr. Bjorn Lomborg

A new "softer" breed of contrarian, called climate delayers[106] or inactivists,[107] began to appear now at congressional hearings with more science-friendly messages. They accepted that climate change was real and anthropogenic, but these scientists were just as resistant to policies to reduce fossil fuel consumption as any other contrarian; the climate scientist Michael E. Mann (1965–) explained[108]:

> As climate change denial shifts from outright denial of the physical science to denial of the scale and magnitude of the threat … fossil fuel front organizations like the Heartland Institute too are evolving in their rhetoric toward the 'kinder, gentler' form of denial.

Or, as Alexandria Ocasio-Cortez tweeted[109]: "Climate delayers aren't much better than climate deniers. With either one if they get their way, we're toast."

An early member of this new group of contrarians was the political scientist Bjorn Lomborg.[110] Giving testimony at the same congressional hearing mentioned above, he praised Mr. Gore[111]: "for making global warming cool again" before criticizing the former vice president[112]:

> Al Gore and the many people he has inspired have goodwill and great intentions. However, he has got carried away and has come to show only worst-case scenarios and I think we need to recognize that. This is unlikely to form the basis of good judgment. The problem is compounded in that if we follow Al Gore's recommendations, we will likely end up choosing very bad policies to solve the many problems that we agree need to be addressed. In short, Mr. Gore's logic, with all its good intentions and sincerity, would in effect present an obstacle to saving millions of lives.

Dr. Lomborg ended his presentation with more condescending remarks[113]:

> We are not likely to make good judgments if we vastly exaggerate the bad consequences of global warming and forget the positive incidences of global warming. And that also means we need smarter solutions. The solutions that are being proffered right now are doing very little good at very high cost. There are much better ways to do it; for instance, investment in research and development … Did you spend it on vast, frivolous projects like the Kyoto Protocol? Or did you actually spend it on a lot of things that would end up doing a lot of good for the world first?

In 2014, during the presidency of Barack Obama, Bjorn Lomborg testified that[114]: "Inaction has costs, but so does action. It is likely that climate action will lead to higher total costs in this century." Dr. Lomborg has been the most frequent non-American contrarian at congressional hearings.

I have questioned Dr. Lomborg on Twitter in #ClimateBrawl about his use of the term "alarmist" in his tweets, to which he answered[115]: "Alarmist: someone who *incorrectly* say millions will die, when right number close to ~zero. Not a complicated definition, I would think." He clarified that he was referring to deaths only from sea level rise and "not from all impacts." When I challenged him for having no training in the climate sciences, Dr. Lomborg replied[116]: "Climate impact on society is — not surprisingly — a social science question (with climate science input). That's what I do." Dr. Lomborg never explained to me, despite my asking, the source of his "climate science input."

During the administration of George W. Bush, contrarians had become a routine fixture at congressional hearings on climate change where they gave their opinions discrediting the peer-reviewed literature creating doubt about the scientific consensus. The Republicans relished their testimony. Disappointingly, there was little pushback from Democrats who seemed at a loss about how to respond to the climate denial that had infected the hearings. The Democrats came better prepared during the Obama administration.

President Barack Obama: 2009–17

Before becoming president, Senator Obama (1961–) pushed for action to avert a climate crisis[117]:

> For decades, we have been warned by legions of scientists and mounds of evidence that global warming is real, that we couldn't just keep burning fossil fuels and contributing to the changing atmosphere without consequence. Yet for decades, far too many have ignored the warnings, either dismissing the science as a hoax … [page]
>
> The climate changes we are experiencing are already causing us harm, but in the end, it will not primarily be us who deal with its most devastating effects. It will be our children and our grandchildren.
>
> This is our generation's chance to protect their futures.

The speeches given by Barack Obama and George W. Bush on climate change stand in sharp contrast to one another, with Senator Obama stressing the importance of urgent action. His passionate interest in this issue continued into his presidency.

At the start of his second term, President Obama again made the case for action on climate change[118]:

> Some may still deny the overwhelming judgment of science, but none can avoid the devastating impact of raging fires and crippling drought and more powerful storms. The path towards sustainable energy sources will be long and sometimes difficult. But America cannot resist this transition, we must lead it. We cannot cede to other nations the technology that will power new jobs and new industries, we must claim its promise.

Unlike his predecessor who only talked about climate change, on March 27, 2012, President Obama introduced bold policy initiatives by issuing an executive order that the EPA would modify the Clean Air Act to include greenhouse gas emissions from new power plants.[119] A major policy that emerged from the amended act was the Clean Power Plan in the summer of 2015[120]:

> Right now, our power plants are the source of about a third of America's carbon pollution. That is more pollution than our cars, our aeroplanes, and our homes generate combined …
>
> There will be critics on what we are trying to do. There will be cynics that say it cannot be done … the special interests and their allies in Congress were already mobilizing to oppose it with everything they got. They will claim that this plan will cost you money… They claim we need to slash our investments in clean energy; it is a waste of money, even though they are happy to spend billions per year in subsidizing oil companies. They claim that this plan will kill jobs …
>
> We're the first generation to feel the impact of climate change. We're the last generation that can do something about it. We only get one home. We only get one planet.

When introducing the Clean Power Plan, President Obama had voiced his concern over the deteriorating relationship with the Republicans who had become a party of political climate deniers. During the 2015 State of the Union Address, President Obama called out Republicans for not answering questions on climate science and concealing their climate denial[121]: "I've heard some folks try to dodge the evidence by saying they're not scientists; that we don't have enough information to act. Well, I'm not a scientist, either. But you know what, I know a lot of really good scientists at NASA, and at NOAA, and at our major universities. And the best scientists in the world are all telling us that our activities are changing the climate." The political discourse on climate change, already deeply partisan, was showing no signs of improving.

The Paris Climate Agreement

After Rio and Kyoto, the next big destinations for climate negotiations were Copenhagen and Paris. The highly anticipated COP (Conference of the Parties of the UNFCCC) in Copenhagen[122] took place at the end of the year President Obama took office. Among the over one hundred world leaders who attended, media attention focussed on the new American president who announced a breakthrough agreement[123]:

> Earlier this evening I had a meeting with the last four leaders I mentioned – from China, India, Brazil, and South Africa. And that's where we agreed to list our national actions and commitments, to provide information on the implementation of these actions through national communications, with

> international consultations and analysis under clearly defined guidelines. We agreed to set a mitigation target to limit warming to no more than 2 degrees Celsius, and importantly, to take action to meet this objective consistent with science.

President Obama had set out to overcome the restrictions imposed by the Byrd-Hagel Resolution – all nations had to have a duty to do something – and the new deal was able to get around this. A formal agreement reached in Paris, on December 12, 2015, stated[124]:

> The Paris Agreement's central aim is to strengthen the global response to the threat of climate change by keeping a global temperature rise this century well below 2 degrees Celsius above pre-industrial levels and to pursue efforts to limit the temperature increase even further to 1.5 degrees Celsius.

The United States formally joined the Paris Agreement on September 3, 2016, and in early October, President Obama held a press conference in the Rose Garden of the White House where he described the historic agreement[125]:

> Ten months ago, in Paris, I said before the world that we needed a strong global agreement to reduce carbon pollution and to set the world on a low-carbon course. The result was the Paris Agreement …
>
> Now, keep in mind, the skeptics said these actions would kill jobs. And instead, we saw – even as we were bringing down these carbon levels – the longest streak of job creation in American history. We drove economic output to new highs. And we drove our carbon pollution to its lowest levels in two decades …
>
> Now, the Paris Agreement alone will not solve the climate crisis. Even if we meet every target embodied in the agreement, we'll only get to part of where we need to go. But make no mistake, this agreement will help delay or avoid some of the worst consequences of climate change … So this gives us the best possible shot to save the one planet we've got.

President Obama had, at last, positioned the United States at the forefront of tackling the global crisis, no minor accomplishment considering the toxic dynamics of American politics. Sadly, the progress was short lived.

Tweets from the Oval Office

Barack Obama had added something new into presidential politics – tweets from the Oval Office; he was the one who introduced @POTUS, as the official Twitter account of the "President of the United States." In his first tweet on this account, on May 18, 2015, he joked[126]: "Hello, Twitter! It's Barack. Really! Six years in, they're finally giving me my own account," which gained over 375K likes. Later that day, Bill Clinton welcomed him to Twitter.[127] The @POTUS account, though, belongs to the sitting president, so it would be inherited later by Donald Trump and then Joe Biden (tweets from President Obama are now archived under @POTUS44).

Senator Obama had been an early member of Twitter. His first account, @BarackObama opened in March 2007, has become the most popular account in Twitter history, until early 2023 when passed by @elonmusk. A twitterverse milestone took place on March 25, 2009, when Barack Obama became the first president of the U.S. to tweet.[128] Another historic Twitter moment had taken place on August 19, 2013, when Barack Obama tweeted Bill Clinton,[129] which was the first tweet between two U.S. presidents.

President Obama began tweeting about climate change on December 8, 2009[130]: "Yesterday we took important steps toward solving the global problem of climate change" (in reference to the Copenhagen conference). He would tweet about climate change frequently. Obama's favourite hashtag is #actonclimate, which he has used 389 times while tweeting under @BarackObama.[131]

Congress

Early during the Obama administration, Steven Chu (1948–) Secretary of Energy was being grilled in a House hearing, to which someone had inserted subtitles in the transcripts (which I had not seen before), including some related to climate denial. One was called "Skepticism of Global Climate Change," where Dana Rohrabacher (1947–) (Republican from California) shared some standard climate denial talking points[132]:

> That is why I guess people have changed the wording to — now it is not global warming, now it is, you know, something to do with climate change. There are prominent scientists, more and more prominent scientists, every day joining the ranks of those who are suggesting that the whole global warming theory is bogus.

Another subtitle was called "More Climate Change Denial," where Paul Broun (1946–) (Republican from Georgia) had made even more extreme claims[133]:

> There are a tremendous number of scientists who would absolutely debunk any human causes of global warming, and I think just for scientific integrity, I ask that you go and look at those things because there is no consensus …
>
> I am an applied scientist, and I believe in scientific integrity. If you will look at a lot of other writings that are peer reviewed, there are many sources of data that show that human-induced global warming is a myth. And I request that you take off your blinders and bias and look because there are many, many scientists who would debunk this whole idea, and it is going to kill our economy and it is going to particularly hurt poor people and people on limited incomes. And to go down this track is going to kill our economy.

A week before, there had been a truly bizarre hearing in the House of Representatives which began sensibly enough when Jim McDermott (1936–) (Democrat from Washington) opened the session[134]:

> Reducing the threat of climate change means reducing the emission of harmful greenhouse gases, such as carbon dioxide. There are various proposals for achieving this goal, but they all ultimately seek to impose some form of limit on those emissions by making the cost reflect their impact on our climate.

But the session quickly deteriorated when his Republican colleague disagreed; John Linder (1942–) (Republican from Georgia) rebutted with a lengthy reply[135]:

> Not so fast. Science is not a democracy. The head count fallacy has been recognized as irrational since Aristotle. Even if science were a democracy, for every scientist who supports the notion of human-caused global warming, there are more than ten who consider that notion pure vanity, and they have made their names public …
>
> Areas are studied for centuries, and then are proven by facts to be correct or incorrect. Both Galileo and Einstein were famous deniers of centuries-old theories. They were right. The consensus was wrong …
>
> Today predictions of future weather calamities are being made by computer games that do not take into consideration scientific observations of the Earth's natural temperature modulations.

The hearing, though, really went off the rails when a surprise witness appeared, the British climate denier Lord Christopher Monckton (1952–), who testified about "a harmless and beneficial trace gas:"[136]

> I should like to warn this honorable House that any proposal to inflict billions of dollars of new taxation on all citizens by charging selectively disfavored industries who arbitrarily rationed permits to emit a harmless and beneficial trace gas [carbon dioxide] that is necessary to all life on Earth and has little effect on its surface temperature, will fall cruelly and disproportionately upon the poor; will threaten their very lives; will gravely diminish the liberty that is the glory of your great nation; will render difficult, if not impossible, the pursuit of happiness.

A month later, at another hearing at the House of Representatives, Lord Monckton returned to Washington.[137] He had been warmly welcomed by the Republican Joe Barton, who said that he was[138]: "one of the most knowledgeable … from a skeptical point of view on this issue of climate change." When Lord Monckton testified that carbon dioxide was "plant food," and we lived in a "carbon-starved planet," John Shimkus (1958–) (Republican from Illinois) was elated.[139] In his opening remarks at this hearing, Mr. Shimkus had started off with verses from the Bible and stated[140]: "The earth will end only when God declares it is time to be over. Man will not destroy this earth … There is a theological debate that this is a carbon starved planet, not too much carbon." These eccentric comments by a senior official in Congress fit well with those made by Lord Monckton who had many supporters among the G.O.P.

The following year, Lord Monckton was back again in Congress for a House hearing called "The Foundation of Climate Science." Ed Markey (1946–) (Democrat from Massachusetts) introduced the session[141]: "Those who deny global warming point to past uncertainties that have been refuted. They ignore the overwhelming observational evidence that the increased levels of heat-trapping pollution are already warming the planet … While the deniers hope to confuse the public, the real-world consequences of inaction mount." Lord Monckton, in his written statement for this hearing, concluded[142]:

> First, it would be orders of magnitude more cost-effective to adapt to any 'global warming' that might occur than to try to prevent it from occurring by trying to tax or regulate emissions of carbon dioxide in any way. Second, there is no hurry.

As with Monckton's other "expert" comments about climate change, these views were completely at odds with the greater scientific community.

Later during the Obama administration, the contrarian Roger Pielke Jr. was called upon to testify where he discussed extreme weather events.[143] Joining him at this session was the contrarian Roy Spencer (1955–)[144]:

> Skeptics generally are skeptical of the view that recent warming is all human-caused, and/or that it is of a sufficient magnitude to warrant immediate action given the cost of energy policies to the poor. They do not claim humans have no impact on climate whatsoever.

Dr. Spencer closed his testimony with[145]: "Thus, the evidence that humans are mostly responsible for either recent warmth or severe weather changes (if such changes exist at all) is equivocal, at best." As with most contrarians, Dr. Spencer was not presenting peer-reviewed science, just his opinion about peer-reviewed science.

Democrats were increasingly pushing back on the use of contrarians by the Republicans. In a House hearing in 2015, Raúl Grijalva (1948–) (Democrat from Arizona) came out swinging against politically driven science[146]:

> Manufacturing doubt can be done in many different ways. You can question the quality of the science that indicates there is a problem with your product. You can find a scientist who agrees with you and fund their research or give them media training. You can take the research to the press and demand that they print both sides of the story as if there was really a controversy.
>
> In a hearing about politically driven science, climate denial is the ultimate case study. Its tactics are well-documented, including paying scientists to produce industry-friendly science …

> There is an ongoing, well-organized, well-funded campaign to deny climate change and the established science behind it. That campaign poses serious risk to the quality of life on this planet and to the American people.

During the Trump administration, at the hearing called "Climate Change: Impacts and the Need to Act," Raúl Grijalva again expressed his frustration with the climate denial espoused by his Republican colleagues[147]: "Today we turn the page on this Committee from climate change denial to climate action … The Trump administration chooses to mock science and mislead the public about what our country will look like if we do nothing." At the end of the session, he mentioned the softer version of climate denial he had picked up on from the contrarian testimonies during the hearing[148]:

> We are into a different phase, which is climate change avoidance. And what can we do to stall, change, tinker with the science, raise issues that are meant to slow any solution-seeking or policies or legislative initiatives to deal with this very urgent problem.

Judith Curry had been a witness at this hearing.

Dr. Judith Curry

During the Obama administration in a Senate hearing, Judith Curry[149] shared the epiphany that had led to her conversion to become a contrarian[150]:

> Prior to 2009, I felt that supporting the IPCC consensus on climate change was a responsible thing to do. I bought into the argument don't trust what one scientist says, trust what an international team of 1,000 scientists have said after years of careful deliberation.
>
> That all changed for me in November 2009, following the leaked "Climategate" e-mails that illustrated the sausage making and even bullying that went into building the consensus. I started speaking out, saying that scientists needed to do better at making the data and supporting information publicly available, being more transparent about how they reach conclusions, doing a better job of assessing uncertainties, and actively engaging with scientists having minority perspectives.
>
> The response of my colleagues to this is summed up by the title of a 2010 article in the Scientific American, "Climate Heretic: Judith Curry Turns on Her Colleagues." I came to the growing realization that I had fallen into the trap of group think …
>
> As a result of my analyses that challenge IPCC conclusions, I have been called a denier by other climate scientists and most recently by Senator Sheldon Whitehouse.

The reason given for her conversion, "climategate" (the public release of illegally hacked emails from some IPCC scientists), makes little, if any, sense, as several investigations found there had been no misconduct on the part of the scientists.

Dr. Curry was a witness at the hearing on "Data or Dogma? Promoting Open Inquiry in the Debate Over the Magnitude of Human Impact on Earth's Climate," where Ted Cruz (1970–) (Republican from Texas) opened the session by labelling mainstream scientists "alarmists" and accused Democratic members of the committee of trying to censor "actual science."[151] The "actual science" that Senator Cruz was protecting from censorship was the testimony of the contrarians, John Christy, Judith Curry, William Happer (1939–)[152] and Mark Steyn (1959–).[153] Ted Cruz also stated[154]:

> And yet an individual named Galileo dared to actually be a scientist and take measurements and stand up to that enforced consensus. And I would note it was the Roman Inquisition that brought heretics before it who dared to say that the Earth rotates around the Sun, and today the global warming alarmists have taken the language of the Roman Inquisition, going so far as labeling anyone who dares point to the actual science as a denier, which is, of course, the language of religion. It is calling someone a heretic.

This hearing, chaired by Senator Cruz, was itself an inquisition. Overall, the presence of so many contrarians as witnesses and Cruz's own climate denial had made a mockery, not of science, but of Congress.

Judith Curry was a witness at the Committee on Environment and Public Works, where she explained[155]: "My research on understanding the dynamics of uncertainty at the climate science-policy interface has led me to question whether these dynamics are operating in a manner that is healthy for either the science or the policy process." She then closed her testimony with the remark[156]: "Motivated by the precautionary principle to avoid dangerous anthropogenic climate change, attempts to modify the climate through reducing CO_2 emissions may turn out to be futile." After questioning by Senator Whitehouse (1955–) (Democrat from Rhode Island), Dr. Curry, who resented being called a contrarian or a denier, defended her "skepticism:"[157]

> Skepticism is one of the norms of science. The way that we test theories and ideas is to challenge them. And a good theory will be able to defend itself against challenges. When people try to defend their theory by calling people who challenge their theory by names, deniers, whatever, that is not a good sign that it is a strong theory … I don't think climate scientists would call me a contrarian.

A year later, Dr. Curry had again objected to being labelled a climate denier[158]:

> As a result of my analyses that challenge IPCC conclusions, I have been called a denier by other climate scientists and most recently by Senator Sheldon Whitehouse. My motives have been questioned by Representative Grijalva in a recent letter sent to the president of Georgia Tech.

During the Trump administration, Dr. Curry was protesting the way contrarians were treated[159]:

> The 'war on science' that I am most concerned about is the war from *within* science – scientists and the organizations that support science who are playing power politics with their expertise and passing off their naïve notions of risk and political opinions as science. When the IPCC consensus is challenged or the authority of climate science in determining energy policy is questioned, these activist scientists and organizations call the questioners 'deniers' and claim 'war on science.' These activist scientists seem less concerned with the integrity of the scientific process than they are about their privileged position and influence in the public debate about climate and energy policy. They do not argue or debate the science – rather, they denigrate scientists who disagree with them. These activist scientists and organizations are perverting the political process and attempting to inoculate climate science from scrutiny – this is the real war on science.

Legitimate criticisms of science are presented in the peer-reviewed literature, but contrarian opinions on accepted science can be met with fierce rejection by other scientists when only presented at congressional hearings (or in a blog).

At a House session in 2010 titled "A Rational Discussion of Climate Change: The Science, The Evidence, The Response," Republican Ralph Hall pointed out that Dr. Curry could not "easily be characterized as a climate change 'denier' or 'skeptic,'" and wanted to know why so many scientists were out to get "skeptics." In reply, she mentioned her blog, "Climate Etc.," and offered the following[160]:

> While free market fundamentalism and 'big oil' may have been a major source of skepticism in the past, the current dominant group of skeptics, enabled by the blogosphere, seeks accountability. Many of these skeptics have professional backgrounds and extensive experience with the practical application of science and regulation, without any particular political motivations and certainly without funding from 'big oil.'

A blog had become Curry's favourite platform for her contrarian views. Blogs are sites, especially popular with climate deniers, where any claims can be published without scrutiny.

At a much later hearing (and at others[161]), Dr. Curry again referenced her blog[162]:

> For the past decade, I have been promoting dialogue across the full spectrum of understanding and opinion on the climate debate through my blog Climate Etc. (judithcurry.com). I have learned about the complex reasons that intelligent, educated and well-informed people disagree on the subject of climate change, as well as tactics used by both sides to try to gain a political advantage in the debate.

In this hearing, Dr. Curry used the term "uncertain" (and derivatives thereof) over two dozen times in her formal statement.[163]

When I asked Dr. Curry on Twitter if she had peer-reviewed studies to support her claims, she replied[164]:

> I abandoned academic 'gamesmanship' 3 yrs ago (~200 peer reviewed publications from my academic era; I still publish a few papers/yr) … I abandoned academia, which has become highly politicized, to pursue science in its purest form. The trappings of academia do not define a scientist or science.

Later in the Twitter thread, Dr. Curry boasted[165]: "Since you are such a stickler for peer review, I checked your citations on Google Scholar … and compared with mine … I don't think you really want to have this debate with me." Dr. Curry does indeed have more citations than I do – many more – but none of her citations disprove the science of climate change.

In another #ClimateBrawl engagement on Twitter, after I referred to Dr. Curry as a "contrarian," she corrected me[166]:

> I hold no opinion that is outside the bounds of IPCC AR6 WGI … Humans are contributing to climate change, through emissions and land use. Sun, volcanoes, long term ocean oscillations & geological processes also contribute to climate change. CO2 emissions should be reduced over the course of 21st C.

Since she holds "no opinion that is outside the bounds of IPCC," I asked Dr. Curry to provide the reference from the AR6 that said reducing carbon dioxide emissions was "not urgent;" she did not reply and later blocked me.

President Donald Trump: 2017–21

The new president mimicked Ronald Reagan; except he had a lot more than solar panels in the White House to rip out. President Donald Trump (1946–) initiated an "everything-Obama-did" purge.[167] He fired many top climate supporters within the bureaucracy and appointed climate deniers to key posts within the federal government. On March 28, 2017, an executive order was issued for the EPA to review the Clean Power Plan,[168] and a watered-down "Affordable Clean Energy" rule replaced it on June 19, 2019.[169]

President Trump even targeted the crown jewels of Obama's climate policies – the Paris Climate Agreement. On June 1, 2017, in the Rose Garden (where President Obama had celebrated his historic achievement less than a year before), he withdrew America from the treaty[170]:

> As President, I can put no other consideration before the wellbeing of American citizens. The Paris Climate Accord is simply the latest example of Washington entering into an agreement that disadvantages the United States to the exclusive benefit of other countries, leaving American workers – who I love – and

> taxpayers to absorb the cost in terms of lost jobs, lower wages, shuttered factories, and vastly diminished economic production.

At the announcement, he also sarcastically quipped: "I was elected to represent the citizens of Pittsburgh, not Paris." This was the second termination of an international agreement on climate change that had been signed by the United States; the previous Republican president had done the same with the Kyoto Protocol, at the beginning of the century. And who benefitted most from the tearing up of the Paris Climate Agreement? It was, of course, the energy-industrial complex.

Some global leaders were more diplomatic about the betrayal,[171] but many were outraged by this decision. Antonio Tajani (1953–), the president of the European parliament, expressed his disappointment[172]: "It is a matter of trust and leadership. This decision will hurt the US and the planet." Stronger words came from French president Emmanuel Macron (1977–) who stated President Trump had[173]: "committed an error for the interests of his country, his people and a mistake for the future of our planet … Don't be mistaken on climate: there is no plan B because there is no planet B." The Prime Minister of Belgium Charles Michel (1975–) tweeted[174]: "I condemn this brutal act against #ParisAccord @realDonaldTrump Leadership means fighting climate change together." Such international scorn is usually reserved for the most disturbing actions on the global stage, and President Trump had garnered more than most.

Other prominent figures also spoke out. The former president of Ireland Mary Robinson (1944–) regarded Trump's withdrawal as an[175]: "egregious act of climate irresponsibility," and the scholar Noam Chomsky (1928–) expressed his own strong views that[176]: "pulling out of the Paris negotiations should be regarded as one of the worst crimes in history." The physicist Stephen Hawking (1942–2018), as well, was deeply disturbed by the action of the president[177]: "By denying the evidence for climate change, and pulling out of the Paris Climate Agreement, Donald Trump will cause avoidable environmental damage to our beautiful planet, endangering the natural world, for us and our children."

President Trump remained steadfast in his climate denial despite all the criticism and was not even swayed by the alarming report from the National Climate Assessment the following year – among its stark conclusions[178]:

> rising temperatures, sea level rise, and changes in extreme events are expected to increasingly disrupt and damage critical infrastructure and property, labor productivity, and the vitality of our communities … is expected to cause substantial net damage to the U.S. economy throughout this century.

After its release, on November 26, 2018, President Trump was asked about the report, and he responded[179]: "Yeah, I don't believe it … I've seen it. I've read some of it, and it's fine."

In his one term as president, Donald Trump demolished the climate change initiatives of the previous administration, including the all-important Paris Climate Agreement. By electing a hard-core climate denier to the highest office,

America had placed the world on a much more dangerous path. There is more to come on President Trump in the following chapter.

Congress

The Trump term witnessed more pushback in Congress against climate denialism by Democrats. Eddie Bernice Johnson became the chair of the Science, Space and Technology Committee and, on February 13, 2019, at the hearing "The State of Climate Science and Why It Matters," she stated[180]: "Though this Administration has regrettably chosen to ignore the findings of its own scientists in regards to climate change, we as lawmakers have a responsibility to protect the public's interest. I plan to do this by making sure this Committee is informed by the most relevant and up to date science as we work to conduct our legislative and oversight responsibilities."

An interesting House hearing took place in April 2019, "Climate Change, Part I: The History of a Consensus and the Causes of Inaction." One of the expert witnesses, economist Jeffrey Sachs (1954–), placed the blame of the failure of Congress to act on funding by Big Oil[181]:

> With the science established, the climate disasters at hand, the future risks evident, the technological solutions available, and the diplomatic framework established, the question remains why Congress has so flagrantly failed?
>
> In my view, it is money. The oil industry supports the campaign funding of much of the Congress, much of this committee, and much of all of the Congress, especially on the Republican side.

A week before, Sean Casten (1971–) (Democrat from Illinois) also chastised Congress for its dismal performance on the issue of climate change[182]:

> You know, as a lot of people have said up here, over the last four decades, we have had a bipartisan consensus to do nothing. And that is shameful. That is bipartisan agreement, but it is an agreement for inaction, at least on the scale that is required.
>
> Now, some of that action may be driven by corruption, some is driven by denial, some is driven by cowardice. But I think in all cases, there are far too many in this town who are content to sit and wait for public opinion to force them to act. And acting in response to polls may make you an effective politician, but it is the opposite of leadership.

An attempt to break out of the congressional malaise on climate change during Trump's presidency came from the House Select Committee on the Climate Crisis, which had been formed by Nancy Pelosi (1940–) (Democrat from California) in January 2019.[183] At a hearing of this committee, a new contrarian, the author Michael Shellenberger (1971–),[184] gave testimony. He provided a lengthy list of all that was wrong with renewable energy and said[185]: "No credible scientific body has ever claimed that climate change threatens the collapse of civilization, much less the extinction of the human species, which is what 'existential' threat means." I had engaged Michael Shellenberger on Twitter in #ClimateBrawl, but only briefly, as he blocked me.

After a series of hearings, on June 30, 2020, the House Select Committee on the Climate Crisis issued an over 500-page plan by the majority (i.e., the Democrats), "Solving the Climate Crisis, The Congressional Action Plan for a Clean Energy Economy and a Healthy, Resilient, and Just America:"[186]

> In this report, the majority staff for the Select Committee lays out a framework for comprehensive congressional action to satisfy the scientific imperative to reduce carbon pollution as quickly and aggressively as possible, make communities more resilient to the impacts of climate change, and build a durable and equitable clean energy economy.

A long overdue recommendation was that Congress push forward climate legislation.[187] The final conclusions of the plan were[188]:

> Confronting the climate crisis requires action across sectors and at all levels of government …
>
> The climate crisis is inextricably linked to the social, economic, and environmental challenges that afflict the nation and world today. But by working together, we can avert the worst impacts of climate change and build a stronger, healthier, and fairer America for everyone. What we choose to do now shapes the future for young people and Americans on the front lines of the climate crisis. Now is the time for Congress to implement these recommendations with urgent and decisive action.

Two months after the House report, a sister-document from the Senate came out, which began[189]:

> The climate crisis threatens our lives and livelihoods. The evidence is clear: we must flatten the warming curve, and fast … We already have the technologies needed to avert catastrophe. We just need the American optimism and the political will to deploy them on an unprecedented scale. To ignite this transition, we need Congress to act.

Two years later, during Biden's administration, Congress did act, but it was only Democrats who pushed legislation through.

John Christy

The last of the five major contrarians being discussed is John Christy,[190] who appeared at congressional hearings under various administrations (as mentioned above). Late in the term of Bill Clinton, Dr. Christy was a witness at a session called "The Science Behind Global Warming." In his opening remarks, the Republican Senator John McCain stated[191]:

> Most importantly, any actions the United States takes in response to claims of global warming must be based on the best science available and not on rhetoric or political expedience …
>
> I do not have a plan. I am sorry to say that I do not have a plan because I do not have, nor do the American people have, sufficient information and knowledge. But I do believe that Americans and we who are policymakers in all branches of government, should be concerned about mounting evidence that indicates that something is happening.

John Christy proceeded to criticize the climate sciences[192]:

> I will close with three questions and a plea. Is the climate changing? Yes, it always has and it always will, but it is very difficult to detect on decadal time scales.
>
> Are climate models useful? Yes, and improving. At this point, their utility is mostly in global average scale, yet there are still some significant shortcomings even there.
>
> Is that portion of climate change due to human factors good, bad, or inconsequential? And that, no one knows, although we do know that the plant world thrives on additional CO_2 in the atmosphere.

His last answer that "no one knows" was clearly not based on the peer-reviewed literature. Science knew that "climate change due to human factors" was, in fact, bad.

In 2006, the chairperson Tom Davis (1949–) (Republican from Virginia) opened a congressional hearing on climate change[193]:

> There are not very many people left these days who would argue global warming isn't happening, per se …
>
> It is our job to ask whether we are responding appropriately, whether a scientific consensus exists, and whether we are facilitating the research and ensuring an unbiased review when there is not.
>
> Knowledge is refined through continuous inquiry and, yes, through skepticism.

Two contrarians, Roger Pielke Jr. and John Christy, testified. The political scientist Roger Pielke Jr. made the comment[194]: "I fully intend that this perspective be viewed as an alternative to the two-sided debate that has been caricatured as climate skeptics versus climate alarmists. Perhaps those holding this third position might be characterized as climate realists." The term "climate alarmists" was used by climate deniers to discredit traditional climate scientists, while the term "climate realists" was used by climate deniers to flatter themselves. He also noted the polarization within Congress, with most Democrats, but only a minority of

Republicans, believing climate change was a serious issue.[195] In 2018 and 2019, I had engagements with Roger Pielke Jr. on Twitter in #ClimateBrawl, but he eventually blocked me. During his turn to testify, John Christy argued[196]: "Think of the poor people out there. If energy costs rise, that does specifically and directly affect them." The claim that any policy which reduces GHG emissions will hurt the poor was a common talking point of climate deniers.

In 2011, Dr. Christy had complained that the "consensus science" and the "climate establishment" were basically censoring contrarians.[197] He went on to state that the consensus was "murky science" and a "political notion." Of course, consensus is the inevitable outcome of all strong theories of science. Dr. Christy, though, was warming up the political crowd for a bold request: Congress should fund contrarians (such as himself) to provide "alternative hypotheses" to those of the IPCC (and other government reviews of the climate sciences) to bypass, what he called, the "gatekeepers." In short, Dr. Christy was seeking financial support for opinions that could not pass peer review.

On September 20, 2012, Dr. Christy took the idea to fight "consensus science" a step further – introducing the "Red Team" concept – and proposed[198]:

> … funds be allocated to a group of well-credentialed scientists to produce an assessment that expresses legitimate, alternative hypotheses that have been (in their view) marginalized, misrepresented or ignored in previous IPCC reports (and thus EPA and National Climate Assessments). Such activities are often called "Red Team" reports and are widely used in government and industry. Decisions regarding funding for "Red Teams" should not be placed in the hands of the current "establishment" but in panels populated by credentialed scientists who have experience in examining these issues …
>
> What this proposal seeks is to provide to the Congress and other policymakers a parallel scientifically-based assessment regarding the state of climate science which addresses issues which here-to-for have been un- or under-represented by previous tax-payer funded, government-directed climate reports. In other words, our policymakers need to see the entire range of findings regarding climate change.

Although it did not garner much attention at the time, Dr. Christy was not deterred.

At a Senate hearing a year later, Dr. Christy was not present, but Jeff Sessions (1946–) (Republican from Alabama) submitted a brief called "Why We Need a Red Team Approach to Climate Science," written by Richard McNider and John Christy, which had concluded[199]:

> Congress should require the National Science Foundation (NSF) to coordinate an interagency process that would allocate 5 percent of the federal climate change research budget to a Red Team effort. The review of the Red Team science or policy research proposals could be carried out by review panels selected from the "skeptic community," with minority representation from the pro-climate change community …

> If we do this, in the future if the nation and the world take action on climate change and it is wrong, at least history will record that we looked at the issue from all sides."

The concept was promoted again in 2015 by Dr. Christy[200]: "Since there are many more proposals than funding allows, a contrarian proposal has essentially no chance of receiving funding because the panel decides these things by votes." At this hearing, he was supported by the contrarian William Happer: "And so I would like to second Dr. Christy's request for a red team. A Team B also would be a good idea to provide a rigorous review of how well is science working."[201]

Two years on, during the Trump administration, Dr. Christy was still advocating for this approach[202]:

> One way to aid congress in understanding more of the climate issue than what is produced by biased "official" panels of the climate establishment is to organize and fund credible 'Red Teams' … I would expect such a team would offer to congress some very different conclusions regarding the human impacts on climate.

This time around, after several years of selling the "Red Team" concept, he had finally found a political champion in Scott Pruitt (1968–), the head of the EPA.[203] The EPA had even reached out to the right-wing think tank Heartland Institute to identify potential members for such a team.[204] After Pruitt's departure, President Trump took on the project himself and looked into forming the "President's Committee on Climate Security," with William Happer, Judith Curry, Richard Lindzen, and John Christy as members.[205] The intent was for the contrarian panel to challenge the scientific consensus and climate science studies, but the scheme never materialized.[206] More recently, the Red Team concept was promoted by another contrarian, Steven Koonin (1951–).[207]

The "Red Team" was a horrible idea for science, since the fringe group of contrarians would have been given unprecedented power over the greater scientific community and the scrutiny of the peer-reviewed literature.

President Joe Biden: 2021–

With the election of Joe Biden (1942–), there were hopes again for climate action. Before he had become vice president, Senator Biden had warned about the dangers of climate change[208]:

> We missed the chance to turn the impending threat of catastrophic climate change into an opportunity to reduce the security threat of our dependence on oil, to reduce the health threat from pollution, to reduce the sheer waste and inefficiency in our economy …
>
> This is a classic tragedy of the commons. We have treated our atmosphere as a costless dump for the waste gases that are the byproduct of our great wealth.

> There was a time when we could plead ignorance. That day is past. The science is now clear.

Climate change was discussed during the presidential debates, including the last one on October 22, 2020, when Mr. Biden stated[209]:

> It's the number one issue facing humanity. And it's the number one issue for me … Climate change is the existential threat to humanity … Unchecked, it is going to actually bake this planet. This is not hyperbole. It's real. And we have a moral obligation.

Joe Biden has used Twitter to spread his message about climate change. In the 2020 campaign, he had tweeted[210]: "Climate change is the challenge that's going to define our American future. And if I have the honor of being elected President, we won't tinker around the edges, we'll seize this opportunity and meet this moment in history." He also attacked Donald Trump for his climate denial[211]: "Donald Trump's climate denial may not have directly caused the record fires we're seeing on the West Coast. But if he gets four more years in the White House, they'll only become more common, devastating, and deadly" (only a few months before, Barack Obama had said the same[212]). And just before the election, on November 1st, Joe Biden tweeted[213]: "I believe climate change is an existential threat to humanity. Donald Trump doesn't even think it exists. It's that simple, folks." Since first tweeting on climate change in 2020, Joe Biden had tweeted about the topic almost fifty times by mid-2022.

Upon becoming president, Joe Biden issued several initiatives on climate change, including rejoining the Paris Climate Agreement.[214] As President Biden approached his 100th day in office, Noam Chomsky, in a rare moment of praise, stated[215]: "Biden's [climate change] program is the best … that has yet been proposed by any president … it is a move in the right direction." Later, Senator Mitt Romney (1947–) (Republican from Utah), too, had some encouraging words for the president on the challenges of policymaking[216]:

> President Joe Biden is a genuinely good man, but he has yet been unable to break through our national malady of denial, deceit, and distrust … Too often, Washington demonstrates the maxim that for evil to thrive only requires good men to do nothing.

A harsh blow to President Biden's climate plans took place on June 30, 2022, when the Supreme Court ruled that the Clean Air Act does not give the EPA the authority to regulate greenhouse gas emissions from power plants.[217] The president was livid about the decision, tweeting[218]:

> The Supreme Court's ruling in West Virginia vs. EPA risks damaging our nation's ability to keep our air clean and combat climate change. We cannot ignore the existential threat the climate crisis poses. Our fight against climate change must carry forward, and it will.

And he issued a formal statement[219]:

> Since the Clean Air Act was passed by a bipartisan majority in Congress in 1970, the landmark law has enabled both Democratic and Republican administrations to protect and improve the air we breathe, cutting air pollution by 78 percent even as our economy quadrupled in size. Yet today's decision sides with special interests that have waged a long-term campaign to strip away our right to breathe clean air …
>
> And we will keep pushing for additional Congressional action ..
>
> Our fight against climate change must carry forward, and it will.

Then a month later, on July 20, 2022, President Biden pushed Congress for action[220]:

> So my message today is this: Since Congress is not acting as it should … this is an emergency.
>
> As President, I'll use my executive powers to combat climate — the climate crisis in the absence of congressional actions …
>
> And so does Congress, which — notwithstanding the leadership of the men and women that are here today — has failed in this duty. Not a single Republican in Congress stepped up to support my climate plan. Not one.

Unexpectedly, after more than 50 years of failing to act on climate change, Congress finally did something – the first legislation on climate change was passed in mid-August 2022, as part of the "Inflation Reduction Act" (see below), which was praised in an editorial[221]:

> When the House of Representatives passed landmark climate legislation on Friday, Joe Biden chalked up one of the surprise successes of his presidency … Ultimately the climate emergency needs a fundamental economic restructuring. Mr Biden's new environmental law is a good start, but there's a very long way to go.

In his 2023 State of the Union Address, President Biden commented on the new legislation[222]:

> Look, the Inflation Reduction Act is also the most significant investment ever to tackle the climate crisis … The climate crisis doesn't care if your state is red or blue. It is an existential threat. We have an obligation to our children and grandchildren to confront it. I'm proud of how America is at last stepping up to the challenge.

Since President Reagan, America's climate initiatives have been on a wild roller coaster ride between Republican and Democratic administrations. President Biden appears to be following the pattern of his Democratic predecessors, but what will the next Republican president do?

Congress

A summary of U.S. government actions on climate change, including those of Congress, were presented by the Congressional Research Service in late 2021. During the first several months of the Biden administration, over 750 legislative proposals addressing climate change had been introduced by Congress, 34 of which had received floor consideration.[223] A dejected Senator Whitehouse stated[224]: "The legislative part has been, so far, a calamity that we've got no serious legislation passed." And the prospects looked bleak for 2022, but surprise, surprise.

In August 2022, an unexpected congressional breakthrough took place, at long last, with the "Inflation Reduction Act" being passed, by the narrowest of margins (Senate 51–50; House 220–207). The name does not sound like legislation on climate change, but for the overly sensitive ears of American politicians, one could not have dared call it the "Greenhouse Gas Reduction Act." The law provides financial incentives on green energy initiatives for industry and the public. However, there are two important aspects still missing: no price on carbon and no targets on GHG reduction. The bill has been hailed as the greatest climate legislation in U.S. history, but that is easy to say since it is the first. In fairness, though, the first steps are always the hardest. Al Gore stated[225]: "In crossing this threshold we have changed history and will never go backwards … I'm extremely optimistic that this will be a critical turning point in our struggle to confront the climate crisis." Then, Mr. Gore tweeted[226]: "Today, the US finally said yes to climate action. Let us embrace this moment as a tipping point for climate action – in America and around the world."

Climate Brawl with the Contrarians

Congressional hearings on climate change, spanning over half a century, showed that the efforts by mainstream scientists to communicate the serious threat of climate change to politicians had been undermined by a handful of fringe scientists – the messengers of climate denialism in the halls of Congress were the contrarians. What motivated them to oppose the accepted science? We can only speculate on their reasons, but it is unlikely that it had much to do with science directly. Many contrarians were associated with right-wing think tanks and may have been financially and/or ideologically motivated. Some of them have become better known among conservatives for being contrarians than for their own scientific achievements; so perhaps, they were enticed by the "celebrity" status.[227] In any case, the contrarians, for all their controversial testimonies, are less to blame for bringing climate denial to Congress than the Republican elites who repeatedly called upon their services,[228] for the sole purpose of providing what they desperately wanted to hear – no need to reduce fossil fuels.

The invitation of contrarians as expert witnesses at congressional hearings on climate change was a dishonest attempt to circumvent the voice of the greater scientific community. These hearings were intended to inform politicians about the latest science, and the contrarians often admitted that they were not there for that purpose. The Republicans welcomed them as witnesses precisely because they were not expert reporters on the science per se, but expert doubters. As the scientific evidence grew and the consensus became only stronger, contrarians remained steadfast in their doubt, to the delight of Republicans who had no interest in ever restricting the use of fossil fuels. The messages of the contrarians contained a few common themes repeated over and over: no scientific consensus, too much doubt about the science, and, above all, no need for politicians to do anything.

For decades, then, the contrarian parade had marched through Washington to the steady beat of the Republican anti-science drum: doubt, doubt, doubt. Advocates of climate denial have had a free reign to shout out their propaganda without being challenged, no matter how bizarre or misleading their claims. Democrats and scientific witnesses seldom called out the contrarians, but there were some notable interactions worth mentioning, and these are discussed in detail below to highlight the climate denial counter movement, known as climate brawl, at congressional hearings.

Jim Cooper – 1989

The climate brawl with contrarians began as soon as they appeared in Washington; in February 1989, Jim Cooper (1954–) (Democrat from Tennessee) questioned the contrarian Patrick Michaels[229]: "The vested interests are so powerful and so overwhelming … Dr. Michaels, I see you squirming. Would you like to say something?" And Patrick Michaels replied:

> The reason I am squirming is because I tend to agree, by and large, with your statement but I have seen others in the government make statements that people who express what they characterize as the contrarian point of view and I guess that is me - are doing a disservice and they are, in fact, impeding our ability to act on this issue.

Jim Cooper then gently scolded Dr. Michaels[230]:

> You also have to realize that in the inevitable cartooning tendency, there are people who want to paint you unfairly as being antigreenhouse effect. You have to resist strongly that tendency and that characterization.
>
> Many people don't realize that in Washington anything can have a lobby representing it, even scientific uncertainty. That is probably one of the biggest and best-funded lobbies if you are talking about tobacco smoke, for example. So please bear in mind that there are people who would like to use your testimony to shield less noble causes …

Afterwards, Dr. Michaels added a twist of contrarian humour, when he stated that he did not "mean to appear as the Darth Vader of the environment."[231]

Credit must be given to Representative Cooper for initiating an impressive, albeit mild, climate brawl by questioning the intent of one of the first contrarian witnesses at a congressional session. Unfortunately, too few of his colleagues would follow in his footsteps.

John Kerry – 2000

The new millennium started off well for the climate brawl. Senator Kerry (1943–) (Democrat from Massachusetts) had set the tone with his opening remarks[232]:

> And I must say, Mr. Chairman, I have been a little bit dumbfounded and somewhat disturbed by the level of skepticism that exists, and has existed over a long period of time, in the U.S. Congress with respect to this issue.

Later in the session, Robert Watson (1948–), the chairman of the IPCC, challenged the propaganda of the climate deniers[233]:

> However, the good news is that if we go beyond political ideology, there are numerous cost-effective ways to mitigate climate change …
>
> And, let me leave you with one simple observation. Many of the global warming skeptics today are the same skeptics who questioned whether human activities were destroying the earth's fragile ozone layer and increasing the level of damaging ultraviolet radiation reaching the Earth's surface. These skeptics argued against national and global action to protect the ozone layer.
>
> We now know that human activities were destroying the ozone layer and thankfully governments from around the world, working with industry, ignored the skeptics and cost-effective solutions were developed, thus protecting all life on Earth from the damaging - damaging ultraviolet radiation.

The mention of "political ideology" by Dr. Watson speaks for itself. As head of the IPCC, he would have observed its workings more than most climate scientists.

This time, John Christy was the contrarian on the hot seat. Dr. Christy knew this might be a lively session. He had begun his presentation by stating[234]: "Considering the varying levels of skepticism represented on this panel, it would be apparent that I am very likely the witness that is most skeptical, but not agnostic, regarding our ability to predict future climate." During questioning, Senator Kerry chided Dr. Christy with some climate brawl humour[235]: "Well, it is hard to find a whole lot of contrarians now … it is hard to fill a room with them." Dr. Watson added: "I would say there are a half a dozen key contrarians, which include Dick Lindsen [sic], Fred Singer and Pat Michaels, but I would say the large majority of the scientists clearly fall on one side." Senator Kerry then asked about the differences between the contrarians and other scientists; a lead author of the IPCC, Kevin Trenberth (1944–) explained[236]:

> I think we need to take into account some of the ideologies that come into play and recognize that there are different views of the world … And we have seen this in the tobacco industry, for instance, where often the strategy is to denigrate the science and to say that there is not a problem and recognize that they do have a vested interest.

Senator Kerry continued his climate brawl: "Well, it is completely fair and, I think, accurate to say, that some of the denigration of science has emanated from specific industries highly vested in fossil fuel. Is that accurate?" And Dr. Trenberth agreed. Kerry's remark "hard to fill a room with them" accurately reflected the small number of contrarians seen in Washington.

Senator Kerry was an early politician to pushback against contrarians and the energy-industrial complex. Few climate brawls involving a politician challenging contrarians in Congress could match this one.

James Hansen – 2009

There was a rare climate brawl between a mainstream scientist and a contrarian. John Christy was again a witness and joining him was the famous climate scientist Jim Hansen who jumped right in with a powerful plea[237]:

> We have a planet in peril …
>
> It is clear that we cannot burn all fossil fuels releasing the waste products into the air without handing our children a situation in which amplifying feedbacks begin to run out of their control, with severe consequences for nature and humanity …
>
> We must move beyond fossil fuels anyhow, so why not do it sooner for the benefit of our children. Not to do so and knowing the consequences is, I think, immoral …
>
> For the sake of our children and grandchildren, we cannot let the special interests win this fight.

When it was his turn, Dr. Christy presented his doubt on the climate models.[238] Dr. Hansen was asked to comment on the contrarian viewpoint during the question period, and he responded[239]:

> It is a tactic of those who want to do nothing to make it sound like there is a debate. In fact, I think that is the wrong road to go down. I think if there is any question about the reality of this, which scientifically, there is not, then you should ask, Congress should ask the National Academy of Sciences, which is the most authoritative scientific body in the world, to deliver a report back to Congress or the President should ask for that.
>
> The science has become crystal clear. There is an issue and we can see it happening. It is not based on climate models. It is looking at what is happening in the real world.

As Dr. Hansen had emphatically testified, there was no debate over the science of climate change.

Richard Somerville – 2011

A congressional hearing in 2011 was a landmark climate brawl with contrarians (which also included the climate denialism of the Republicans). Setting the stage, the chair Ed Whitfield (1943–) (Republican from Kentucky) began the hearing with climate denial sarcasm[240]:

> I might say that I only brought one of my many books that questions global warming and the science on global warming. I am delighted to see that at least one member brought a number of books. I couldn't get all mine in the car. Anyway, that is the reason we have these hearings, to hear both sides of the issue.
>
> I might say also that we have had 24 hearings in the House of Representatives over the past 4 years relating to the science for climate change and/or global warming. One thing that really stuck out to me is that these computer models seem to have difficulty making seasonal or yearly forecasts and they certainly, according to many scientists, have great difficulty trying to forecast 100 years down the road …
>
> Of course, the rhetoric coming from the White House is that the sky is falling and carbon emissions are going through the roof.

Bobby Rush (1946–) (Democrat from Illinois) countered and called out the propaganda of the contrarians[241]:

> On the other side, you have a very small, less than 5 percent, of the scientists in the community, who range from straight-out climate change denial to those who would dispute the certainty that the claims that human behavior is contributing to climate change …
>
> So I ask you, Mr. Chairman, and all my Republican colleagues, to remember to listen to what the science is telling us … by showing the American people that we can be serious about finding solutions and that we are not just here for political infighting and scorekeeping.

Jay Inslee (1951–) (Democrat from Washington) also lashed out at his Republican colleagues[242]:

> I want to thank all of you for being here today, but I have to express some degree of embarrassment that a Nation that went to the Moon, mapped the human genome, established the best software companies in the world, does now have one of its great parties adopt a chronic anti-science syndrome; one of its great parties that has decided to have an allergy to consensus science instead of respect for science and scientists. And that is embarrassing.

These exchanges display climate brawls by Democrats against the climate denialism of their Republican peers.

But what was really special about this hearing from a climate brawl perspective was the testimony of Richard Somerville (1941–) of the Scripps Institution of Oceanography who came out with a masterful rebuke of climate denial and the contrarians[243]:

> It is a standard tactic of many climate "skeptics" or "contrarians" (terms commonly used to denote those who reject central findings of mainstream climate change science) to try to frame this issue in terms of the whole edifice of modern climate science hanging from some slender thread. Thus, if a given scientist uses intemperate language, or a particular measurement is missing from an archive, or a published paper has a minor mistake in it, the whole unstable scientific structure comes tumbling down, or so the skeptics would have people believe …
>
> In my opinion, many people need to learn more about the nature of junk or fake science, so they will be better equipped to recognize and reject it. There are a number of warning signs that can help identify suspicious claims. One is failure to rely on and cite published research results from peer-reviewed journals. Trustworthy science is not something that appears first on television or the Internet. Reputable scientists first announce the results of their research by peer-reviewed publication in well-regarded scientific journals. Peer review is not a guarantee of excellent science, but the lack of it is a red flag. Peer review is a necessary rather than a sufficient criterion.
>
> Another warning sign is a lack of relevant credentials on the part of the person making assertions …
>
> Finally, we must be alert for any hint of delusions of grandeur on the part of those who would insist that they are correct, while nearly everyone else in the entire field of climate science is badly mistaken.

Dr. Somerville perfectly exposed the nature of contrarians and other climate deniers, before the term "climate deniers" was even popular (he used the term climate skeptics).

Overall, these climate brawls reveal what should have been done at every climate hearing. The contrarians, with their worn-out message of doubt, are still appearing at congressional hearings, but their testimonies seem to have lost their luster – not because of climate brawls (which are still too few to have much of an impact) or because the science is winning per se, but because Republicans no longer depend on the contrarians to legitimize their anti-science position, having completely surrendered to climate denialism.

Congressional Climate Chatter

Congress had been warned that the burning of fossil fuels[244] "could cause a remarkable change in climate," and "we are making perhaps the greatest geophysical experiment in history" – this testimony was provided in Congress by Roger Revelle back in 1956! Since then, leading climate scientists have come to Washington sharing the message of the peer-reviewed literature that climate change was a serious threat and had to be addressed. Science has never wavered from this message, except for emphasizing more and more the urgency for political action. During the 1980s, the science of climate change was reaching a stage where political action was warranted, but the following decade, due to the release of a massive propaganda campaign by the energy-industrial complex and the arrival of the contrarians, the political will for climate action was stifled and has remained so.

After examining over half a century of testimonies at congressional hearings, the evolution of American politics on climate change can be mapped out as follows:

1950s: Weather – politicians were interested in the climate because of weather control and modification, especially for military purposes (such weather hearings continued into the 1970s), and the modern age of the science of climate change was underway.
1960s: Environment – climate experts were routinely invited to general environmental hearings but received lukewarm interest from politicians, and the focus was on funding for more research; and a president of the United States first took direct interest in the science of climate change.
1970s: Global Cooling – politicians were distracted by global cooling versus global warming, and hearings specifically on the science of climate change were held for the first time as the scientific evidence grew stronger.
1980s: Paradigm Shift – a transition took place as climate denialism was gaining over climate skepticism and sound science, and political polarization on climate change was underway (and contrarian op-eds started to appear in public media).
1990s: Contrarians – Republican politicians invited contrarians to hearings to cast doubt on the science of climate change to block legislative action (which continues today), and the Kyoto negotiations were the primary focus of hearings.
2000s: Denial Cabal – members of the Republican elite were censoring climate science and persecuting leading climate scientists (discussed in chapter four).
2010s: Denial Machine – the energy-industrial complex had achieved undue influence over policymaking on climate change suggesting a level of political interference characteristic of a deep state (discussed in chapter five).

During these hearings, climate change was discussed under various terms in the Congressional Record; Table 3.1 lists them and their frequency, illustrating the popularity of the subject.

Congress was in a chatty mood about climate change as the above table illustrates. In addition, between 1996 and 2015, there were 2,727 speeches in the Congressional Record that mentioned "climate change," "global warming," or "greenhouse gas" at least twice, but the discourse styles of Democrats and Republicans had been different[245]: "Democrats communicated in ways that are message-based, emphasizing the weight of scientific evidence, while Republicans tended towards a softer, cue-based narrative based on anecdotes and storytelling."

In 2011, the Congressional Research Service issued a study, "Climate Change: Conceptual Approaches and Policy Tools," which recognized that[246]: "Over time, the consequences of climate change for the United States and the globe will be influenced by choices made or left to others by the U.S. Congress." Despite all the congressional chatter, unfortunately the choices were "left to others by the U.S. Congress," especially the energy-industrial complex, for yet another decade.

Table 3.1 Popular Climate Change Terms Used in Congress

Term	*Records*	*Top User*
Climate Change	25,057	Richard Durbin
Greenhouse Gas	13,992	Barbara Boxer
Global Warming	11,299	Harry Reid
Paris (Climate) Accord/Agreement	3,054	Sheldon Whitehouse
Climate (Change) Science	2,257	Sheldon Whitehouse
Greenhouse Effect	1,857	Barbara Boxer
Climate (Change) Action	1,782	Sheldon Whitehouse
Climate Crisis	1,654	Sheldon Whitehouse
Kyoto Protocol	1,572	Barbara Boxer
IPCC	1,308	James Inhofe
Climatic Change	618	Byron Dorgan
Climate (Change) Denial	327	Sheldon Whitehouse
Climate (Change) Denier(s)	312	Sheldon Whitehouse
Global Cooling	233	James Inhofe
Act on Climate	214	Sheldon Whitehouse
Climate Justice	193	Adriano Espaillat
Climategate/Climate Gate	169	James Inhofe
Climate Catastrophe	109	Sheldon Whitehouse

Source: https://www.govinfo.gov/ (full Congressional Record to July 21, 2022; for >100 records; bracketed items were with and without that item; some records may contain more than one of the above terms and are counted separately).

The odd bill had previously passed one of the houses,[247] and the best-known climate initiative of Congress was the Byrd-Hagel Resolution (passed by the Senate) which crippled U.S. negotiations at Kyoto. The only law responsible for federal climate initiatives is over 50 years old – the 1970 Clean Air Act.[248] It became relevant after the Supreme Court decision of 2007[249] declared carbon dioxide and other greenhouse gases to be pollutants, but then, in 2022, the Supreme Court limited the scope of that ruling; Chief Justice John Roberts (1955–) delivered the opinion of the court, which ended with[250]:

> But the only interpretive question before us, and the only one we answer, is more narrow: whether the "best system of emission reduction" identified by EPA in the Clean Power Plan was within the authority granted to the Agency in Section 111(d) of the Clean Air Act. For the reasons given, the answer is no.
>
> Capping carbon dioxide emissions at a level that will force a nationwide transition away from the use of coal to generate electricity may be a sensible "solution to the crisis of the day." New York v. United States, 505 U. S. 144, 187 (1992). But it is not plausible that Congress gave EPA the authority to adopt on its own such a regulatory scheme in Section 111(d). A decision of such magnitude and consequence rests with Congress itself…

In other words, the EPA could not mandate the "best system of emission reduction" to climate change in the power sector – that authority resided with Congress. The solution to climate change in the United States always points to Congress.

There was a glimmer of hope in February 2016, when the Climate Solutions Caucus, a bipartisan caucus in the House was formed (a similar group formed three years later in the Senate) to act on climate change by Carlos Curbello (1980–) (Republican from Florida) and Ted Deutch (1966–) (Democrat from Florida). Mr. Curbello stated[251]:

> The establishment of the Climate Solutions Caucuses in both chambers was a sign that the polarization was finally starting to ease; that Democrats and Republicans could at least open a dialogue and explore ways to collaborate. Modest bipartisan collaboration began soon thereafter. Maybe it was a turning point.

And he added that during the 115th Congress, the Climate Solution Caucus rallied to defeat anti-climate amendments for the first time under a GOP majority. While this was an encouraging development, there is no further evidence that climate denialism is softening within Republican ranks. We are so desperate to see political progress on the climate front that if Republicans even admit climate change is real, it becomes headline news; Americans are basically victims of the Stockholm syndrome, as they yearn for any reason to hope. The apparent softening of views is mainly political greenwashing – we are just seeing the G.O.P. adopt other shades of climate denialism.

Finally, after decades of chatting about climate change, in August 2022, Congress passed the Inflation Reduction Act which is an important start to dealing with the climate crisis; significantly, not one single Republican supported the legislation. Climate denialism has corrupted the Republican Party, especially the political elite of which some have become members of a denial cabal (discussed in the next chapter).

Notes

1 George H. W. Bush Library 1989, p. 6; this is one of the earliest records of a U.S. president using the term "climate change."
2 Bush 1990a.
3 Kutney 2014, pp. 138–142.
4 Bush 1990b; for the closing statements of President Bush at the conference see Bush 1990c.
5 Hudson 2018.
6 Shabecoff 1990.
7 Gore 1990.
8 See Hudson 2018; also see Waldman and Hulac 2018.
9 Nitze in Mintzer and Leonard 1994, p. 192.
10 For a review of the Rio Conference see Kutney 2014, pp. 8–13.
11 Hecht and Tirpak 1995, p. 393.
12 Shabecoff 1989.
13 Bush 1989.
14 Shabecoff 1989.
15 Kjellen in Mintzer and Leonard 1994, p. 162; also see p. 160.
16 Dasgupta in Mintzer and Leonard 1994, p. 134.

17 Bodansky in Mintzer and Leonard 1994, pp. 55–56; Borione and Ripert in Mintzer and Leonard 1994, p. 80; Kjellin in Mintzer and Leonard 1994, pp. 151, 158, 164; Tariq in Mintzer and Leonard 1994, p. 209; Rahman and Roncerel in Mintzer and Leonard 1994, p. 268; Cass 2006, p. 69.
18 Rich 2018.
19 Rahman and Roncerel in Mintzer and Leonard 1994, p. 254.
20 George H. W. Bush Library 1990a, p. 5.
21 George H. W. Bush Library 1990b, p. 3.
22 George H. W. Bush Library 1990c, p. 3.
23 George H. W. Bush Library 1991c, pp. 7–8.
24 George H. W. Bush Library 1991a, pp. 4–6.
25 George H. W. Bush Library 1991b, p. 10.
26 George H. W. Bush Library 1991b, p. 11.
27 George H. W. Bush Library 1992a, pp. 6–7.
28 Sebenius in Mintzer and Leonard 1994, p. 281.
29 Goldemberg in Mintzer and Leonard 1994, p. 182.
30 Bush 1992a.
31 George H. W. Bush Library 1992b, p. 1.
32 George H. W. Bush Library 1992c.
33 Bush 1992b; also see George H. W. Bush Library 1992d.
34 National Academy of Sciences 1991, p. vii.
35 National Academy of Sciences 1991, p. 72.
36 United States Senate 1991, p. 45.
37 United States Senate 1991, p. 3.
38 United States Senate 1991, p. 29.
39 National Academy of Sciences 1991, p. x.
40 A referenced biography of contrarian scientists (and politicians) is available at DeSmog, "Climate Disinformation Database;" also see Skeptical Science 2020. The tobacco industry had used similar types of scientists, who they called the "Whitecoats" (Diethelm and McKee 2009).
41 United States Senate 1985, p. 1.
42 United States House of Representatives 1988, p. 24.
43 United States Senate 1989b, pp. 1, 3.
44 United States House of Representatives 1989, p. 1.
45 United States Senate 1989a, p. 1.
46 DeSmog, "Patrick J. Michaels;" for an early article see Michaels 1989.
47 United States House of Representatives 1989, pp. 78–81.
48 United States House of Representatives 1989, pp. 89–90.
49 United States Senate 1989a, p. 254; in his testimony, Dr. Michaels ended his testimony with a quote from Reid Bryson (p. 257).
50 DeSmog, "S. Fred Singer"; Mann 2012; Schwartz 2020.
51 United States Senate 1989a, p. 276.
52 United States House of Representatives 1991, p. 1.
53 United States House of Representatives 1991, p. 156. Dr. Michaels included a copy of a letter sent to President Bush on February 1, 1991, on a symposium – Global Climatic Change: A New Vision for the 1990s – of contrarian scientists that questioned the dangers of global warming (pp. 133, 136–137).
54 United States House of Representatives 1995, p. 114.
55 United States House of Representatives 1995, p. 25.
56 United States House of Representatives 2002, p. 1.
57 DeSmog, "Roger A. Pielke Sr."
58 United States House of Representatives 2002, p. 77.
59 United States House of Representatives 2002, p. 78.

60 Waldman 2022.
61 Kerr 1992, p. 1138.
62 DeSmog, "Richard Lindzen"; also see Mann 2012, pp. 68–69; Mann and Toles 2016, pp. 58–61.
63 See Koch Docs 1991; Patrick Michaels was also at the conference. An early climate denial conference had been in Phoenix, AZ, in October 1990 (United States House of Representatives 1991, pp. 136–137).
64 United States Senate 1992, p. 1.
65 United States Senate 1992, p. 2; also see pp. 3–4 for other comments on the Marshall Institute report.
66 United States Senate 1992, p. 51; also giving testimony was William Happer, who would be known as a contrarian but, at the time, was working for the Department of Energy (pp. 135–137).
67 United States House of Representatives 1996, p. 333.
68 United States Senate 2001a, p. 24.
69 United States Senate 2001b, p. 325.
70 United States Senate 2001b, p. 322.
71 United States Senate 2001b, p. 323.
72 United States Senate 2001b, pp. 366–367; for a review of the Clinton actions on climate change see Royden 2002.
73 Cass 2006, p. 97.
74 William J. Clinton Library 1993, p. i.
75 Cass 2006, p. 98.
76 William J. Clinton Library 1993.
77 Clinton 2000.
78 See Kutney 2014, pp. 142–151.
79 Inglis 2015.
80 United States Senate 1997.
81 See United States Department of State 1998.
82 United States Senate 2001a, p. 5.
83 United States House of Representatives 2019e, p. 12.
84 United States House of Representatives 2019e, p. 38.
85 McGreal 2022b.
86 See Kutney 2014, pp. 142–151, and references therein.
87 See Kutney 2014, p. 152.
88 For an analysis and commentary see Cohen and Egelston 2003.
89 National Academy of Sciences 2001, Appendix A.
90 National Academy of Sciences 2001, p. 3.
91 Bush 2001.
92 Bush 2002.
93 Bush 2007.
94 United States Senate 2007a, p. 52.
95 Bush 2008.
96 Supreme Court of the United States 2007, pp. 1443–1444.
97 EPA 2009; also see United States Senate 2007b.
98 United States Senate 2007a, p. 10.
99 United States Senate 2007a, pp. 139–1053.
100 United States Senate 2007a, p. 119.
101 United States House of Representatives 2007b, p. 9.
102 United States House of Representatives 2007b, p. 24.
103 United States House of Representatives 2007b, p. 25.
104 United States House of Representatives 2007b, pp. 29–30.
105 United States House of Representatives 2007b, p. 41.

106 Romm 2007; also see Begley 2010; Mann 2015; Cantor 2019; Montlake 2019; Lamb et al. 2020; Levantesi and Corsi 2021; King, Janulewicz, and Arcostanzo 2022; White House 2022.
107 Mann 2021; also see Dutton and Mann 2019.
108 See Montlake 2019.
109 Ocasio-Cortez 2019.
110 DeSmog, "Bjorn Lomborg".
111 United States House of Representatives 2007b, p. 52.
112 United States House of Representatives 2007b, p. 53.
113 United States House of Representatives 2007b, pp. 58–59, also see p. 126.
114 Lomborg 2014, p. 1; also see United States Senate 2014b.
115 Lomborg 2019a.
116 Lomborg 2019b.
117 United States Senate 2007a, pp. 87–89; for earlier comments on climate change by Senator Obama see United States Senate 2005, p. 20.
118 Obama 2013a; for remarks on the following State of the Union Address, see Obama 2013b.
119 EPA 2012.
120 Obama 2015c.
121 Obama 2015a.
122 See Kutney 2014, pp. 154–155.
123 Obama 2009c.
124 UNFCCC 2020, "Key Aspects of the Paris Agreement."
125 Obama 2016; by May 2023, 195 out of 198 nations had ratified the Paris Agreement.
126 Obama 2015b; the historic tweet was even marked by an official White House announcement (White House 2015).
127 Clinton 2015.
128 Obama 2009a.
129 Obama 2013c.
130 Obama 2009b.
131 TweetBinder Blog undated.
132 United States House of Representatives 2009c, p. 43.
133 United States House of Representatives 2009c, pp. 47–48.
134 United States House of Representatives 2009b, p. 4.
135 United States House of Representatives 2009b, pp. 5–6.
136 United States House of Representatives 2009b, p. 40.
137 United States House of Representatives 2009d, p. 117.
138 United States House of Representatives 2009d, p. 15.
139 United States House of Representatives 2009d, p. 237.
140 United States House of Representatives 2009d, p. 9; also see Samuelsohn 2010; for later comments by Mr. Shimkus on climate change, see United States House of Representatives 2019b, pp. 5–6.
141 United States House of Representatives 2010a, pp. 1–2.
142 United States House of Representatives 2010a, p. 61.
143 United States Senate 2013, p. 302; for more, see DeSmog, "Roger Pielke Jr."
144 United States Senate 2013, p. 317.
145 United States Senate 2013, p. 322.
146 United States House of Representatives 2015, p. 6; Raúl Grijalva was known for his challenging of contrarians; see, for example, Maize 2015.
147 United States House of Representatives 2019a, pp. 1–2.
148 United States House of Representatives 2019a, pp. 113–114.
149 DeSmog, "Judith Curry."
150 United States Senate 2015.

151 United States Senate 2015.
152 Dr. Happer discussed the benefits of CO2 in the atmosphere.
153 Mr. Steyn was particularly aggressive, as he was complaining about the persecution of climate deniers, especially himself by Michael E. Mann, whom he lambasted throughout his testimony.
154 United States Senate 2015.
155 United States Senate 2014a, p. 171.
156 United States Senate 2014a, p. 183.
157 United States Senate 2014a, p. 210.
158 United States Senate 2015.
159 United States House of Representatives 2017, p. 32.
160 United States House of Representatives 2010b, p. 209.
161 And she had done so at previous Congressional hearings; see, for example, United States Senate 2015.
162 United States House of Representatives 2019a, p. 85. At another hearing a few months later, Dr. Currie dismissed the association of climate change with extreme weather events, such as hurricanes; Dr. Mann was also a witness (see United States House of Representatives 2019f).
163 United States House of Representatives 2019a, pp. 85–93.
164 Curry 2019a.
165 Curry 2019b.
166 Curry 2022a,b.
167 For a summary of deregulation of climate measures during the Trump administration, see Columbia Law School 2022, which listed 176 items.
168 Trump 2017a.
169 Nuccitelli 2019.
170 Trump 2017b.
171 Ward 2017; Watts and Connolly 2017.
172 See Watts and Connolly 2017.
173 See Watts and Connolly 2017.
174 Michel 2017.
175 See Carrington 2019.
176 See Hackett 2019.
177 Ghosh 2017.
178 USGCRP 2018, “Chapters”, “Summary Findings”, “2. Economy”.
179 Cama 2018.
180 Johnson 2019.
181 United States House of Representatives 2019d, p. 3.
182 United States House of Representatives 2019c, p. 51.
183 House Select Committee on the Climate Crisis 2020, p. 17.
184 DeSmog, “Michael Shellenberger”.
185 Shellenberger 2020; also see Shellenberger 2022.
186 House Select Committee on the Climate Crisis 2020, p. 2.
187 House Select Committee on the Climate Crisis 2020, p. 25.
188 House Select Committee on the Climate Crisis 2020, p. 535.
189 Senate Democrats’ Special Committee on the Climate Crisis 2020, p. 1.
190 DeSmog, “John Christy”; Mann 2012, pp. 171–173.
191 United States Senate 2000, p. 2.
192 United States Senate 2000, pp. 37–38.
193 United States House of Representatives 2006, pp. 1–2.
194 United States House of Representatives 2006, p. 154.
195 United States House of Representatives 2006, p. 192.
196 United States House of Representatives 2006, p. 197.

197 United States House of Representatives 2011, pp. 73–74. Accusations of censorship against contrarians were common at congressional hearings, such as by Dr. Michaels (United States House of Representatives 1995, p. 25), Republican Tom Davis (United States House of Representatives 2007a, p. 10), Republican Dana Rohrabacher (United States House of Representatives 2007c, pp. 4–5), and Dr. Curry (United States House of Representatives 2017, p. 32).
198 See United States House of Representatives 2017, pp. 48–49.
199 United States Senate 2013, p. 658.
200 United States Senate 2015.
201 United States Senate 2015.
202 United States House of Representatives 2017, p. 37.
203 Holden 2018.
204 Lavelle 2018.
205 D'Angelo 2019a, b; Davenport and Landler 2019.
206 Waldman 2019.
207 Koonin 2021, pp. 197–201.
208 United States Senate 2007a, pp. 124–125.
209 Newburger 2020; for more on Biden's climate plans, see Aronoff 2020; Irfan 2020.
210 Biden 2020a.
211 Biden 2020b.
212 Obama 2020.
213 Biden 2020c.
214 White House 2021.
215 Hasan 2021.
216 Romney 2022.
217 Supreme Court of the United States 2022.
218 Biden 2022a.
219 Biden 2022b.
220 Biden 2022c.
221 Editorial 2022.
222 Biden 2023.
223 Lattanzio et al. 2021, pp. 16–17, 33–37. As of November 19, 2021, the Congressional Research Service had issued 363 reports that mention "climate change."
224 McGreal 2022a.
225 Milman 2022.
226 Gore 2022.
227 For an interesting review of contrarians see Schwartz 2019.
228 For example, Dr. Christy appeared as a witness in Congressional hearings about 20 times (Schwartz 2019), and Dr. Curry ten times (United States House of Representatives 2019f, "WITNESSES, Dr. Judith Curry").
229 United States House of Representatives 1989, p. 101.
230 United States House of Representatives 1989, pp. 101–102.
231 United States House of Representatives 1989, p. 109.
232 United States Senate 2000, p. 21.
233 United States Senate 2000, pp. 55–56.
234 United States Senate 2000, p. 36.
235 United States Senate 2000, p. 73.
236 United States Senate 2000, pp. 73–74.
237 United States House of Representatives 2009a, pp. 6–7.
238 United States House of Representatives 2009a, pp. 24–25, 36–37.
239 United States House of Representatives 2009a, p. 38.
240 United States House of Representatives 2011, pp. 1–2.
241 United States House of Representatives 2011, pp. 4–5.

242 United States House of Representatives 2011, p. 178.
243 United States House of Representatives 2011, pp. 43–47.
244 United States House of Representatives 1956, p. 473.
245 Guber, Bohr, and Dunlap 2020; also see United States Senate 2007a, pp. 139–1053.
246 Leggett 2011.
247 C2ES undated a. The only major bill to have had some success was the American Clean Energy and Security Act written by Henry Waxman and Edward Markey, which passed the House of Representatives in June 2009, but was not even passed on to the Senate (also see Nilsen 2021).
248 C2ES undated b.
249 Supreme Court of the United States 2007, pp. 1443–1444.
250 Supreme Court of the United States 2022, p. 31.
251 Curbello 2021.

References

Aronoff, Kate (2020). "Why Is Biden Clinging to the Dream of Green Factories?" *The New Republic*, July 15; https://newrepublic.com/article/158515/biden-clinging-dream-green-factories

Begley, Sharon (2010). "Debunking Lomborg, the Climate-Change Skeptic." *Newsweek*, February 21; https://www.newsweek.com/debunking-lomborg-climate-change-skeptic-75173

Biden, Joe (2020a). Tweet, Twitter, July 14; https://twitter.com/JoeBiden/status/1283168143867707392

Biden, Joe (2020b). Tweet, Twitter, September 16; https://twitter.com/JoeBiden/status/1306236396378546179

Biden, Joe (2020c). Tweet, Twitter, November 1; https://twitter.com/JoeBiden/status/1322989285377667073

Biden, Joe (2022a). Tweet, Twitter, June 30; https://twitter.com/JoeBiden/status/1542647170192625664

Biden, Joe (2022b). *Statement by President Joe Biden on Supreme Court Ruling on West Virginia v. EPA*. The White House, June 30; https://www.whitehouse.gov/briefing-room/statements-releases/2022/06/30/statement-by-president-joe-biden-on-supreme-court-ruling-on-west-virginia-v-epa/

Biden, Joe (2022c). *Remarks by President Biden on Actions to Tackle the Climate Crisis*. The White House, July 20; https://www.whitehouse.gov/briefing-room/speeches-remarks/2022/07/20/remarks-by-president-biden-on-actions-to-tackle-the-climate-crisis/

Biden, Joe (2023). *Remarks of President Joe Biden – State of the Union Address as Prepared for Delivery*. The White House, February 7; https://www.whitehouse.gov/briefing-room/speeches-remarks/2023/02/07/remarks-of-president-joe-biden-state-of-the-union-address-as-prepared-for-delivery/

Bush, George H. W. (1989). *Remarks at the Annual Meeting of the Boards of Governors of the International Monetary Fund and World Bank Group*. George H.W. Bush Presidential Library & Museum, September 27; https://bush41library.tamu.edu/archives/public-papers/958

Bush, George H. W. (1990a). *Remarks to the Intergovernmental Panel on Climate Change*. The American Presidency Project, February 5; https://www.presidency.ucsb.edu/documents/remarks-the-intergovernmental-panel-climate-change

Bush, George H. W. (1990b). *Remarks at the Opening Session of the White House Conference on Science and Economics Research Related to Global Change*. The American

Presidency Project, April 17; https://www.presidency.ucsb.edu/documents/remarks-the-opening-session-the-white-house-conference-science-and-economics-research

Bush, George H. W. (1990c). *Remarks at the Closing Session of the White House Conference on Science and Economics Research Related to Global Change*. The American Presidency Project, April 18; https://www.presidency.ucsb.edu/documents/remarks-the-closing-session-the-white-house-conference-science-and-economics-research

Bush, George H. W. (1992a). *Message to the Congress on Environmental Goals*. George H.W. Bush Presidential Library & Museum, March 24; https://bush41library.tamu.edu/archives/public-papers/4102

Bush, George H. W. (1992b). *Address to the United Nations Conference on Environment and Development in Rio de Janeiro, Brazil*. George H.W. Bush Presidential Library & Museum, June 12; https://bush41library.tamu.edu/archives/public-papers/4417

Bush, George W. (2001). *President Bush Discusses Global Climate Change*. White House, President George W. Bush, June 11; https://georgewbush-whitehouse.archives.gov/news/releases/2001/06/20010611-2.html

Bush, George, W. (2002). *President Announces Clear Skies & Global Climate Change Initiatives*. The White House, President George W. Bush, February 14; https://georgewbush-whitehouse.archives.gov/news/releases/2002/02/20020214-5.html

Bush, George W. (2007). *President Bush Delivers State of the Union Address*. The White House, President George W. Bush, January 23; https://georgewbush-whitehouse.archives.gov/news/releases/2007/01/20070123-2.html

Bush, George W. (2008). *President Bush Discusses Climate Change*. The White House, President George W. Bush, April 16; https://georgewbush-whitehouse.archives.gov/news/releases/2008/04/20080416-6.html

C2ES (undated a). *Congress Climate History*. Center for Climate and Energy Solutions; https://www.c2es.org/content/congress-climate-history/

C2ES (undated b). *Congress and Climate Change*. Center for Climate and Energy Solutions; https://www.c2es.org/content/congress-and-climate-change/

Cama, Timothy (2018). "Trump on Dire Warnings in Climate Report: 'I Don't Believe It.'" *The Hill*, November 26; https://thehill.com/policy/energy-environment/418289-trump-on-dire-warnings-in-climate-report-i-dont-believe-it

Cantor, Matthew (2019). "Could 'Climate Delayer' Become the Political Epithet of Our Time?" *Guardian*, March 1; https://www.theguardian.com/environment/2019/mar/01/could-climate-delayer-become-the-political-epithet-of-our-times?CMP=share_btn_tw

Carrington, Damian (2019). "Climate Change Denial Is Evil, says Mary Robinson." *Guardian*, March 26; https://www.theguardian.com/environment/2019/mar/26/climate-change-denial-is-evil-says-mary-robinson

Cass, Loren R. (2006). *The Failures of American and European Climate Policy*. Albany: State University of New York Press; https://books.google.ca/books/about/The_Failures_of_American_and_European_Cl.html?id=ZQXFtYdw7nsC&redir_esc=y

Clinton, Bill (2000). *Address before a Joint Session of the Congress on the State of the Union*. The American Presidency Project, January 27; https://www.presidency.ucsb.edu/documents/address-before-joint-session-the-congress-the-state-the-union-7#axzz1y9IhNx00

Clinton, Bill (2015). Tweet, Twitter, May 18; https://twitter.com/BillClinton/status/600389785769881600

Cohen, Maurie J., and Egelston, Anne (2003). "The Bush Administration and Climate Change: Prospects for an Effective Policy Response." *Journal of Environmental Policy &. Planning* 5 (4), p. 315, December; https://www.tandfonline.com/doi/abs/10.1080/1523908032000171611

Columbia Law School (2022). *Climate Deregulation Tracker*. Columbia University in the City of New York, Columbia Law School, Columbia Climate School, Sabin Center for Climate Change Law; https://climate.law.columbia.edu/climate-deregulation-tracker

Curbello, Carlos (2021). Personal Communications by email. May 18.

Curry, Judith (2019a). Tweet, Twitter, March 9; https://twitter.com/curryja/status/1104420855952498688

Curry, Judith (2019b). Tweet, Twitter, March 9; https://twitter.com/curryja/status/1104517312487469056

Curry, Judith (2022a). Tweet, Twitter, November 29; https://twitter.com/curryja/status/1597692704132997120

Curry, Judith (2022b). Tweet, Twitter, November 29; https://twitter.com/curryja/status/1597696641028730881

D'Angelo, Chris (2019a). "Meet the Ostriches under Consideration for Trump's Anti-Science Climate Panel." *HuffPost*, February 27; https://www.huffingtonpost.ca/entry/trump-climate-change-panel_us_5c7470c6e4b0bf16620250d4

D'Angelo, Chris (2019b). "Trump's Attack on Climate Science Echoes Big Oil's 1998 Denial Campaign." *HuffPost*, June 27; https://www.huffingtonpost.ca/entry/trump-climate-panel-american-petroleum-institute-1998-campaign_n_5d129d6fe4b04f059e4b3d18

Davenport, Coral, and Landler, Mark (2019). "Trump Administration Hardens Its Attack on Climate Science." *New York Times*, May 27; https://www.nytimes.com/2019/05/27/us/politics/trump-climate-science.html?edf=78

DeSmog (undated). "Bjorn Lomborg;" https://www.desmogblog.com/bjorn-lomborg

DeSmog (undated). "Climate Disinformation Database;" https://www.desmogblog.com/global-warming-denier-database

DeSmog (undated). "John Christy;" https://www.desmogblog.com/john-christy

DeSmog (undated). "Judith Curry;" https://www.desmogblog.com/judith-curry

DeSmog (undated). "Michael Shellenberger;" https://www.desmogblog.com/michael-shellenberger

DeSmog (undated). "Patrick J. Michaels;" https://www.desmogblog.com/patrick-michaels

DeSmog (undated). "Richard Lindzen;" https://www.desmogblog.com/richard-lindzen

DeSmog (undated). "Roger A. Pielke Sr.;" https://www.desmogblog.com/roger-pielke-sr

DeSmog (undated). "Roger Pielke Jr.;" https://www.desmogblog.com/roger-pielke-jr

DeSmog (undated). "S. Fred Singer;" https://www.desmogblog.com/s-fred-singer

Diethelm, Pascal, and McKee, Martin (2009). "Denialism: What Is It and How Should Scientists Respond?" *European Journal of Public Health* 19 (1), p. 2, January; https://academic.oup.com/eurpub/article/19/1/2/463780

Dutton, Andrea, and Mann, Michael E. (2019). "A Dangerous New Form of Climate Denialism Is Making the Rounds." *Newsweek*, August 22; https://www.newsweek.com/dangerous-new-form-climate-denialism-making-rounds-opinion-1455736

Editorial (2022). "The Guardian View on Biden's Green Deal: Leadership after Trump's Denialism." *Guardian*, August 14; https://www.theguardian.com/commentisfree/2022/aug/14/the-guardian-view-on-bidens-green-deal-leadership-after-trumps-denialism

EPA (2009). *Endangerment and Cause or Contribute Findings for Greenhouse Gases under the Section 202(a) of the Clean Air Act*. U.S. EPA, Climate Change; https://www.epa.gov/climate-change/endangerment-and-cause-or-contribute-findings-greenhouse-gases-under-section-202a

EPA (2012). *EPA Proposes First Carbon Pollution Standard for Future Power Plants*. March 27; EPA's Web Archive; https://archive.epa.gov/epapages/newsroom_archive/newsreleases/9b4e8033d7e641d9852579ce005ae957.html

George H. W. Bush Library (1989). *Memorandum of Conversation, First Expanded Bilateral Session with Chairman Gorbachev of the Soviet Union.* George H.W. Bush Presidential Library and Museum, December 2; https://bush41library.tamu.edu/files/memcons-telcons/1989-12-02--Gorbachev%20Malta%20First%20Expanded%20Bilateral%20Meeting%20GB%20and%20Gorbachev%20in%20Malta.pdf

George H. W. Bush Library (1990a). *Memorandum of Conversation, Meeting with Prime Minister Margaret Thatcher of Great Britain.* George H.W. Bush Presidential Library and Museum, April 13; https://bush41library.tamu.edu/files/memcons-telcons/1990-04-13--Thatcher.pdf

George H. W. Bush Library (1990b). *Memorandum of Conversation, Telephone Call to Prime Minister Ruud Lubbers of the Netherlands.* George H.W. Bush Presidential Library and Museum, April 17; https://bush41library.tamu.edu/files/memcons-telcons/1990-04-17--Lubbers.pdf

George H. W. Bush Library (1990c). *Memorandum of Conversation, Bilateral Meeting with German Chancellor Helmut Kohl.* George H.W. Bush Presidential Library and Museum, July 9; https://bush41library.tamu.edu/files/memcons-telcons/1990-07-09--Kohl.pdf

George H. W. Bush Library (1991a). *Memorandum of Conversation, Meeting with Fernando Collor de Mello President of Brazil.* George H.W. Bush Presidential Library and Museum, June 18; https://bush41library.tamu.edu/files/memcons-telcons/1991-06-18--Collor%20de%20Mello.pdf

George H. W. Bush Library (1991b). *Memorandum of Conversation, Opening Session of the London Economic Summit.* George H.W. Bush Presidential Library and Museum, July 15; https://bush41library.tamu.edu/files/memcons-telcons/1991-07-15--Opening%20Session%20[1].pdf

George H. W. Bush Library (1991c). *Memorandum of Conversation, Second Plenary, London Economic Summit.* George H.W. Bush Presidential Library and Museum, July 16; https://bush41library.tamu.edu/files/memcons-telcons/1991-07-16--Mitterrand%20[2].pdf

George H. W. Bush Library (1992a). *Memorandum of Conversation, Meeting with Chairman Helmut Kohl of Germany.* George H.W. Bush Presidential Library and Museum, March 21; https://bush41library.tamu.edu/files/memcons-telcons/1992-03-21--Kohl.pdf

George H. W. Bush Library (1992b). *Memorandum of Telephone Conversation, Telecon with Brian* [sic] *Prime Minister Mulroney, Prime Minister of Canada.* George H.W. Bush Presidential Library and Museum, May 1; https://bush41library.tamu.edu/files/memcons-telcons/1992-05-01--Mulroney.pdf

George H. W. Bush Library (1992c). *Memorandum of Telephone Conversation, Telecon with Chancellor Kohl of Germany.* George H.W. Bush Presidential Library and Museum, May 6; https://bush41library.tamu.edu/files/memcons-telcons/1992-05-06--Kohl.pdf

George H. W. Bush Library (1992d). *The President's News Conference with Prime Minister John Major of the United Kingdom at Camp David.* George H.W. Bush Presidential Library and Museum, June 7; https://bush41library.tamu.edu/archives/public-papers/4401

Ghosh, Pallab (2017). "Hawking Says Trump's Climate Stance Could Damage Earth." *BBC News*, July 2; https://www.bbc.com/news/science-environment-40461726

Gore, Al (1990). "To Skeptics on Global Warming …" *New York Times*, April 22; https://www.nytimes.com/1990/04/22/opinion/to-skeptics-on-global-warming.html

Gore, Al (2022). Tweet, Twitter, August 12; https://twitter.com/algore/status/1558207319065989122

Guber, Deborah Lynn, Bohr, Jeremiah, and Dunlap, Riley E. (2020). "'Time to Wake Up:' Climate Change Advocacy in a Polarized Congress, 1996–2015." *Environmental Politics*,

July 7; https://www.tandfonline.com/doi/abs/10.1080/09644016.2020.1786333?needAccess=true&journalCode=fenp20

Hackett, Robert (2019). "Noam Chomsky: 'In a Couple of Generations, Organized Human Society May Not Survive.'" *National Observer*, February 12; https://www.nationalobserver.com/2019/02/12/features/noam-chomsky-couple-generations-organized-human-society-may-not-survive-has-be

Hasan, Mehdi (2021). "Legendary Activist Noam Chomsky on Biden's Presidency and the Modern GOP." The Mehdi Hasan Show, *MSNBC*, April 18; https://www.msnbc.com/mehdi-on-msnbc/watch/legendary-activist-noam-chomsky-on-biden-s-presidency-and-the-modern-gop-110429765669

Hecht, Alan D., and Tirpak, Dennis (1995). "Framework Agreement on Climate Change: A Scientific and Policy History." *Climatic Change* 29, p. 371; https://link.springer.com/article/10.1007/BF01092424

Holden, Emily (2018). "Scott Pruitt Never Gave Up EPA Plans to Debate Climate Science, Records Show." *Guardian*, December 22; https://www.theguardian.com/environment/2018/dec/21/scott-pruitt-epa-climate-change-science-red-team-blue-team-debate

House Select Committee on the Climate Crisis (2020). *Solving the Climate Crisis, the Congressional Action Plan for a Clean Energy Economy and a Healthy, Resilient, and Just America*. Select Committee on the Climate Crisis, Majority Staff Report, 116th Congress, June; https://climatecrisis.house.gov/report

Hudson, Marc (2018). *George Bush Sr Could Have Got in on the Ground Floor of Climate Action – History Would Have Thanked Him*. The Conversation, December 5; https://theconversation.com/george-bush-sr-could-have-got-in-on-the-ground-floor-of-climate-action-history-would-have-thanked-him-108050

Inglis, Bob (2015). *Bob Inglis – Acceptance Speech*. John F. Kennedy Presidential Library and Museum; https://www.jfklibrary.org/events-and-awards/profile-in-courage-award/award-recipients/bob-inglis-2015

Irfan, Umair (2020). "We Asked Joe Biden's Campaign 6 Key Questions about His Climate Change Plans." *Vox*, October 22; https://www.vox.com/21516594/joe-biden-climate-change-covid-19-president

Johnson, Eddie Bernice (2019). *The State of Climate Science and Why It Matters*. House Science, Space, and Technology Committee, United States House of Representatives, February 13; https://docs.house.gov/meetings/SY/SY00/20190213/108915/HHRG-116-SY00-MState-J000126-20190213.pdf

Kerr, Richard A. (1992). "Greenhouse Science Survives Skeptics." *Science* 256 (5060), p. 1138, May 22; https://science.sciencemag.org/content/256/5060/1138.2

King, Jennie, Janulewicz, Lukasz, and Arcostanzo, Francesca (2022). *Deny, Deceive, Delay: Documenting and Responding to Climate Disinformation at COP26 & Beyond – Executive Summary*. Institute for Strategic Dialogue, June 9; https://www.isdglobal.org/isd-publications/deny-deceive-delay-documenting-and-responding-to-climate-disinformation-at-cop26-and-beyond/

Koch Docs (1991). *1991 CATO Climate Denial Conference Flyer*. Koch Docs; https://kochdocs.org/2019/08/12/1991-cato-climate-denial-conference-flyer-and-schedule/#annotation/a442303

Koonin, Steven E. (2021). *Unsettled: What Climate Science Tells Us, What It Doesn't, And Why It Matters*. Dallas: BenBella Books; https://www.google.ca/books/edition/Unsettled/uVMAEAAAQBAJ?hl=en&gbpv=1&dq=Unsettled:+What+Climate+Science+Tells+Us,+What+It+Doesn%E2%80%99t,+And+Why+It+Matters&printsec=frontcover

Kutney, Gerald W. (2014). *Carbon Politics and the Failure of the Kyoto Protocol*. London: Routledge.

Lamb, William F., Mattioli, Giulio, Levi, Sebastian, Roberts, Timmons, and others (2020). "Discourses of Climate Delay." *Global Sustainability* 3, e17, July 1; https://www.cambridge.org/core/journals/global-sustainability/article/discourses-of-climate-delay/7B11B722E3E3454BB6212378E32985A7

Lattanzio, Richard K., Leggett, Jane A., Procita, Kezee, Ramseur, Jonathan L., Clark, Corrie E., Croft, Genevieve K., and Miller, Rena S. (2021). *U.S. Climate Change Policy*. Congressional Research Service, October 28; https://crsreports.congress.gov/product/pdf/R/R46947

Lavelle, Marianne (2018). "Trump EPA Sued Over Refusal to Release Heartland Institute Communications." *Inside Climate News*, March 15; https://insideclimatenews.org/news/15032018/epa-documents-foia-heartland-institute-scott-pruitt-red-team-climate-change-denial-lawsuit

Leggett, Jane A. (2011). *Climate Change: Conceptual Approaches and Policy Tools*. Congressional Research Service, August 29; https://crsreports.congress.gov/product/details?prodcode=R41973

Levantesi, Stella, and Corsi, Giulio (2021). *Climate Deniers Are Using These Four Major Scare Tactics to Stop Climate Action*. DeSmog, November 16; https://www.desmog.com/2021/11/16/four-major-climate-denial-scare-tactics-twitter/

Lomborg, Bjorn (2014). *Examining the Threats Posed by Climate Change: The Effects of Unchecked Climate Change on Communities and the Economy*. U.S. Senate Committee on Environment and Public Works, Hearings, Subcommittee on Clean Air and Nuclear Safety, July 29; https://www.epw.senate.gov/public/index.cfm/hearings?ID=EBCD8F70-CC3F-978D-C157-58C503716D6F

Lomborg, Bjorn (2019a). Tweet, Twitter, June 3; https://twitter.com/BjornLomborg/status/1135643116717969409

Lomborg, Bjorn (2019b). Tweet, Twitter, June 3; https://twitter.com/BjornLomborg/status/1135638855233744896

Maize, Kennedy (2015). "Climate McCarthyism or Just Stupidity?" *Power*, March 3; https://www.powermag.com/blog/climate-mccarthyism-or-just-stupidity/

Mann, Michael E. (2012). *The Hockey Stick and the Climate Wars*. New York: Columbia University Press.

Mann, Michael E. (2015). "The Assault on Climate Science." *New York Times*, December 8; https://www.nytimes.com/2015/12/08/opinion/the-assault-on-climate-science.html

Mann, Michael E. (2021). *The New Climate War*. New York: Public Affairs.

Mann, Michael E., and Toles, T. (2016). *The Madhouse Effect*. New York: Columbia University Press.

McGreal, Chris (2022a). "Interview, Climate Action Has Been 'a Calamity', Says Senate Democrat Sheldon Whitehouse." *Guardian*, March 25; https://www.theguardian.com/environment/2022/mar/25/senator-sheldon-whitehouse-climate-crisis

McGreal, Chris (2022b). "'What We Now Know … They Lied': How Big Oil Companies Betrayed Us All." *Guardian*, April 21; https://www.theguardian.com/tv-and-radio/2022/apr/20/what-we-now-know-they-lied-how-big-oil-companies-betrayed-us-all

Michaels, Patrick J. (1989). "The Greenhouse Climate of Fear." *Washington Post*, January 8; see United States Senate (1989), p. 56.

Michel, Charles (2017). Tweet, Twitter, June 1; https://twitter.com/CharlesMichel/status/870365886233837569

Milman, Oliver (2022). "Al Gore Hails Biden's Historic Climate Bill as 'a Critical Turning Point.'" *Guardian*, August 12; https://www.theguardian.com/us-news/2022/aug/12/al-gore-climate-turning-point-inflation-reduction-act

Mintzer, Irving M., and Leonard, J. Amber (ed.) (1994). *Negotiating Climate Change*. Cambridge: Cambridge University Press.

Montlake, Simon (2019). "What Does Climate Change Have to Do with Socialism?" *Christian Science Monitor*, August 5; https://www.csmonitor.com/Environment/2019/0805/What-does-climate-change-have-to-do-with-socialism

National Academy of Sciences (1991). *Policy Implications of Greenhouse Warming*. National Academy of Sciences, Washington, DC: National Academy Press; https://www.nap.edu/catalog/1794/policy-implications-of-greenhouse-warming

National Academy of Sciences (2001). *Climate Change Science: An Analysis of Some Key Questions*. National Research Council. Washington, DC: National Academy Press; https://www.nap.edu/catalog/10139/climate-change-science-an-analysis-of-some-key-questions

Newburger, Emma (2020). "Joe Biden calls Climate Change the 'Number One Issue Facing Humanity.'" *CNBC*, October 24; https://www.cnbc.com/2020/10/24/joe-biden-climate-change-is-number-one-issue-facing-humanity.html

Nilsen, Ella (2021). "Democrats Think Now Is Their Last, Best Chance to Pass a Big Climate Bill." *Vox*, June 23; https://www.vox.com/22537509/democrats-climate-bill-biden-waxman-markey

Nuccitelli, Dana (2019). *The Trump EPA Strategy to Undo Clean Power Plan*. Yale Climate Connection, June 21; https://www.yaleclimateconnections.org/2019/06/the-trump-epa-strategy-to-undo-the-clean-power-plan/

Obama. Barack (2009a). Tweet. Twitter, March 25; https://twitter.com/BarackObama/status/1389362776

Obama, Barack (2009b). Tweet, Twitter, December 8; https://twitter.com/BarackObama/status/6470396547

Obama, Barack (2009c). *President Obama at Copenhagen Climate Change Conference*. The White House, President Barack Obama, Video; December 18; https://obamawhitehouse.archives.gov/photos-and-video/video/president-obama-copenhagen-climate-change-conference

Obama, Barack (2013a). *Inaugural Address by President Barack Obama*. The White House, President Barack Obama, January 21; https://obamawhitehouse.archives.gov/the-press-office/2013/01/21/inaugural-address-president-barack-obama

Obama, Barack (2013b). *Remarks by the President in the State of the Union Address*. The White House, President Barack Obama, February 12; https://obamawhitehouse.archives.gov/the-press-office/2013/02/12/remarks-president-state-union-address

Obama, Barack (2013c). Tweet, Twitter, August 19; https://twitter.com/BarackObama/status/369540473553055744

Obama, Barack (2015a). *Remarks by the President in State of the Union Address | January 20, 2015*. The White House, President Barack Obama, January 20; https://obamawhitehouse.archives.gov/the-press-office/2015/01/20/remarks-president-state-union-address-january-20-2015

Obama, Barack (2015b). Tweet, Twitter, May 18; https://twitter.com/POTUS44/status/600324682190053376

Obama, Barack (2015c). *President Obama Announces the Clean Power Plan*. The White House, President Barack Obama, Video, August 3; https://obamawhitehouse.archives.gov/photos-and-video/video/2015/08/03/president-obama-announces-clean-power-plan;

for a summary of the Clean Power Plan see *Climate Change and President Obama's Action Plan*. The White House, President Barack Obama, December 12; https://obamawhitehouse.archives.gov/president-obama-climate-action-plan

Obama, Barack (2016). *Remarks by the President on the Paris Agreement*. The White House, President Barack Obama, October 5; https://obamawhitehouse.archives.gov/the-press-office/2016/10/05/remarks-president-paris-agreement

Obama, Barack (2020). Tweet, Twitter, March 31; https://twitter.com/BarackObama/status/1245007713387544576

Ocasio-Cortez, Alexandria (2019). Tweet, Twitter, February 24; https://twitter.com/AOC/status/1099854840145031168

Rich, Nathaniel (2018). "Losing Earth: The Decade We Almost Stopped Climate Change." *New York Times*, August 1; https://www.nytimes.com/interactive/2018/08/01/magazine/climate-change-losing-earth.html

Romm, Joe (2007). *Debating Bjorn Lomborg, Global Warming Delayer*. ThinkProgress, September 12; https://thinkprogress.org/debating-bj-rn-lomborg-global-warming-delayer-ecd3c314cd45/

Romney, Mitt (2022). "America Is in Denial." *The Atlantic*, July 4; https://www.theatlantic.com/ideas/archive/2022/07/mitt-romney-republican-denial-biden-election/661468/

Royden, Amy (2002). "U.S. Climate Change Policy under President Clinton: A Look Back." *Golden Gate University Law Review* 32 (4), p. 415; https://digitalcommons.law.ggu.edu/cgi/viewcontent.cgi?referer=https://www.google.com/&httpsredir=1&article=1842&context=ggulrev;US

Samuelsohn, Darren (2010). "Shimkus Cites Genesis on Climate." *Politico*, November 10; https://www.politico.com/story/2010/11/shimkus-cites-genesis-on-climate-044958

Schwartz, Dan (2019). *The Last of the Climate Deniers Hold On*. Typeinvestigations, November 18; https://www.typeinvestigations.org/investigation/2019/11/18/the-last-of-the-climate-deniers-hold-on/

Schwartz, John (2020). "S. Fred Singer, a Leading Climate Change Contrarian Dies at 95." *New York Times*, April 11; https://www.nytimes.com/2020/04/11/climate/s-fred-singer-dead.html

Senate Democrats' Special Committee on the Climate Crisis (2020). *The Case for Climate Action, Building a Clean Economy for the American People*. Senate Democrats, August 25; https://www.democrats.senate.gov/newsroom/press-releases/08/25/2020/senate-democrats-climate-committee-releases-new-report-on-climate-action-plan-to-build-clean-economy-for-american-people

Shabecoff, Philip (1989). "E.P.A. Chief Says Bush Will Not Rush into a Treaty on Global Warming." *New York Times*, May 13; https://www.nytimes.com/1989/05/13/us/epa-chief-says-bush-will-not-rush-into-a-treaty-on-global-warming.html

Shabecoff, Philip (1990). "Bush Denies Putting Off Action on Averting Global Climate Shift." *New York Times*, April 19; http://www.nytimes.com/1990/04/19/us/bush-denies-putting-off-action-on-averting-global-climate-shift.html?scp=57&sq=philip+shabecoff&st=nyt

Shellenberger, Michael (2020). *Testimony of Michael D. Shellenberger, Founder and President of Environment Progress, for the House Select Committee on the Climate Crisis*. July 28; https://climatecrisis.house.gov/vibrantandjusthearing

Shellenberger, Michael (2022). *Michael Shellenberger Testimony to the House Committee on Oversight & Reform For a Hearing on: "Fueling the Climate Crisis: Examining Big Oil's Prices, Profits, and Pledges."* September 15; https://oversight.house.gov/sites/democrats.oversight.house.gov/files/Shellenberger%20Testimony.pdf

Skeptical Science (2020). *Climate Misinformation by Source*; https://skepticalscience.com/misinformers.php

Supreme Court of the United States (2007). *127 S.Ct. 1438 (2007) 549 U.S. 497 Massachusetts et al., Petitioners, v. Environmental Protection Agency et al. No. 05-1120.* April 2; https://scholar.google.ca/scholar_case?case=18363956969502505811&q=Massachusetts+v.+EPA,+549+U.S.+497&hl=en&as_sdt=2006&as_vis=1

Supreme Court of the United States (2022). *West Virginia et al. v. Environmental Protection Agency et al. No. 20-1539*, June 30; https://www.supremecourt.gov/opinions/21pdf/20-1530_n758.pdf

Trump, Donald (2017a). *Presidential Executive Order on Promoting Energy Independence and Economic Growth.* White House, March 28; https://trumpwhitehouse.archives.gov/presidential-actions/presidential-executive-order-promoting-energy-independence-economic-growth/

Trump, Donald (2017b). *Statement by President Trump on the Paris Climate Accord.* White House, June 1; https://trumpwhitehouse.archives.gov/briefings-statements/statement-president-trump-paris-climate-accord/

TweetBinder Blog (undated). *Barack Obama on Twitter – 2008/2021 Analysis*; https://www.tweetbinder.com/blog/barack-obama-twitter/

UNFCCC (2020). *The Paris Agreement.* United Nations Climate Change; https://unfccc.int/process-and-meetings/the-paris-agreement/the-paris-agreement

United States Department of State (1998). *Statement by Vice President Gore on the United States' Signing of the Kyoto Protocol.* U.S. Department of State Archives, November 12; https://1997-2001.state.gov/global/global_issues/climate/981112_gore_climate.html

United States House of Representatives (1956). *National Science Foundation, International Geophysical Year.* Hearings before the Subcommittees of the Committee on Appropriations, House of Representatives, 84th Congress, Second Session, March 8, p. 426; https://babel.hathitrust.org/cgi/pt?id=umn.31951d007420l6t;view=1up;seq=476

United States House of Representatives (1988). *Implications of Global Warming for Natural Resources.* Oversight Hearings before the Subcommittee on Water and Power Resources of the Committee on Interior and Insular Affairs, House of Representatives, 100th Congress, Second Session, September 27, October 17; https://babel.hathitrust.org/cgi/pt?id=umn.31951d00283282e;view=1up;seq=3

United States House of Representatives (1989). *Global Warming.* Hearings before the Subcommittee on Energy and Power of the Committee on Energy and Commerce, House of Representatives, 101st Congress, First Session, February 21 and May 4; https://babel.hathitrust.org/cgi/pt?id=uc1.31210007429986;view=1up;seq=1

United States House of Representatives (1991). *Global Climate Change and Greenhouse Emissions.* Hearings before the Subcommittee on Health and the Environment of the Committee on Energy and Commerce, House of Representatives, 102nd Congress, First Session, February 21 and August 1; https://babel.hathitrust.org/cgi/pt?id=pst.000019267487;view=1up;seq=1

United States House of Representatives (1995). *Scientific Integrity and Public Trust: The Science behind Federal Policies and Mandates: Case Study 2 – Climate Models and Projections of Potential Impacts of Global Climate Change.* Hearing before the Subcommittee on Energy and Environment of the Committee on Science, U.S. House of Representatives, 104th Congress, First Session, November 16; https://babel.hathitrust.org/cgi/pt?id=uc1.31210012872972&view=1up&seq=1

United States House of Representatives (1996). *U.S. Global Change Research Programs: Data Collection and Scientific Priorities.* Hearing before the Committee on Science,

U.S. House of Representatives, 104th Congress, Second Session, March 6; https://babel.hathitrust.org/cgi/pt?id=pst.000025977899&view=1up&seq=3

United States House of Representatives (2002). *The U.S. National Climate Change Assessment: Do the Climate Models Project A Useful Picture of Regional Climate?* Hearing before the Subcommittee on Oversight and Investigations of the Committee on Energy and Commerce, House of Representatives, 107th Congress, Second Session, July 25; https://babel.hathitrust.org/cgi/pt?id=pst.000049657166&view=1up&seq=1

United States House of Representatives (2006). *Climate Change: Understanding the Degree of the Problem*. Hearing before the Committee on Government Reform, House of Representatives, 109th Congress, Second Session, July 20; https://babel.hathitrust.org/cgi/pt?id=pst.000058948934;view=1up;seq=1

United States House of Representatives (2007a). *Allegations of Political Interference with the Work of Government Climate Change Scientists*. Hearing before the Committee on Oversight and Government Reform, House of Representatives, 110th Congress, First Session, January 30; https://babel.hathitrust.org/cgi/pt?id=pst.000061491786&view=1up&seq=1

United States House of Representatives (2007b). *Perspectives on Climate Change*. Hearing before the Subcommittee on Energy and Air Quality of the Committee on Energy and Commerce and the Subcommittee on Energy and Environment of the Committee on Science and Technology, House of Representatives, 110th Congress, First Session, March 21; https://babel.hathitrust.org/cgi/pt?id=umn.31951d02737035q&view=1up&seq=5

United States House of Representatives (2007c). *Shaping the Message, Distorting the Science: Media Strategies to Influence Science Policy*. Hearing before the Subcommittee on Investigations and Oversight, Committee on Science and Technology, House of Representatives, 110th Congress, First Session, March 28; https://babel.hathitrust.org/cgi/pt?id=pst.000061515956&view=1up&seq=3

United States House of Representatives (2009a). *Hearing on Scientific Objectives for Climate Change Legislation*. Hearing before the Committee on Ways and Means, U.S. House of Representatives, 111th Congress, First Session, February 25; https://babel.hathitrust.org/cgi/pt?id=mdp.39015090388078&view=1up&seq=1

United States House of Representatives (2009b). *Protecting Lower-Income Families While Fighting Global Warming*. Hearing before the Subcommittee on Income Security and Family Support of the Committee on Ways and Means, U.S. House of Representatives, 111th Congress, First Session, March 12; https://babel.hathitrust.org/cgi/pt?id=pst.000066755180&view=1up&seq=1

United States House of Representatives (2009c). *New Directions for Energy Research and Development at the U.S. Department of Energy*. Hearing before the Committee on Science and Technology, House of Representatives, 111th Congress, First Session, March 17; https://babel.hathitrust.org/cgi/pt?id=mdp.39015090407142&view=1up&seq=1&skin=2021&q1=denial

United States House of Representatives (2009d). *Preparing for Climate Change: Adaptation Policies and Programs*. Hearing before the Subcommittee on Energy and Environment of the Committee on Energy and Commerce, House of Representatives, 111th Congress, First Session, March 25; https://babel.hathitrust.org/cgi/pt?id=umn.31951d03586253x&view=1up&seq=5

United States House of Representatives (2010a). *The Foundation of Climate Science*. Hearing before the Select Committee on Energy Independence and Global Warming, House of Representatives, 111th Congress, Second Session, May 6; https://www.govinfo.gov/content/pkg/CHRG-111hhrg62590/pdf/CHRG-111hhrg62590.pdf

United States House of Representatives (2010b). *A Rational Discussion of Climate Change: The Science, the Evidence, the Response*. Hearing before the Subcommittee on Energy and Environment, Committee on Science and Technology, House of Representatives, 111th Congress, Second Session, November 17; https://www.govinfo.gov/content/pkg/CHRG-111hhrg62618/pdf/CHRG-111hhrg62618.pdf

United States House of Representatives (2011). *Climate Science and EPA's Greenhouse Gas Regulations*. Hearing before the Subcommittee on Energy and Power of the Committee on Energy and Commerce, House of Representatives, 112th Congress, First Session, March 8; https://babel.hathitrust.org/cgi/pt?id=umn.31951d03457933v&view=1up&seq=3

United States House of Representatives (2015). *Zero Accountability: The Consequences of Politically Driven Science*. Oversight Hearing before the Subcommittee on Oversight and Investigations of the Committee of Natural Resources, U.S. House of Representatives, 114th Congress, First Session, April 29; https://www.govinfo.gov/content/pkg/CHRG-114hhrg94404/pdf/CHRG-114hhrg94404.pdf

United States House of Representatives (2017). *Climate Science: Assumptions, Policy Implications, and the Scientific Method*. Hearing before the Committee on Science, Space, and Technology, House of Representatives, 115th Congress, First Session, March 29; https://www.govinfo.gov/content/pkg/CHRG-115hhrg25098/pdf/CHRG-115hhrg25098.pdf

United States House of Representatives (2019a). *Climate Change: The Impacts and the Need to Act*. Oversight Hearing before the Committee on Natural Resources, U.S. House of Representatives, 116th Congress, First Session, February 6; https://www.govinfo.gov/content/pkg/CHRG-116hhrg34954/pdf/CHRG-116hhrg34954.pdf

United States House of Representatives (2019b). *Time for Action: Addressing the Environmental and Economic Effects of Climate Change*. Hearing before the Subcommittee on Environment and Climate Change of the Committee on Energy and Commerce, House of Representatives, 116th Congress, First Session, February 6; https://www.govinfo.gov/content/pkg/CHRG-116hhrg35330/pdf/CHRG-116hhrg35330.pdf

United States House of Representatives (2019c). *Generation Climate: Young Leaders Urge Climate Action Now*. Hearing before the Select Committee on the Climate Crisis, 116th Congress, First Session, April 4; https://www.govinfo.gov/content/pkg/CHRG-116hhrg36812/pdf/CHRG-116hhrg36812.pdf

United States House of Representatives (2019d). *Climate Change, Part I: The History of a Consensus and the Causes of Inaction*. Hearing before the Subcommittee on Environment of the Committee on Oversight and Reform, House of Representatives, 116th Congress, First Session, April 9; https://www.govinfo.gov/content/pkg/CHRG-116hhrg36637/pdf/CHRG-116hhrg36637.pdf

United States House of Representatives (2019e). *The Need for Leadership to Combat Climate Change and Protect National Security*. Hearing before the Committee on Oversight and Reform, House of Representatives, 116th Congress, First Session, April 9; https://www.govinfo.gov/content/pkg/CHRG-116hhrg36439/pdf/CHRG-116hhrg36439.pdf

United States House of Representatives (2019f). *Recovery, Resiliency and Readiness – Contending with Natural Disasters in the Wake of Climate Change*. Hearing before the Subcommittee on Environment of the Committee on Oversight and Reform, House of Representatives, 116th Congress, First Session, June 25; https://oversight.house.gov/legislation/hearings/recovery-resiliency-and-readiness-contending-with-natural-disasters-in-the-wake

United States Senate (1985). *Global Warming*. Hearing before the Subcommittee on Toxic Substances and Environmental Oversight of the Committee on Environment and Public Works, United States Senate, 99th Congress, First Session, December 10; https://babel.hathitrust.org/cgi/pt?id=umn.31951d00275504i;view=1up;seq=1

United States Senate (1989a). *International Environmental Agenda for the 101st Congress*. Hearing of the Committee on Foreign Relations, United States Senate, 101st Congress, First Session, April 20; https://babel.hathitrust.org/cgi/pt?id=umn.31951d002918167&view=1up&seq=1

United States Senate (1989b). *Climate Surprises*. Hearing before the Subcommittee on Science, Technology, and Space of the Committee on Commerce, Science, and Transportation, United States Senate, 101st Congress, First Session, May 8; https://babel.hathitrust.org/cgi/pt?id=uc1.b5176796&view=1up&seq=1

United States Senate (1991). *Policy Implications of Greenhouse Warming*. Hearing before the Committee on Commerce, Science, and Transportation, 102nd Congress, First Session, April 25; https://babel.hathitrust.org/cgi/pt?id=pst.000018818604&view=1up&seq=1

United States Senate (1992). *Global Climate Change*. Hearings before the Committee on Energy and Natural Resources, United States Senate, 102nd Congress, Second Session, May 6; https://babel.hathitrust.org/cgi/pt?id=uc1.b5124509&view=1up&seq=3

United States Senate (1997). *S.Res.98 – A Resolution Expressing the Sense of the Senate Regarding the Conditions for the United States Becoming a Signatory to Any International Agreement on Greenhouse Gas Emissions under the United Nations Framework Convention on Climate Change*. Congress.gov. Agreed to in Senate, 105th Congress, First Session, July 25; https://www.congress.gov/bill/105th-congress/senate-resolution/98/text

United States Senate (2000). *The Science behind Global Warming*. Hearing before the Committee on Commerce, Science, and Transportation, United States Senate, 106th Congress, Second Session, May 17; https://babel.hathitrust.org/cgi/pt?id=uc1.b5183343&view=1up&seq=1

United States Senate (2001a). *Intergovernmental Panel on Climate Change (IPCC) Third Assessment Report*. Hearing before the Committee on Commerce, Science, and Transportation, United States Senate, 107th Congress, First Session, May 1; https://babel.hathitrust.org/cgi/pt?id=uc1.b5183414&view=1up&seq=1

United States Senate (2001b). *Clean Air Act Oversight Issues, Science of Global Climate Change and Greenhouse Gas Emissions*. Hearings before the Subcommittee on Clean Air, Wetlands, Private Property, and Nuclear Safety, and the Committee on Environment and Public Works, United States Senate, 107th Congress, First Session, May 2, p. 309; https://babel.hathitrust.org/cgi/pt?id=uc1.b5139047;view=1up;seq=5

United States Senate (2005). *Kyoto Protocol: Assessing the Status of Efforts to Reduce Greenhouse Gases*. Hearing before the Committee on Environment and Public Works, United States Senate, 109th Congress, First Session, October 5; https://babel.hathitrust.org/cgi/pt?id=pst.000063527162;view=1up;seq=1

United States Senate (2007a). *Senators' Perspectives on Global Warming*. Hearing before the Committee on Environment and Public Works, United States Senate, 110th Congress, First Session, January 30; https://babel.hathitrust.org/cgi/pt?id=uc1.31822034529552;view=1up;seq=1

United States Senate (2007b). *The Implications of the Supreme Court's Decision Regarding EPA's Authorities with Respect to Greenhouse Gases under the Clean Air Act*. Hearing before the Committee on Environment and Public Works, United States Senate, 110th Congress, First Session, April 24; https://babel.hathitrust.org/cgi/pt?id=uc1.31822038354130;view=1up;seq=1

United States Senate (2013). *Climate Change: It's Happening Now*. Hearing before the Committee on Environment and Public Works, United States Senate, 113th Congress, First Session, July 18; https://www.govinfo.gov/content/pkg/CHRG-113shrg95976/pdf/CHRG-113shrg95976.pdf

United States Senate (2014a). *Review of the President's Climate Action Plan*. Hearing before the Committee on Environment and Public Works, United States Senate, 113th Congress, Second Session, January 16; https://www.govinfo.gov/content/pkg/CHRG-113shrg97581/pdf/CHRG-113shrg97581.pdf

United States Senate (2014b). *Examining the Threats Posed by Climate Change*. Hearing before the Subcommittee on Clean Air and Nuclear Safety of the Committee on Environment and Public Works, United States Senate, 113th Congress, Second Session, July 29; https://www.govinfo.gov/content/pkg/CHRG-113shrg98184/html/CHRG-113shrg98184.htm

United States Senate (2015). *Data or Dogma? Promoting Open Inquiry in the Debate over the Magnitude of Human Impact on Earth's Climate*. Hearing before the Subcommittee on Space, Science, and Competitiveness of the Committee on Commerce, Science, and Transportation, United States Senate, 114th Congress, First Session, December 8; https://www.govinfo.gov/content/pkg/CHRG-114shrg21644/html/CHRG-114shrg21644.htm

USGCRP (2018). *Impacts, Risks, and Adaptation in the United States: Fourth National Climate Assessment, Volume II* [Reidmiller, D. R., C. W. Avery, D. R. Easterling, K. E. Kunkel, K. L. M. Lewis, T. K. Maycock, and B. C. Stewart (eds.)]. Washington, DC: U.S. Global Change Research Program; https://nca2018.globalchange.gov/chapter/front-matter-about/

Waldman, Scott (2019). "Emails: Trump Aide Had Blueprints to Unravel Climate Science." *E&E News*, December 10; https://www.eenews.net/stories/1061769389

Waldman, Scott (2022). "Patrick Michaels, Influential Climate Denier, Dies at 72." *E&E News*, July 19; https://www.eenews.net/articles/patrick-michaels-influential-climate-denier-dies-at-72/

Waldman, Scott, and Hulac, Benajmin (2018). "This Is When the GOP Turned Away from Climate Policy." *E&E News*, December 5; https://www.eenews.net/stories/1060108785/

Ward, Alex (2017). "Foreign Leaders to Trump: There Is No 'Better Deal' Than Paris." *Vox*, June 1; https://www.vox.com/world/2017/6/1/15726602/paris-agreement-trump-foreign-leaders-laugh

Watts, Jonathan, and Connolly, Kate (2017). "World Leaders React after Trump Rejects Paris Climate Deal." *Guardian*, June 2; https://www.theguardian.com/environment/2017/jun/01/trump-withdraw-paris-climate-deal-world-leaders-react

White House (2015). *Introducing @POTUS: President Obama's Twitter Account*. The White House, President Barack Obama, May 18; https://obamawhitehouse.archives.gov/blog/2015/05/17/introducing-potus-presidents-official-twitter-account

White House (2021). *FACT SHEET: President Biden Takes Executive Actions to Tackle the Climate Crisis at Home and Abroad, Create Jobs, and Restore Scientific Integrity across Federal Government*. The White House, January 27; https://www.whitehouse.gov/briefing-room/statements-releases/2021/01/27/fact-sheet-president-biden-takes-executive-actions-to-tackle-the-climate-crisis-at-home-and-abroad-create-jobs-and-restore-scientific-integrity-across-federal-government/

White House (2022). *Readout of White House Climate Science Roundtable on Countering "Delayism" and Communicating the Urgency of Climate Action*. The White House, February 25; https://www.whitehouse.gov/ostp/news-updates/2022/02/25/readout-of-white-house-climate-science-roundtable-on-countering-delayism-and-communicating-the-urgency-of-climate-action/

William J. Clinton Library (1993). *Climate Change Action Plan 9-22-93 1:00-2:00*. Clinton Digital Library, September 22; https://clinton.presidentiallibraries.us/items/show/19991

4 The Denial Cabal

The evolution of political thought on climate change in the White House and in Congress was covered in the previous two chapters. Climate science had, at first, gained the attention of politicians but that gave way to climate denialism, at least among Republicans, beginning with the presidency of George H. W. Bush. During the term of President George W. Bush, a "denial cabal" of hard-core climate deniers had formed among the G.O.P. elite which has been a barrier for the United States to enact policies to deal with the climate crisis.

Political Climate Denialism

Climate denialism had spread like a plague during the 1990s, and by the start of the new millennium, climate denialism was already rife within the Republican ranks and its elite. Climate deniers would be found chairing congressional committees, in cabinet and other senior posts, and even in the Oval Office – the most outspoken and zealous on climate denial among the political elite, I have designated as members of a denial cabal.

Secretary James Schlesinger

Who was the first to deny the accepted science of climate change from within the ranks of the American political elite? A strong candidate for the first political climate denier would be the economist James Schlesinger (1929–2014), who had been Secretary of Defense under Richard Nixon and Gerald Ford, and Secretary of Energy under Jimmy Carter. He was discussed in chapter two as Director of the CIA in 1973, just before the strangely unbalanced report on global cooling had been issued, which first raised my suspicions about Dr. Schlesinger.

During a cabinet meeting, an extraordinary exchange took place when Dr. Schlesinger corrected President Carter[1]:

> Mr. Bergland [Secretary of Agriculture] asked whether the burning of coal will create a change in climatic conditions – an allegation in recent press reports. Dr. Schlesinger reponded [sic] that there is some speculation concerning a "greenhouse effect" … The President said that the trend during

DOI: 10.4324/9781003455417-5

the first part of the century was toward a slight increase in temperature, but Dr. Schlesinger noted that there is some evidence to support the opposite conclusion.

As Secretary of Energy during the Carter administration, he had used "uncertainty" as an excuse for inaction on climate change. On July 7, 1977, President Carter had been sent a memorandum from Frank Press with the subject line: "Release of Fossil CO_2 and the Possibility of a Catastrophic Climate Change," and James Schlesinger immediately dismissed the warning, responding the next day that[2]: "The policy implications of this issue are still too uncertain to warrant presidential involvement or policy initiatives." His note emphasizing uncertainty indicated that the president should not be involved. Two years later, the Department of Energy again reacted dismissively to the warnings from climate experts, this time by James Speth.[3] President Carter sided with the Department of Energy and declared his support for liquid fuels from coal, which caused outrage among some within the scientific community. At the time, James Schlesinger was just finishing his tenure as the first Secretary of Energy.

In 2003, no longer in politics, James Schlesinger wrote an opinion piece for the *Washington Post*, in which his denial leanings were on full display[4]: "The CO_2/climate-change relationship has hardened into orthodoxy – always a worrisome sign – an orthodoxy that searches out heretics and seeks to punish them … A premature commitment to a fixed policy can only proceed with fear and trembling." He continued to author opinion pieces, including an especially derogatory one in the *Wall Street Journal*, titled "The Theology of Global Warming."[5] Dr. Schlesinger was reported to be[6]: "probably the nation's leading voice when it comes to promoting a contrarian interpretation of the science in these venues [mainstream media]," and had been called out for being on the board of the world's largest private-sector coal company. Schlesinger's op-eds appeared during the days of the George W. Bush administration, just as the denial cabal was taking shape.

President Ronald Reagan: 1981–9

President Reagan inspired a new paradigm in politics, and climate change was caught up in the movement of regulation minimalism. His dogmatic free-market stance erased climate change from the Republican agenda, and the polarization of climate change politics was underway.[7] At the time there was growing evidence that anthropogenic climate change was a threat to humanity; so, the 1980s could have been a breakout decade for political action, but President Reagan ensured this would not happen under his watch.[8] Senator Kerry (1943–) (Democrat from Massachusetts) summed it up well when he stated[9]: "In 1980, before President Reagan arrived in Washington, we were the world's leader in alternatives and renewables … in 1981 the funding was cut completely."

So was Ronald Reagan the first president to be a political climate denier? Maybe not, as he was simply apathetic about climate change (which for the time was less outrageous than it would be today) and considered the issue better handled by business than government. Reagan's laissez-faire politics and the rise of neoliberalism cleared a path for climate denialism to march unopposed into Washington, leading to the creation of the denial cabal (and the denial machine discussed in the next chapter).

President George H. W. Bush: 1989–93

Ronald Reagan opened the door, but it was George H. W. Bush who welcomed climate denialism into the Oval Office. The latter expressed views common to climate deniers,[10] which can be seen in the following speech[11]:

> We acknowledge a broad spectrum of views on these issues, but our respect for a diversity of perspective does not diminish our recognition of our obligation or soften our will to produce policies that work. Some may be tempted to exploit legitimate concerns for political positioning. Our responsibility is to maintain the quality of our approach, our commitment to sound science, and an open mind to policy options.

There was only a "broad spectrum of views" if you included contrarians and other climate deniers. Making matters worse, this speech was addressed to the IPCC which had just issued the consensus view on the science. While politicians should have "an open mind to policy options," as he pointed out, doing nothing on climate change was not an appropriate policy choice. Then a few months later, on July 9, 1990, President Bush met with Helmut Kohl, Chancellor of Germany and made the remark[12]: "Regarding CO_2, only 4% of the CO_2 in the atmosphere comes from human sources; the other 96% comes from natural sources. On global warming, there still are questions about the data." Like many in his own party, President Bush displayed climate denial views and vehemently opposed global environmental initiatives, but he and other members of his administration were still less extreme than John Sununu.[13]

White House Chief of Staff John Sununu

President George H. W. Bush had appointed as his White House Chief of Staff John H. Sununu (1939–)[14] who believed he was guarding President Bush from the "environmental cabal" that was threatening America's prosperity, and he had a powerful grip over the administration's climate change agenda until he resigned in early 1992.[15] Bush's climate policy and Sununu's role were described as follows[16]: "It was based instead on a volatile mixture of ideology and politics … These men believed that the climate change issue was being used by environmentalists to impose their 'anti-growth agenda' on the US economy."

John Sununu, the czar of climate change within the administration, gave the order to his staff[17]: "I don't want anyone in this administration without a scientific background using 'climate change' or 'global warming' ever again." Even senior governmental officials had to bow to his authority, including Jim Baker, Secretary of State (who officially had removed himself from the portfolio because of his former oil ties),[18] and William Reilly of the EPA, who also reported that Dr. Sununu told a Cabinet meeting[19]: "that the models on which climate change were premised were fundamentally flawed and the best atmospheric scientists had yet to become involved in climate research." The reference to "the best atmospheric scientists" likely referred to the contrarians.

Had John Sununu been operating rogue within the administration? No, for after William Reilly went to George H. W. Bush for direction on the climate file, the president referred him back to Dr. Sununu,[20] and William Reilly was forced into a "marginal" role.[21] Clearly, John Sununu was tasked with doing the dirty work on climate change for the president.

His influence on climate discussions was not restricted to internal White House matters either. On April 13, 1990, during a meeting between President Bush and Prime Minister Thatcher, who brought up the new IPCC report that warned about the threats of climate change, Dr. Sununu challenged the British prime minister[22]: "The body of the text for that report doesn't support the summary of conclusions. It appears that some who are anti-growth use the environmental issue for anti-growth objectives." Much later, reflecting on this period, he commented on the posturing of world leaders on climate change before the Rio conference[23]:

> Frankly, the leaders in the world at that time were at a stage where they were all looking how to seem like they were supporting the policy without having to make hard commitments that would cost their nations serious resources.

A DNC (Democratic National Committee) report critiqued the environmental performance of the Bush administration[24]:

> But the White House effect quickly devolved into the Whitewash effect. The Bush administration that pledged its allegiance to combatting the greenhouse effect and other environmental issues has consistently shown itself to be a tool of industrial interests and a foe of taking any real action,

and then quoted the *Boston Globe*:

> In May 1989, 'genius' John Sununu found flaws in the testimony of greenhouse expert James Hansen, head of NASA's Goddard Institute, who planned to tell Congress that immediate action was in order to prevent drought-inducing global warming from progressing. Sununu rewrote the Hansen testimony to read that the jury was still out on global warming.

John Sununu had promoted a report from the George C. Marshall Institute, "Scientific Perspectives of the Greenhouse Problem," written by the contrarians William Nierenberg, Robert Jastrow and Frederick Seitz. The report claimed that there was no problem caused by greenhouse gases, and that the world was heading towards a cooling phase. Climate scientists protested; Stephen Schneider bitterly complained[25]: "Sununu is holding the report up like a cross to a vampire, fending off greenhouse warming." And earlier, Jim Hansen characterized Sununu's overall negative influence on the Bush administration by saying[26]: "He was being pushed in a very bad direction by the elder Sununu, a real jackass from my perspective as a scientist."

The denier credentials of John Sununu were no secret; in the nomenclature of the day, he was called[27]: "the single biggest skeptic in the White House." And an article critical of him in the *New York Times* reported[28]: "Dr. Sununu seems to find most specialists in energy and environmental policy to be hysterics, bolstered by bad scientists and unreliable economists," which prompted the contrarian Richard Lindzen to protest to the editor.[29] As recently as 2015, Sununu's support for climate denial had not dampened.[30]

President George W. Bush: 2001–9

While a few senior officials in the federal government had been climate deniers, a full-fledged denial cabal did not get established until the administration of George W. Bush. Early in his term, Republicans had hired a consultant to prepare a strategy paper to defend against policies on climate change in which the following talking points were recommended[31]:

1 The scientific debate remains open …
2 Americans want a free and open discussion …
3 Technology and innovation are the key in arguments on both sides …
4 The "international fairness" issue is the emotional home run …
5 The economic argument should be secondary.

The introduction to this strategy paper began with[32]: "A compelling story, even if factually inaccurate, can be more emotionally compelling than a dry recitation of the truth."

In the same year as this propaganda guide was prepared, the EPA issued the Climate Action Report which essentially bowed to Bush's climate agenda[33]:

> Climate change is a long-term problem, decades in the making, that cannot be solved overnight. A real solution must be durable, science-based, and economically sustainable. In particular, we seek an environmentally sound approach that will not harm the U.S. economy, which remains a critically important engine of global prosperity. We believe that economic development is key to protecting the global environment. In the real world, no one will forego meeting basic family needs to protect the global commons. Environmental protection is neither achievable nor sustainable without opportunities for continued development and greater prosperity.

The report recommended a series of soft initiatives – such as carbon intensity reductions, voluntary measures, innovation and research, energy efficiency improvements – and then used the standard climate denier argument that the "future use of fossil fuel" could not be addressed without first "reducing uncertainty." The contents of the Climate Action Report from the EPA could have just as easily come from the API (American Petroleum Institute). Initiatives like these became popular strategies on climate change for conservative governments across the globe. Moreover, the EPA report paid extraordinary tribute to President George W. Bush for his climate change initiatives, including his "innovative approaches," although no material climate action had been adopted by him. The president was lauded throughout, being directly named 16 times and by the title "President" 54 times. Nevertheless, George W. Bush disliked the report and distanced himself from it because of the pushback by right-wing critics.[34] Apparently, for the G.O.P. elite, even such a spiritless climate report was still unacceptable.

The Chairman of the IPCC Robert Watson (1948–) had complained that[35]: "The only person who doesn't believe the science is President Bush." So radical was Bush's anti-science climate agenda that sociologists Aaron McCright and Riley Dunlap wrote that it was[36]: "a level of politicization of science reminiscent of the Soviet Union's Lysenko era."[37] Throughout his time in office, President Bush derided the Kyoto Protocol[38] and will always be remembered for withdrawing the United States from the treaty.[39] Publicly, he would deliver rhetoric supporting climate change legislation, but as reported in *Rolling Stone* in June 2007[40]: "Even when Bush proposes what looks like a plan, it's designed to stall real progress on global warming." The article, also, implicated his vice president Dick Cheney.

Vice President Dick Cheney

As George H. W. Bush had relinquished the climate change portfolio to John Sununu, George W. Bush did the same with Dick Cheney (1941–). The Administrator of the EPA, Christine Todd Whitman (1946–), realized that Mr. Cheney had gained control of the climate portfolio, and she objected to the "political science" that had been enthusiastically adopted by him and the administration[41]: "What disturbed me most was the administration's record of taking the most extreme of the science – what I call the 'political science' – and giving it the same weight as the real science." Her term "political science" likely referred to contrarian science as discussed in the last chapter.

In an interview in February 2007, the vice president publicly admitted to not accepting the science on climate change[42]: "We're going to see a big debate on it going forward … the extent to which it is part of a normal cycle versus the extent to which it's caused by man." And his hard-core views have hardly diminished over time; Mr. Cheney was still promoting climate denial more than a decade later (November 2018)[43]:

> we are getting to the point where everyone has to admit there is climate change. There is still a debate on how it is caused. That is not surprising. But it is going to be a major challenge of our time and for this generation and future generations.

There is no "debate on how it is caused," except in the minds of climate deniers.

Senator James Inhofe

Senator James Inhofe (1934–) (Republican from Oklahoma)[44] initially claimed that he[45]: "was a believer that manmade anthropogenic gases actually affected climate change" and only switched sides after seeing reports on how expensive it would be for the U.S. to meet its Kyoto targets, as the "hysteria was setting in." He joined the denial cabal after he became chair of the Environment and Public Works Committee in 2003.

His speech on the floor of the Senate – which I have dubbed the "Denial Manifesto" – was a 12,000-word, two-hour diatribe that targeted the IPCC, promoted the views of contrarians, and mocked the science of climate change; below are some of the ideas he presented[46]:

> Much of the debate over global warming is predicated on fear, rather than science. Global warming alarmists see a future plagued by catastrophic flooding, war, terrorism, economic dislocations …
>
> Thus far no one has seriously demonstrated any scientific proof that increased global temperatures would lead to the catastrophes predicted by alarmists. In fact, it appears that just the opposite is true: that increases in global temperatures may have a beneficial effect on how we live our lives …
>
> It is my fervent hope that Congress will reject prophets of doom who peddle propaganda masquerading as science in the name of saving the planet from catastrophic disaster …
>
> Let me be very clear: alarmists are attempting to enact an agenda of energy suppression that is inconsistent with American values of freedom, prosperity, and environmental progress …
>
> With all of the hysteria, all of the fear, all of the phony science, could it be that man-made global warming is the greatest hoax ever perpetrated on the American people?

Many of these ideas have since become embedded in the doctrine of climate denialism, and the following terms mentioned by him in this speech are routinely used by the denial cabal and other climate deniers:

- "alarmist" – to insult climate scientists
- "catastrophic" – to exaggerate the warnings of science
- "settled" or "proof" – to undermine the validity of science
- "skeptic" – to legitimize climate deniers and contrarians
- "hoax" – to demean the climate sciences (Inhofe's trademark insult[47]).

The day after this extraordinary speech, Senator Inhofe chaired a hearing of the Committee on Environment and Public Works, where he set out to define a new framework for his congressional hearings[48]:

> I announced three very outrageous things that we were going to do in this committee that have not been done before. No. 1, we are going to try to base our decisions, things that we do, on sound science. No. 2, we are going to be

> looking at the costs of some of these regulations, some of these policies that we have, and determine what they are going to be. And No. 3, we are going to try to reprogram the attitudes of the bureaucracy so that they are here not to rule, but to serve.

At this hearing, he criticized the "hockey-stick" graph (discussed later) of Michael E. Mann (1965–) who was present as a witness; here is an example of one exchange between them[49]:

> Senator INHOFE. I would like to call your attention to the recent op/ed in the *Washington Post* by Dr. James Schlesinger, who was Energy Secretary under President Carter. In it, he wrote, "There is an idea among the public that the science is settled. That remains far from the truth." He has also acknowledged the Medieval Warming Period and the Little Ice Age. Do you question the scientific integrity of Dr. Schlesinger?
>
> Dr. MANN. I do not think I have questioned scientific integrity. I have questioned scientific expertise in the case of Drs. Willie Soon and David Legates with regard to issues of paleoclimate. As far as Schlesinger is concerned, I am not familiar with any peer-reviewed work that he has submitted to the scientific literature, so I would not be able to evaluate his comments in a similar way.

Senator Inhofe argued that Dr. Mann had "impugned the integrity" of his colleagues.[50] There were several testy exchanges that day, or as Senator Hillary Clinton (1947–) (Democrat from New York) remarked "the questioning and the testimony has been somewhat lively, if not controversial and contested."[51] Even after the hearing, Senator Inhofe was not done badgering the climate scientist, sending him over 50 questions in writing.[52] Dr. Mann later summarized his view of the hearing[53]:

> I felt good about how the hearing went, Science seemed to have prevailed that day ... Among other things, I point out that the intellectual bankruptcy of the climate change denial campaign had been in full display in that hearing.

Dr. Mann and his famous graph were attacked by Senator Inhofe at every opportunity; the noted climate scientist is mentioned in the records of Congress 141 times of which 40 were from Senator Inhofe. In late September 2019, James Inhofe awarded Dr. Mann[54] one of his "Climate Hypocrite Awards;" other "winners" he announced on Twitter were Al Gore, Alexandria Ocasio-Cortez, a joint one to U.N. climate officials and Hollywood elites, and the Obamas.[55]

On October 11, 2004, he was back on the Senate floor for another two-hour marathon speech, where he again attacked the IPCC and Dr. Mann, and mentioned alarmist, catastrophic, settled, proof, and hoax, numerous times. The "Senator of Denial" was even quoting Vladimir Putin's economic advisor Andrei Illarionov (1961–) who had claimed that the Kyoto Protocol[56]: "would deal a powerful blow on the whole humanity similar to the one humanity experienced when Nazism and communism flourished." Senator Inhofe, then, introduced his own conspiracy theory[57]: "Those who are afraid of the newest and best science are usually the same people who are afraid that the more the public actually knows, the more it will interfere with their grand geopolitical plans to ration America's energy." The "newest and best science" was undoubtedly a reference to the contrarians.

In early 2005, he called catastrophic global warming nothing more than a fund-raising tool for environmental extremists who viewed it as "an article of religious faith."[58] At this hearing, he promoted a novel by popular fiction writer Michael Crichton (1942–2008), *State of Fear*, about evil eco-terrorists who were associated with climate activism. Among the "experts" at a later Senate hearing in the same year, Senator Inhofe had invited Dr. Crichton, taking great pains to emphasize his credentials (which were, however, unrelated to the climate sciences).[59]

Early in 2007, with Senator Barbara Boxer (1940–) (Democrat from California) as chair of the Committee on Environment and Public Works (and James Inhofe then as the ranking member[60]), there was an unusual hearing with Al Gore being invited to give his views on global warming. Senator Inhofe turned it into an interrogation, demanding simple yes or no answers from the former vice president which elicited warnings from Senator Boxer.[61] In 2013, Senator Boxer and Senator Inhofe were sparring again which prompted her to say[62]: "I do not know what it will take to convince you and the deniers of what is out the window."

In 2015, the Senator of Denial had a busy year on the climate front. On January 20th, Sheldon Whitehouse (1955–) (Democrat from Rhode Island) proposed a largely symbolic amendment "to express the sense of the Senate that climate change is real and not a hoax." From the floor of the Senate, Mr. Whitehouse stated[63]:

> if we can find a few Republican Senators who will say publicly that climate change is real. We can then go on to if it is real, let's have a conversation about what we do about it, because recklessly continuing to dump megatons of carbon into the atmosphere every year is not a solution.

As the "hoax" claim was one of Inhofe's trademarks, the vote was likely an indictment of him, and the sly senator was not about to fall into that political trap. The next day, he was on the Senate floor agreeing with Whitehouse's amendment that the climate was changing but the hoax is that it was caused by man,[64] and the amendment passed 98 to 1 with no harm done to the reputation of Senator Inhofe.

In a colourful stunt on the Senate floor a month later, the 80-year-old senator talked about the "hysteria on global warming" and, holding a snowball, declared: "It's a snowball from just outside here. So it is very, very cold out." He maintained that the unusually cold weather proved the world was not warming, even though he acknowledged that "2014 [had] been the warmest year on record."[65] Not long afterwards, Senator Inhofe attacked the Clean Power Plan of President Obama to control carbon emissions from power plants, quoting a noted contrarian[66]: "This proposal is about expanding the government's control into every aspect of American lives. As MIT climate scientist Richard Lindzen has said, 'Controlling carbon is a bureaucrat's dream. If you control carbon, you control life.'" Inhofe's busy year ended with him criticizing Pope Francis (1936–) for supporting climate action.[67]

Near the end of his last term as senator, on July 20, 2022, he tweeted against the climate initiatives of President Biden[68]: "First, Dems used cow farts to justify their climate agenda. Now, according to the Biden admin, the sky is falling and the only way to solve it is to throw more money at liberal climate policies." Then, on November 16, 2022, in his farewell speech on the Senate floor, he gave one last parting shot at climate change[69]:

> … I pushed back against the Obama administration's far-left policies that sought to upend the lives of Oklahomans like the Paris Climate Agreement,[70] the Waters of the U.S. Rule, the Clean Power Plan, and many others. These policies were really about giving Washington bureaucrats sweeping control over the lives of millions of Americans. We are debating a lot of these same issues today, and I expect these disagreements will continue long into the future.

Over the years, the senator has been met by a variety of monikers for his climate denial which he had himself listed in a Senate speech[71]: "The 'noisiest climate skeptic,' 'the Senate's resident denier bunny,' 'traitor,' 'dumb,' 'crazy man,' 'science abuser,' 'Holocaust denier,' 'villain of the month,' 'hate filled,' 'war mongering,' 'Neanderthal,' 'Genghis Khan.'" Senator Inhofe had also been called "one of the world's most vociferous climate skeptics"[72] and "the original climate-denier in chief."[73] Furthermore, he has been criticized for his "unabashed climate idiocy."[74]

The Denial Manifesto established the gold-standard spins and talking points of climate deniers, which I see several times per day on Twitter. Senator Inhofe's views became popular within the Republican Party and spread through its ranks. Sociologist Robert Brulle has written[75]: "Just as in a theatrical show, there are stars in the spotlight. In the drama of climate change, these are often prominent contrarian scientists or conservative politicians, such as Senator James Inhofe." Although many Republican politicians have followed Inhofe's lead on climate change, none have matched his flair and zeal – not even Donald Trump.

The House of Denial

There were climate deniers in the House of Representatives as well, including Joe Barton, Ralph Hall, Dana Rohrabacher, and Lamar Smith, who were among the members of the denial cabal. Below are some examples of their climate denial.

Joe Barton

Joe Barton (1949–) (Republican from Texas) chaired the Committee on Energy and Commerce from 2004 to 2007. He began harassing Michael E. Mann in 2005 – demanding details of his work and personal correspondence – and his baseless accusations have continued. A review of Mr. Barton's antics concluded[76]: "But what's going on in Congress is not evidence-based query, it's a politically-driven attack on science."

In 2007, there was a hearing with the title "Climate Change: Are Greenhouse Gas Emissions from Human Activities Contributing to the Warming of the Planet?" where Joe Barton presented typical climate denial rhetoric[77]:

> I said then that I accept that the science on this matter is uneven, uncertain, and evolving. That certainly hasn't changed, but now we seem to be pressuring ourselves, or someone is pressuring us, to legislate first and get the facts later. I hope we won't do that. I want to make sure we get the best information available so we have a full and accurate definition of the problem before we start making decisions that will be among the toughest of our careers ...
>
> Discussion of capping CO_2 often misses an essential fact. Carbon dioxide, unlike carbon monoxide and other compounds ending in "oxide," is not toxic. It is not a pollutant. It is not only natural, it is indispensable for life on this planet.

Among the witnesses at this hearing was the well-known contrarian John Christy who offered his own doubts about the climate sciences.[78]

Mr. Barton turned to religion to debunk climate change; in 2009, he said: "You can't regulate God,"[79] and in 2013, he argued[80]: "I would point out that if you are a believer in the Bible, one would have to say the great flood is an example of climate change. And that certainly wasn't because mankind had overdeveloped hydrocarbon energy." In editorials, Joe Barton had been labelled a "notorious global warming denier"[81] and a "long-time denier of global warming."[82]

Ralph Hall

At a hearing from 2007, a House committee had been discussing the findings of the latest IPCC report (AR4). Among the members of the committee was Ralph Hall (1923–2019) (Republican from Texas), who stated[83]:

> Rather than focusing on ways to raise energy prices for Americans, we ought to be discussing what the United States could accomplish with the right investments in energy research and energy development. We have to consider the enormous benefits and the cost of adaptation, and we must not lose sight of other pressing national priorities and understand the overall burden of all the national needs to the average citizen.

Mr. Hall was promoting the typical delaying tactic that the solution to the problem of climate change was more research. In addition, he endorsed adaptation; however, adaptation is not a solution, but it can help you survive until there is one. Moreover, even with adaptation, mitigation must still be carried out, or the climate crisis will keep worsening, and you will be doing adaptation forever with diminishing returns.

There was an interesting hearing on climate change from the perspective of coal-fired power utilities, where Mr. Hall appeared to play to the audience when he advocated for American energy[84]:

> And to me, other than the word of prayer, the word energy is probably the most important word in the dictionary for our children and grandchildren, who we'll send overseas to take some energy away from people, when if we could just use energy we have right here at home, could prevent a war for them.

Later, Mr. Hall criticized the National Academy of Sciences and praised the book *Fed Up*, by Governor Rick Perry (1950–) (Republican from Texas), which proposed that climate science was a conspiracy to get funding. Like Joe Barton, he was best known in the denial cabal for playing the god card, even uttering[85]: "I don't think we can control what God controls … I'm really more fearful of freezing. And I don't have any science to prove that."

Dana Rohrabacher

One of the early members of the denial cabal in Congress had been Dana Rohrabacher (1947–) (Republican from California). In 1995, during the Clinton administration, for example, he chaired a session – "Scientific Integrity and Public Trust" – where he proclaimed[86]:

> In this atmosphere I believe this subcommittee has a duty to continue to present balanced panels. And I hope we will produce a useful dialogue on controversial scientific issues of the day …
>
> Well, as long as I am Subcommittee chairman, we are going to try our best to have both sides of every issue presented, and side by side, and promote dialogue between the expert witnesses.

In response to these opening remarks, Tim Roemer (1956–) (Democrat from Indiana) identified a disturbing trend within the Republican Party[87]: "It is ironic that systematic dismantling of environmental research and environmental statutes has now been held up by the Republican leadership as a litmus test for ideological purity."

At a hearing in 2007, during the term of George W. Bush, Mr. Rohrabacher defended contrarians as "very respectable scientists" and dismissed the scientific consensus on climate change[88]:

> What I see happening more and more in this debate over global warming is that those people who are advocating this position end up not answering the charges of very respectable scientists, and again, one need only look at my website to find the names of hundreds of these prominent scientists from major universities who are not part of this so-called consensus …

> Now, I will have to tell you, that is about as arrogant and about as anti-scientific an attitude, and it is prevailing in this debate. I mean, I don't want to hear about consensus anymore, proving that someone is right.

His speech attacking the scientific consensus was typical among political climate deniers who recognized that climate denialism would collapse, if the public accepted the consensus.

At a later congressional hearing that same year, the funding for contrarians by the energy-industrial complex was raised in a climate brawl by the witness Kassie Siegel of the Center of Biological Diversity[89]:

> I would like to point out that the paper was funded by the Charles G. Kotch [sic] Charitable Fund, American Petroleum Institute, and ExxonMobil, and that that authors include discredited climate deniers Willie Soon, David Legates, Sally Baliunas, and others. I would also like to point out that it was published as a viewpoint in the journal *Ecological Complexity*, not as a peer-reviewed science article. This is essentially an op-ed masquerading as a peer-reviewed science journal.

When it was Rohrabacher's turn to ask questions, he falsely accused Dr. Jim Hansen of being financed by George Soros (1930–)[90] and absurdly asked Kassie Siegel to respond to this spurious allegation, when she had not even spoken about the famous NASA scientist. He then again challenged the scientific consensus by presenting quotes from "respected scientists" who were mainly contrarians.[91]

His denial comments continued into the presidency of Barack Obama; for example, on December 8, 2011[92]:

> In my lifetime, there has been no greater example of this threat, which Eisenhower warned us about [scientific-technological elite], than the insidious coalition of research science and political largesse—a coalition that has conducted an unrelenting crusade to convince the American people that their health and their safety and— yes—their very survival on this planet is at risk due to manmade global warming …
>
> The manmade global warming theory is being pushed by people who believe in global government. They have been looking for an excuse for an incredible freedom-busting centralization of power for a long time, and they've found it in the specter of manmade global warming.
>
> For the past 30 years, the alarmists have been spouting "Chicken Little" climate science.

In 2013, Mr. Rohrabacher was questioning Ernest Moniz (1944–), the Secretary of Energy. When the testimony veered into the greenhouse effect, he became aggressive[93]:

> But there are other people with credentials, like Richard Lindzen from MIT, who are very skeptical of some of the research that has been going

> on and have articulated that, yet we have just a few weeks ago, an offshoot of President Obama's reelection campaign listed climate deniers. The only other use of that term is a Holocaust denier. Do you use the term denier for those people who disagree with you on climate science, and do you think that term is appropriate in engaging in a civil discourse over a scientific issue?

His mention of "climate deniers" was unexpected, as Dr. Moniz had not mentioned the term in his testimony, and making matters worse, Mr. Rohrabacher associated the term with the Holocaust which is a nasty tactic of climate deniers who shamelessly exploit the term "Holocaust" in an attempt to smear those defending science (discussed in chapter one). Of course, there is no connection between a Holocaust denier and a climate denier (except in the minds of the latter), but the intent is to provoke public outrage against those using the term climate denier. Such toxic comments appear on a weekly basis on Twitter, especially against the friends of #ClimateBrawl.

Lamar Smith

Lamar Smith (1947–)[94] (Republican from Texas) gained notoriety as a climate denier in 2009 for his conspiracy theory about corruption among climate scientists[95]:

> We now know that prominent scientists were so determined to advance the idea of human-made global warming that they worked together to hide contradictory temperature data. But for 2 weeks, none of the networks gave the scandal any coverage on their evening news programs, and when they finally did cover it, their reporting was largely slanted in favor of global warming alarmists.

He became the chair of the House Committee on Science, Space and Technology after Ralph Hall, in 2013, during Obama's second term. "Subpoena-happy" Smith has been accused of practising a "campaign of intimidation."[96] A 2015 op-ed by Michael E. Mann in the *New York Times* exposed his egregious abuse of political power[97]: "Let's end the McCarthy-like assault on science led by the Lamar Smiths of the world. Our nation is better than that." The following year, Mr. Smith chaired a session to justify his excessive use of subpoenas, where Eddie Bernice Johnson (1935–) (Democrat from Texas) challenged his misuse of power[98]:

> This hearing appears to be the culmination of a politically motivated oversight agenda that has been applauded by oil, gas, and mining interests and broadly condemned by the public, the media and the independent scientific community across the country and around the world. The Committee Majority has abused the Committee's oversight powers to harass NOAA climate scientists, going so far as to threaten former NASA astronaut, and current NOAA Administrator Kathy Sullivan, with contempt, all in an attempt to undercut the notion of human-caused climate change.

A brief mention must be made of Mo Brooks (1954–) (Republican from Alabama) who declared at this session a popular climate-denial talking point[99]:

> I know of no person on Earth who denies that climate change occurs. Anyone who knows anything about Earth's history knows the Earth's climate has always changed to hotter, to colder, to wetter, to dryer, and the like. And the Earth's climate always will change.

One sees these misleading comments daily by climate deniers on Twitter but rarely in Congress.

Smith's favourite target was the EPA. In one of his sessions on "Ensuring Sound Science at EPA," Eddie Bernice Johnson lamented[100]: "Congress is one of the few places where you can find the rhetoric of climate change denial alive and well." At a later hearing, he was still on the EPA warpath[101]:

> Sound science should be at the core of the EPA's mission. Legitimate science should underlie all actions at the Agency, from research to regulations, and be an integral part of justifying their actions. Unfortunately, over the last eight years, the EPA has pursued a political agenda, not a scientific one [under the Obama administration] …
>
> With the transition to a new Administration [Donald Trump], there is now an opportunity to right the ship at the EPA and steer the Agency in the right direction. The EPA should be open and accountable to the American people and use legitimate science.

The session had the Trump-like title: "Making EPA Great Again;" a scathing review of this hearing was humorously named "The Congressman Who's Trying to Make the Environment Worse, Again."[102]

In 2018, Lamar Smith was heading up another hearing about climate change, and his climate denial was as strong as ever which the following comments demonstrate[103]:

> The climate is always changing, but what remains uncertain is the extent to which humans contribute to that change. What is certain is that human ingenuity will play a significant role in resolving future environmental issues …
>
> In the field of climate science, there is legitimate concern that scientists are biased in favor of reaching predetermined conclusions. This inevitably leads to alarmist findings that are wrongfully reported as facts. Anyone who then questions the certainty of these findings is wrongly labeled a denier.

Using the term "alarmist" is typical climate denial talk and, in no time at all, Lamar Smith was using it again.

Upon his announcement not to seek re-election, in an article called "Friend of Science," Lamar Smith defended his congressional record on climate change[104]:

> We started to rein in alarmist predictions that were not based on good science and discovered that sometimes they cherry-picked data, or no data existed at all. We'll follow the science wherever it takes us, but it has to be science, not science fiction.

This is how many members of the denial cabal portray themselves – as a "friend" of science. In fact, climate deniers, political or otherwise, are foes of science.

President Donald Trump: 2017–21

On December 6, 2009, American business leaders signed an open letter to President Obama and Congress in the *New York Times* which read in part[105]: "As business leaders we are optimistic that President Obama is attending Copenhagen with emissions targets. Additionally, we urge you, our government, to strengthen and pass United States legislation, and lead the world by example." Among the signatories was: "Donald J. Trump, Chairman … The Trump Organization." Within two years, however, Donald Trump had shown himself to be a climate denier and, like many of them, expressed such views frequently on Twitter.[106]

The "Tweeter-in-Chief"

Unlike previous Republican presidents, Donald Trump did not stand behind a Sununu or a Cheney curtain to hide his denial agenda; he openly used his Twitter account to show off his climate denial.[107] Why did Donald Trump tweet? His own explanation was[108]:

> I use Social Media not because I like to, but because it is the only way to fight a VERY dishonest and unfair 'press,' now often referred to as Fake News Media. Phony and non-existent 'sources' are being used more often than ever. Many stories & reports a pure fiction!

Aside from attempting to undermine mainstream media (or as he would say "Fake News Media"), his tweets were a favourite communication tool, used, from what I have seen, for self-promotion, as a testing ground for potential policy initiatives (once he became president), and/or as a public forum weapon to crucify and humiliate his opponents.

Social media are an echo chamber for propaganda, which Donald Trump was able to take full advantage of, as the reach of disinformation is influenced by the number of followers of those tweeting; President Trump had exceptional reach, with 88 million followers on Twitter. And his tweets were not just amplified by his millions of regular followers but also by human trolls and automated-machine bots. An expert on climate communications had found, for example, that bots had played a major role when President Trump announced that the U.S. was withdrawing from the Paris Accord.[109]

Long before he was elected president, he had launched the first of his climate denial tweets back on November 1, 2011[110]: "It snowed over 4 inches this past weekend in New York City. It is still October. So much for Global Warming;" a rather standard tweet by a climate denier, deliberately conflating weather and climate. Cold weather was a common theme of many of his tweets, as was the "hoax" of climate science.[111] While his first tweet on global warming garnered 220 "likes," as president, he routinely attracted more than 100,000 likes from the twitterverse.

One of his more imaginative tweets appeared the following year[112]: "The concept of global warming was created by and for the Chinese in order to make U.S. manufacturing non-competitive." The comment, at the time, from the reality TV personality did not receive much attention. But, on January 17, 2016, at a Democratic presidential debate, Bernie Sanders (1941–) (Independent from Vermont) reminded everyone of the absurd China conspiracy theory. The following day on "Fox & Friends," Mr. Trump replied that his tweet had been a joke[113]:

> I know much about climate change. I'd be—received environmental awards. And I often joke that this is done for the benefit of China. Obviously, I joke. But this is done for the benefit of China, because China does not do anything to help climate change.

Later, in a presidential debate, he denied making the China accusation.[114]

After a total of over 100 climate denial tweets before becoming president,[115] he stopped tweeting about climate change for the next three years while in office with one exception, on December 28, 2017, when a severe cold wave hit eastern North America.[116] President Trump resumed his regular tweeting on climate denial at the end of 2018.[117] The reason for this Twitter hiatus on climate denial during his presidency is a mystery. Tweeting once again routinely on climate change, on September 23, 2019, he mocked the climate activist Greta Thunberg (2003–)[118]: "She seems like a happy young girl looking forward to a bright and wonderful future. So nice to see." But Ms. Thunberg had the last laugh; on Trump's final day in office, she tweeted[119]: "He seems like a very happy old man looking forward to a bright and wonderful future. So nice to see!" Her tweet went viral, receiving over 1 million likes in under 24 hours. President Trump could not reply, as tweeting by him had come to an abrupt halt roughly two weeks prior, when his personal account had been suspended by Twitter on January 8, 2021.[120] Twelve days later, his term as president also had ended.

On November 15, 2022, Donald Trump announced he would run again for president in 2024. A few days later, Elon Musk (1971–), as the new owner of Twitter, carried out a Twitter poll to "Reinstate former President Trump"[121] on the platform. After over 15 million votes, the results were very close at 51.8% yes, and then Mr. Musk reinstated @realDonaldTrump and tweeted *Vox Populi, Vox Dei* ("the voice of the people is the voice of God").[122] The process was flawed as a Twitter poll is strongly biased by the followers of the host of the poll and certainly does not represent the voice of the people or even the voice of all twittizens. At the time of this writing, Donald Trump, himself, had not yet returned to Twitter.

The "Denier-in-Chief"

President Trump took climate denial to a radically dark and dangerous political level. His administration was packed full of fossil fuel supporters who did not accept the science of climate change (the following list includes examples of denial quotes, not necessarily made while in office, from his top officials)[123]:

- Vice President – Mike Pence[124]:

  > Global warming is a myth. The global warming treaty is a disaster. There, I said it. Just like the 'new ice age' scare of the 1970's, the environmental movement has found a new chant for their latest 'chicken little' attempt to raise taxes and grow centralized governmental power.

  In a CNN interview, Jake Tapper (1969–) pushed the climate change issue but the vice president was evasive in his answers.[125] And during the VP debate on October 7, 2020, Mike Pence used a common climate denier line[126]: "The climate is changing … But the issue is, what's the cause and what do we do about it? President Trump has made it clear that we're going to listen to science."
- CIA – Director – Mike Pompeo[127]: "There's some [scientists] who think we're warming, there's some who think we're cooling, there's some who think that the last 16 years have shown a pretty stable climate environment." He became Secretary of State on March 13, 2018.
- DOE (Department of Energy) – Secretary – Rick Perry[128]: "Calling CO_2 a pollutant is doing a disservice to the country, and I believe a disservice to the world."
- DOI (Department of the Interior) – Secretary – Ryan Zinke[129]: "It's not a hoax, but it's not proven science either." Mr. Zinke was replaced by David Bernhardt, who has been dismissive about the climate change issue[130] and had consulted with high profile contrarians.[131]
- DOJ (Department of Justice) – Attorney General – Jeff Sessions[132]: "Carbon pollution, CO2, is really not a pollutant. It is a plant food and it does not harm anybody except that it might include temperature increases."
- EPA – Administrator – Scott Pruitt[133]: "Scientists continue to disagree about the degree and extent of global warming and its connection to the actions of mankind." Mr. Pruitt was replaced by Andrew Wheeler[134]: "I believe that man has an impact on the climate, but what's not completely understood is what the impact is."
- HHS (Department of Health and Human Services) – Administrator – Tom Price[135]: "This decision [EPA plans to regulate carbon dioxide] goes against all common sense, especially considering the many recent revelations of errors and obfuscation in the allegedly 'settled science' of global warming."
- Homeland Secretary – Secretary – Kirstjen Nielsen[136]: "I believe that climate change exists. I am not prepared to determine causation."
- HUD (Housing and Urban Development) – Secretary – Ben Carson[137]: "There is no overwhelming science that the things that are going on are man-caused and not naturally caused."

- NASA – Administrator – Jim Bridenstein[138]: "Global temperatures stopped rising 10 years ago. Global temperature changes, when they exist, correlate with Sun output and ocean cycles."
- State Department – Secretary – Rex Tillerson[139]: "The world is going to have to continue using fossil fuels, whether they like it or not." Mike Pompeo (see above) replaced Rex Tillerson as Secretary of State on March 13, 2018.
- USDA (U.S. Department of Agriculture) – Secretary – Sonny Perdue[140]: "I don't know that [human activity is contributing to climate change], nor that I think that it has been proven to be that. There are scientists on both sides of that."

Deeper infiltration of climate deniers occurred within the EPA as well – the contrarian John Christy had even been appointed to the EPA Advisory Board. At the announcement of his appointment, Dr. Christy stated[141]: "There's a benefit, not a cost, to producing energy from carbon," and that it was nature, not people, that caused the rising temperatures. His new boss, Andrew Wheeler, affirmed the "unbiased" process in the selection.[142] The appointment of climate deniers into the government bureaucracy continued until the final days of Trump's administration.[143] This level of climate denialism within the senior government hierarchy was unprecedented for any democracy in the world.

President Trump was queried about his views on climate change in an interview with Piers Morgan (1965–) during a state visit to the UK. Mr. Morgan asked the president if he accepted the science on climate change, to which he responded[144]:

> I believe there's a change in weather, and I think it changes both ways. Don't forget, it used to be called global warming, that wasn't working, then it was called climate change. Now it's actually called extreme weather, because with extreme weather you can't miss.

At a later meeting to discuss the California wildfires, when state officials pleaded with him not to ignore the science, President Trump similarly replied[145]: "It will start getting cooler; you just watch … I don't think science knows." Even when he was no longer president, in an interview on March 21, 2022, he was asked if people were responsible for climate change, and he answered[146]:

> into the 1920s, they were talking about global freezing. OK? In other words, the globe was going to freeze. And then they go global warming, and then they couldn't use that because the temperatures were actually quite cool, and it's many different things … the climate's always been changing.

The denial machine had proudly acknowledged its influence over Donald Trump; the right-wing think tank Heartland Institute, for example, revealed that they had helped the president "understand the truth about climate change."[147] In other words, a climate denial organization was teaching a climate denier about climate denial; it doesn't get more ridiculous than that.

Governor Ron DeSantis

As this manuscript was being completed in mid-2023, Donald Trump was leading in the polls to become the G.O.P. nominee for president in the 2024 election. His major rival at this time was Ron DeSantis (1974–), governor of Florida.

Governor DeSantis has not said a lot about climate change, but opened up to a reporter's question in late 2021[148]:

> I've found is, people when they start talking about things like global warming, they typically use that as a pretext to do a bunch of left-wing things that they would want to do anyways. We're not doing any left-wing stuff ... Be very careful of people trying to smuggle in their ideology.

Then, after the official launch of his presidential run on May 24, 2023, during an interview on *Fox News*, Mr. DeSantis mentioned that he "always rejected the politicization of the weather" when questioned about the impact of climate change to his state.

Pete Maysmith of the League of Conservation Voters has warned that[149]:

> The cost of taking his anti-climate record to the national stage as president would be catastrophic ... DeSantis has already made clear he would unleash his war on climate science, clean energy jobs, and strong pollution safeguards against clean air and clean water.

Of course, Ron DeSantis, as governor does not meet the criteria to be a member of the denial cabal as defined in this study, but if he becomes president, he certainly will.

Senator Joe Manchin

A somewhat special member of the denial cabal is Joe Manchin (1947–) (Democrat from West Virginia) who has been a senator since 2010 (when he succeeded Robert Byrd of Byrd-Hagel Resolution fame) and was chair of the Senate Energy and Natural Resources Committee. What makes Senator Manchin so special is that he is a Democrat. One of his first tweets on "climate change" in mid-2014 summarized his views[150]: "We must strike balance that acknowledges climate change exists & that fossil fuels are a vital part of our energy mix." This became his standard climate denial speak; he supported climate change as real, but not real enough to reduce the use of fossil fuels.

In a hearing, on March 5, 2019, Senator Manchin stated[151]:

> First, man-made climate change is real and it is a serious threat to our citizens, to our economy, to our environment and to our national security ...
>
> Second, all communities, including those in energy producing states like West Virginia and Alaska, are experiencing the harmful impacts of the climate crisis. These impacts are often felt disproportionately in West Virginia

communities which are already suffering from the downturn in the coal production, resulting unemployment and negative effects of coal company bankruptcies on retirement and health care benefits. Therefore, the path to a climate solution must offer states like West Virginia opportunities, not additional economic burdens …

And third, the solutions must be grounded in reality which requires a recognition that fossil fuels are not going anywhere anytime soon.

More recently, he tweeted about his plan for handling climate change[152]: "We can meet America's energy needs while addressing climate change responsibly. That means innovation – not elimination;" a catchy phrase for don't touch fossil fuels, especially coal. Senator Manchin is all for action on climate change, so long as it does not negatively impact his state's coal production.

Michael E. Mann offered this assessment of Senator Manchin[153]: "a modern-day villain, who drives a Maserati, lives on a yacht, courtesy of the coal industry, and is willing to see the world burn as long as it benefits his near-term investment portfolio." And John Podesta of the Centre for American Progress voiced a similar opinion[154]: "It seems odd that Senator Manchin would choose as his legacy to be the one man who single-handedly doomed humanity. But we can't throw in the towel on the planet."

His notoriety hinged on his single vote deciding the fate of climate legislation in the Senate during the early term of Joe Biden; in the end, he had received enough incentives to go along with it. The real culprits, though, are not this one renegade Democratic senator but the 50 Republican senators, none of whom supported the Inflation Reduction Act. All the angst and attention directed at the wayward Democrat were really just evidence of a bigger frustration with the G.O.P. as the party of climate denial.

Climate Brawl with the Denial Cabal

The discussion, so far, has exposed the members of the denial cabal and other political climate deniers; now, we will take a look at their most important climate brawls. Again, as was the case with the contrarians, climate brawls were rare; Democrats and scientific witnesses seldom called out the political climate deniers, but a few notable exchanges are discussed in detail below to highlight the climate denial counter movement, known as climate brawl.

Climate Brawls of the Presidents

George W. Bush

Back in 1989, Al Gore had mentioned government censorship of climate scientists during the term of George H. W. Bush[155]:

> I believe it is unfortunate that this administration actively seeks to censor the scientific evidence concerning global warming and ignore the facts. It is

inexcusable and represents a dangerous pattern of environmental fraud that is becoming all too routine.

The promises of environmental activism are being replaced by environmental indifference and, worse, by concerted efforts to discredit the very best researchers, to ignore the very issues emerging with new urgency around the world, and to push the United States into the shadows when we should be in the lead.

The censorship of climate scientists, though, reached dizzying heights during the administration of George W. Bush. The suppression of science reached a breaking point in 2007, and there was a flurry of congressional climate brawls on the censoring of climate scientists by the White House; the first was a hearing of the House of Representatives, titled "Allegations of Political Interference with the Work of Government Climate Change Scientists." The session was chaired by Henry Waxman (1939–) (Democrat from California) who, in his opening remarks, summarized the charges against the administration[156]:

> For several years, there have been allegations that the research of respected climate scientists was being distorted and suppressed by the Bush administration. Some of these reports claim that Phil Cooney, a former lobbyist for the American Petroleum Industry, was put in charge of the Council on Environmental Quality and imposed his own views on the reports scientists had submitted to the White House …
>
> We are simply seeking answers to whether the White House's political staff is inappropriately censoring impartial Government scientists …
>
> According to the documents we have reviewed, administration officials sought to edit an EPA report: First, to add "balance" by emphasizing the "beneficial effects" of climate change. Second, they tried to delete a discussion of the human health and environmental effects of climate change. Third, to strike any discussion of atmospheric concentrations of carbon because carbon levels are not a "good indicator of climate change," and four, to remove the statement that "changes observed over the last several decades are likely mostly the result of human activities."

Predictably, the ranking member, Tom Davis (1949–) (Republican from Virginia), saw things differently and went on to defend contrarians[157]: "I am no denier, but I am troubled by stories of scientists unable to publish or even complete their research because they are perceived as having the wrong answers or being on the wrong side of the science." His comment side-stepped the major topic of this hearing, as the treatment of the contrarians was not under discussion. In the witness testimonies, most scientists gave details of political interference and censorship by the Bush administration, with the single exception of Roger Pielke Jr. – a contrarian – who blamed the scientists instead, accusing them of political advocacy[158]; in other words, the scientists had only themselves to blame for the censorship by the Bush administration, according to the contrarian.

A week later, a similar hearing took place in the Senate, this time John Kerry laid out the evidence for censorship[159]:

> What they do is they take the science, and they tailor it to reflect their political goals. The interference is stunning—from deleting key words, deleting words, this is George Orwell at its best—deleting "warming climate," deleting "global climate change," deleting "climate change" from press releases, changing agency mission statements, de-emphasizing climate research, denying media access to prominent climate scientists. It's absolutely stunning, what's been going on. And it has to stop.

Other dirty tricks played by groups associated with climate denial were heard at this Senate hearing. The right-wing think tank, the American Enterprise Institute, had offered $10,000 to climate scientists for a negative commentary on the Fourth Assessment Report of the IPCC[160]; Rick Piltz of Climate Science Watch described the serious nature of these disgraceful acts of climate denial[161]:

> And, I do think we have seen the beginnings of an orchestrated political effort to undermine the perception of the IPCC. Because the IPCC's conclusions about global climate change, and its implications, raise questions that could cause pressure for a stronger policy. And people who don't find that politically congenial do have an interest in somehow making it look like the IPCC is somehow controversial.
>
> So, I was concerned, when I saw this. As to whether that was part of this sort of, denialist or contrarian or skeptic effort.

Another House session was held two months later, and Henry Waxman again opened the hearing[162]:

> The document production is not yet complete, but some of the information the committee has already obtained is disturbing. It suggests that there may have been a concerted effort, directed by the White House, to mislead the public about the dangers of global climate change …
>
> Mr. Cooney and his staff made hundreds of separate edits to the government's strategic plan for climate change research. These changes injected doubt in place of certainty, minimized the dangers of climate change, and diminished the human role in causing the planet to warm …

At this hearing, Jim Hansen shared his own thoughts and experience with government censorship[163]:

> The growth in political interference coincides with a growth in power of the executive branch. It seems to me that this growth of power violates principles upon which our democracy is based, especially separation of powers and checks and balances …

> They should not be forcing scientists to parrot propaganda. Indeed during the current administration, NASA scientific press releases have been sent to the White House for editing, as I discuss in my written testimony. If public affairs officers are left under the control of political appointees, it seems to me that inherently they become officers of propaganda.

A week later, the chair of another climate brawl hearing, Brad Miller (1953–) (Democrat from North Carolina), had offered more disturbing allegations against the administration of George W. Bush[164]:

> To the average citizen it looked like a real debate between scientific peers. In fact, the skeptics were in the indirect employ of the oil and gas industry and that obviously conflict of interest was rarely disclosed. Few paid skeptics did any original research, many were not even in the fields in which they claimed expertise, and most simply specialized in attacking as "junk science" the careful, legitimate research that was published in journals and tested by rigorous peer review.
>
> According to the testimony we will hear today, since 2001, the Bush Administration has been part of the effort to manipulate the public debate about climate change. The Bush Administration, at the urging also of the oil and gas industry, muzzled Government scientists whose research supported the consensus view of climate scientists, adding to the public impression that there was substantial doubt among scientists. Press officers whose experience was in politics, not science, editor-suppressed press releases about government research, acted as monitors for government scientists during press interviews, and required that politically reliable scientists speak to the press for each agency.

As with the other 2007 hearings, Brad Miller had accused the Bush administration of censoring the scientific community and using contrarians, but he went further in placing the blame on the energy-industrial complex.

The overwhelming evidence directly from climate scientists of censorship by the Bush administration had sparked a series of climate brawls against the president and his administration which are notable for the sheer number of such congressional hearings in less than a year.

Barack Obama

One politician, especially, has become an honoured member of the climate brawl, Barack Obama (1961–). Senator Obama (Democrat from Illinois) had challenged the climate "skeptics" who held public office[165]:

> Even those who are skeptical about climate change, and still dispute the pace with which climate change is taking place, or are still disputing the causes of climate change, have to acknowledge that something out there is happening

that is disturbing; that it is potentially an enormous problem; and that if we can take intelligent steps now to assure that this problem is dealt with, why wouldn't we do so?

Once in the Oval Office, President Obama ceased calling members of the denial cabal "skeptics" and used the term "deniers" instead. He was the first president to call out "climate deniers" directly; in 2011, for example[166]:

> There are climate change deniers in Congress and when the economy gets tough, sometimes environmental issues drop from people's radar screens. But I don't think there's any doubt that unless we are able to move forward in a serious way on clean energy that we're putting our children and our grandchildren at risk.

President Obama continued to express his frustration with their presence in government[167]: "So unfortunately, inside of Washington we've still got some climate deniers who shout loud, but they're wasting everybody's time on a settled debate. Climate change is a fact." Later, the president added[168]: "The climate change deniers suggest that there's still a debate over the science. There is not." At a conference the following year, he stressed[169]: "the political will – finally to get moving. So the time to heed the critics and the cynics and the deniers is past." Climate denial was mentioned at a White House screening of the documentary film about climate change called "Before the Flood."[170] He also aggressively challenged the propaganda of climate deniers found within the ranks of the G.O.P. on Twitter, tweeting about climate deniers (or similar terms) over 50 times during his second term in office.[171]

Campaigning for Hillary Clinton, President Obama was again criticizing climate denial in Washington[172]:

> You can vote with the climate deniers who want to tear up the agreements we've crafted, and doom our kids to a more dangerous world, or you can vote to keep putting people back to work building a cleaner energy future for all of us.

At another campaign stop, he attacked Donald Trump[173]:

> If you care about the environment and climate change, you can vote for somebody who says it's a Chinese plot, and puts a climate denier in charge of hiring folks at the EPA, or you can vote for somebody who believes in science and will keep America a world leader in fighting to protect our planet.

Despite Obama's support for Hillary Clinton, on November 8, 2016, Donald Trump won the election.

Donald Trump

Not surprisingly, given Trump's overt climate denial and propensity for controversy, climate brawls picked up during his presidency. A textbook example of a climate brawl took place on October 15, 2018, in an interview on *60 Minutes* by Lesley Stahl (1941–), when President Trump was asked directly about his climate denial[174]: "Do you think climate change is a hoax?" to which he answered: "I don't think it's a hoax, I think there's probably a difference. But I don't know that it's manmade." As she was explaining that the administration's own scientists at NOAA and NASA had stated that climate change was human-made, he interrupted her: "We have scientists that disagree with that … I'm not denying climate change, but it could very well go back." Ms. Stahl's interview offers a good lesson on how climate brawls should be done, and she refused to let the president off the hook, telling him he was "denying it."

Many protested the anti-science views of the denier-in-chief of the United States – the former Prime Minister of Australia Malcolm Turnbull (1954–)[175]: "Trump is the leading climate denier in the world," Al Gore[176]: "[Donald Trump is] the face of climate denial," Alexandria Ocasio-Cortez (1989–) (Democrat from New York)[177]: "Trump doesn't even believe climate change is real," and the former head of the EPA, Gina McCarthy (1954–)[178]: "The Trump administration has done everything they can to deny the science and denigrate scientists." At a House hearing, Lloyd Doggett (1946–) (Democrat from Texas) mocked President Trump and his administration[179]:

> The Trump Administration always prefers political fantasy to science and scientific fact. They have questioned and harassed so many scientists across this country, one agency after another, that you have to begin to wonder if they believe in gravity.
>
> Of course, willful ignorance of climate change is not a laughing matter …
>
> Of course, the Trump Administration would say that our years and years of record-breaking heat isn't a dangerous sign of climate change. I guess they would call it alternative climate.

Those outside of political circles were also highly critical of Trump's climate denial. Carol Muffett, CEO of the Center for International Environmental Law, commented[180]: "'Having a climate denying president on the order of Donald Trump simply demonstrates' the success of the industry's campaigns 'to not simply foment doubt, but to foment paranoia, about climate science.'" The scholar Noam Chomsky said[181]: "On issues like the climate, they [the Republicans] are the worst in the world … the Trump administration wants to race towards maximizing the use of fossil fuels." And, according to the noted economist Paul Krugman (1953–)[182]:

> While Donald Trump is a prime example of the depravity of climate denial, this is an issue on which his whole party went over to the dark side years ago. Republicans don't just have bad ideas; at this point, they are, necessarily, bad people.

In 2020, a group of major environmental and public health organizations made the following pronouncement[183]:

> Donald Trump's administration has unleashed an unprecedented assault on our environment and the health of our communities. His policies threaten our climate, air, water, public lands, wildlife, and oceans; no amount of his greenwashing can change the simple fact: Donald Trump has been the worst president for our environment in history.

Further commentary came from an expert on the denial machine, Naomi Oreskes (1958–), who pointed out the dangerous problems created by this president[184]:

> It is deeply problematic if the leadership of the U.S. government is rejecting science, because it sends a signal to the American people and to business leaders that it is fine to reject science, and even to ride roughshod over scientists.

President Trump faced more climate brawls than any other president, or world leader for that matter, as he was repeatedly discredited for his climate denial. While the two Bush presidents had openly displayed climate denial but attempted to appear concerned about the climate crisis at least, Donald Trump never bothered to pretend.

Climate Brawls of Congress

An executive order by the White House is a weak substitute for legislation by Congress; what one president giveth, the next can taketh away. The ultimate responsibility for controlling greenhouse gas emissions rests with Congress, but climate denialism has corrupted the legislative branch of the United States government, at least on the Republican side. Democrats and climate scientists have undertaken a few memorable climate brawls aimed at the denial cabal and the climate denial of the Republican Party overall.

The Gas and Oil Party

Senator Markey (1946–) (Democrat from Massachusetts) has been especially critical of the denial cabal for over a decade. The "Final Staff Report" of the Select Committee of Energy Independence and Global Warming was issued by him on the Bush administration's failures in the handling of the climate crisis,[185] and then, at a hearing the same year, Mr. Markey argued[186]: "The Bush administration's approach to climate change policy has been to deny the science, delay the regulation, and dismiss the critics."

Later, Senator Markey was in another climate brawl at a hearing, called "Data or Dogma? Promoting Open Inquiry in the Debate Over the Magnitude of Human Impact on Earth's Climate, where he was critical of the large number of contrarian witnesses"[187]:

> So this panel that we have in this committee, this last redoubt of denial on the planet, of all the countries on the planet, this last place, you know, has the flip of witnesses that have every other place. We have four here who deny it and one who believes in the science.
>
> And so we clearly here are at a historic moment, and there will be a day when you get your apology, Dr. Titley, for being kind of the sacrificial lamb here, like Galileo, standing up for actual science.

At another hearing, Senator Markey ridiculed President Trump, calling him the "Denier-in-Chief"[188]:

> So, like many other things, Trump is just going to side with the Koch Brothers, side with the oil industry in general. It is all part of a pathological pattern where he is a climate denier, gives his State of the Union Address for an hour and 20 minutes, doesn't mention climate change, names a fossil fuel lobbyist to be the head of the EPA …
>
> We are telling other countries please reduce your greenhouse gases, and Trump is the Denier-in-Chief, and it is just sending the wrong message to the rest of the world …

Climate denialism is endemic within the Grand Old Party, or the "Gas and Oil Party"[189] which became Markey's trademark phrase and was reported in the Congressional Record a dozen times.

The Flat Earth Society

Another derogatory label was preferred by Henry Waxman who accused the House of Representatives for being members of the "Flat Earth Society"[190]:

> … to end the conspiracy of silence in this body surrounding the issue of our time: the growing threat posed by climate change.
>
> We have a moral obligation to be responsible stewards of the environment for our children and future generations. History will not judge the House of Representatives kindly if we continue to ignore the mounting danger and act like the last refuge of the Flat Earth Society.

A few months later, Henry Waxman repeated the comparison[191]:

> The President was right. We don't have time for another meeting of the Flat Earth Society. Saying no to every solution is not a plan. Doing nothing is not a plan. If all the Republicans on this committee do today is criticize, they are either denying the science or ignoring it. No one can accept what the scientists are telling us and fail to support a plan of action … [later in the session]
>
> House Republicans have voted against climate change legislation, they voted against climate regulation, they have voted against climate research and development, and they voted against international climate efforts.

His statement that the "the President was right" was in reference to President Obama who had called the climate deniers in Congress "charter members of the Flat Earth Society."[192] Also, Jay Inslee had earlier used the label,[193] and later, Chuck Schumer (1950–) (Democrat from New York) accused President Trump of being a member.[194] However, Mr. Waxman was best known for the use of the term "Flat Earth Society," having used it 29 times in the Congressional Record.

Graphic Denial

The inquisitions by the denial cabal, under the guise of congressional hearings, had most harshly targeted the climate scientist Michael E. Mann and his hockey-stick graph. The graph had appeared in the "TAR Climate Change 2001" from the IPCC[195]: "New analyses of proxy data for the Northern Hemisphere indicate that the increase in temperature in the 20th century is likely to have been the largest of any century during the past 1,000 years." This graph was a milestone for the science of climate change, for which Dr. Mann was both lauded by the scientific community and demonized by the climate denial community. Although Congress had no right to judge any peer-reviewed science, Republicans viciously attacked the hockey-stick graph, and Dr. Mann became the recipient of vehement harassment and persecution by the denial cabal that was so extreme it shook the scientific community.[196]

An extraordinary example of graphic denial within Congress was an entire hearing on this graph alone in July 2006, "Questions Surrounding the 'Hockey Stick' Temperature Studies: Implications for Climate Change Assessment." Joe Barton, in his written statement, said[197]:

> Congress is in the business of making policy decisions that affect the lives of real people. Science provides us with the answers to many policy questions, and we need to trust it. I do trust science, and I trust it most when it is transparent, open to question, and eager to explain. When research is secretive, automatically and aggressively defensive, and self-reinforcing, it becomes easy to distrust.
>
> As Chairman of the Energy and Commerce Committee, which holds a key role in any policy making relating to climate change, I believe it is incumbent on this Committee must ensure that the very best information is available to make its decisions.

Senator Jay Inslee (1951–) (Democrat from Washington) saw through his gamesmanship[198]:

> America is fully capable of dealing with global warming but not if Congress engages in snipe hunts, arguments about how many statisticians can dance on the head of a pin rather than figuring out what our energy policy should be to get a handle on global warming.

> Now, why are we in this exercise for doubt? I refer you to the first slide I have, which is a memo from the tobacco industry when they were fighting the clear, unalloyed science that tobacco was bad for you. Here is a memo from one of their people: "Doubt is our product." And those who decide that America should stay quiescent, do nothing about global warming, doubt is their product.

Similarly, Henry Waxman revealed Barton's "witch hunt"[199]:

> Chairman Barton began this dubious investigation in June of 2005 when he sent a letter demanding the funding for every study that had ever been conducted by Dr. Michael Mann, demanding he turn over all of the data for all their research and made over burdensome and intrusive requests. The Washington Post accused our Chairman of conducting a witch hunt. The Chairman of the Science Committee, Sherwood Boehlert, called the investigation "misguided" and "illegitimate" …
>
> So I have to submit that I don't find this hearing to be one about truth. It is about sowing doubt and spreading disinformation …

The denial cabal waged a relentless war of disinformation against the science and deliberately targeted the work of Dr. Mann and his famous graph.

The Grand Climate Brawl

An even more egregious hearing against Dr. Mann took place over a decade later, "Climate Science: Assumptions, Policy Implications, and the Scientific Method," chaired by Lamar Smith.[200] A few days before, Mr. Smith had given a presentation at a conference hosted by the right-wing think tank Heartland Institute, where he talked about stacking the deck for his upcoming hearing by inviting three contrarians – John Christy, Roger Pielke Jr., and Judith Curry – to join Michael E. Mann, and smirked[201]: "That's why this hearing is going to be so much fun." His swagger drew enthusiastic applause from the pro-climate denial audience.

The House hearing began with Mr. Smith giving a lengthy lecture on the scientific method[202]:

> Far too often, alarmist theories on climate science originate with scientists who operate outside of the principles of the scientific method …
>
> Alarmist's predictions amount to nothing more than wild guesses. The ability to predict far into the future is impossible. Anyone stating what the climate will be in five hundred years or even at the end of the century is not credible.
>
> All too often, scientists ignore the basic tenants of science to justify their claims. Their ultimate goal appears to be to promote a personal agenda, even if the evidence doesn't support it …

Only when scientists follow the scientific method can policymakers be confident that they are making the right decisions. Until then the debate should continue.

Eddie Bernice Johnson (1935–) (Democrat from Texas) gave a powerful rebuttal and raised the issue of contrarians as witnesses (who were disproportionally represented at the hearing)[203]:

> Republicans in Congress overwhelmingly reject or minimize the scientific consensus on climate change. In this, they follow the leader of the Republican Party, President Trump, who once claimed that climate change was a hoax perpetrated by China. Even on this Committee on Science, Republican Members have postulated sometimes unique theories about climate change, some of which have become punchlines on late night television.
>
> It saddens me, really, that Majority Members of Congress and of this Committee in particular, consistently ignore the thousands of scientists around the world who maintain mainstream climate science views, instead repeatedly calling a handful of preferred witnesses who are here today over and over again to testify. For instance, the three witnesses [the contrarians John Christy, Roger Pielke Jr., and Judith Curry] called by the Majority today have collectively appeared in front of Congress at least 20 times over the past decade.

And the climate brawl picked up during the question period when Dr. Mann went after the political climate deniers for their intimidating tactics[204]:

> I think the attacks against scientists by individuals, groups, many of them allied with fossil fuel interests and fossil fuel front groups, are aimed at several goals. One of them is to silence climate scientists. If you get attacked every time you publish an article that demonstrates the reality and threat of human-caused climate change, if that causes you to become subject to Congressional inquiries and Freedom of Information Act requests, obviously that's very stifling, and I think the intention is to cause scientists to retreat.

But things really got heated when the chair, Lamar Smith, came under fire by Dr. Mann who questioned Smith's attendance at the Heartland Institute conference[205]:

> Well, let me first say the scientific method—we've heard this term quite a bit. The Chairman keeps using this term. I do not think it means what he thinks it means. According to an article that came out a few days ago in the Journal of Science, Chairman Smith was on record at the Heartland Institute. This is a climate change-denying Koch Brothers-funded outlet that has a climate change denier conference every year, and Chairman Smith spoke at that conference …

Mr. Smith immediately cried foul when Dr. Mann called the Heartland conference "a climate change denier conference." Dana Rohrabacher also criticized the use of this term[206]:

> People should be ashamed of yourselves for people who continue to attack other people because they disagree. You call people deniers all you want. You can use every kind of name you want …
>
> Whether or not man's use of CO2 is now creating a warming cycle is what this is all about, and for scientists on either side to try to call names and try to beat somebody into submission, that's a Stalinist tactic. Those using the words "denier" are using a Stalinist tactic.

On the other hand, Paul Tonko (1949–) (Democrat from New York) supported Dr. Mann and attacked the energy-industrial complex and their denial tactics[207]:

> The scientific community thrives on skepticism and uncertainty but denial is something different. Unlike healthy scientific skepticism, climate change denial stands today as one of the great pillars in the pantheon of political manipulation. Decades ago, major players in the fossil fuel industry saw the issue of climate change gaining popular attention. They also realized that any serious effort to reduce carbon pollution and greenhouse gases could be a death blow for their industry. So instead of embracing the clear evidence in front of them that fossil fuels contribute to climate change, they launched one of the most successful misinformation campaigns in our American history, right up there with the tobacco industry lying about cancer risks.

This is how a climate brawl should be done in Washington with both scientists and politicians actively engaged in challenging climate denial. The impact of the climate brawl was visible in how some of the Republicans appeared to be agitated when they lost control of the discussion. The term "denier" appeared 23 times in the minutes of the hearing, which must be a record, making this the "grand climate brawl." As Lamar Smith had promised at the Heartland Institute conference, this contrarian-stacked session would prove "to be so much fun" (just not for him and the other members of his party as it turned out).

In summary, climate denial had infiltrated the Republican Party, and some of its elite members were hard-core climate deniers. A report from the DNC from June 2022 had been called[208] "Fact Check, The GOP is Still the Party of Climate Deniers" which was then born out by not one single Republican voting for the Inflation Reduction Act two months later. Frank Pallone (1951–) (Democrat from New Jersey) had called the leaders of the Republican party "all climate deniers,"[209] and earlier, Noam Chomsky (1928–) had made a similar observation[210]:

It is an astonishing fact about the current era that in the most powerful country in world history, with a high level of education and privilege, one of the two political parties virtually denies the well-established facts about anthropogenic climate change.

A survey in 2021 from the Center for American Progress found[211]:

> … there are still 139 elected officials in the 117th Congress, including 109 representatives and 30 senators, who refuse to acknowledge the scientific evidence of human-caused climate change. All 139 of these climate-denying elected officials have made recent statements casting doubt on the clear, established scientific consensus that the world is warming—and that human activity is to blame. These same 139 climate-denying members have received more than $61 million in lifetime contributions from the coal, oil, and gas industries.

Moreover, the 139 members included the majority of the Republicans in Congress.

Behind the hard-core climate deniers of the political elite – Schlesinger, Bush (GHW), Sununu, Bush (GW), Cheney, Inhofe, Barton, Rohrabacher, Hall, Smith, Trump, and Manchin who ultimately had formed a "denial cabal" – stood an incredibly powerful vested interest group, the energy-industrial complex, which is the topic of the next chapter.

Notes

1 Jimmy Carter Library 1977b.
2 Jimmy Carter Library 1977a.
3 Woodwell et al. 1979, pp. 1–2.
4 Schlesinger 2003; many of his references referred back a quarter of a century to the period of the global cooling discussions.
5 Schlesinger 2005.
6 Mooney 2005.
7 A summary of the president's actions acknowledged this transition (Howe 2014, p. 119; ch. 5).
8 For an insider look at the Reagan administration see Rich 2018, *Part Two, 1983–1989*.
9 United States Senate 2000, p. 78.
10 For the policies and tactics of Reagan and Bush to stall action on climate change, see Kutney 2014, pp. 136–154.
11 Bush 1990.
12 George H. W. Bush Library 1990b.
13 Hoffman 1990.
14 Nitze in Mintzer and Leonard 1994, pp. 192–194; Sebenius in Mintzer and Leonard 1994, p. 288; Gelb 1991; Biello 2019.
15 Nitze in Mintzer and Leonard 1994, p. 193.
16 Nitze in Mintzer and Leonard 1994, p. 189.
17 Rich 2018; the article gives many examples of Sununu's antics.
18 Gelb 1991.
19 Hecht and Tirpak 1995, p. 402.

20 Rich 2018.
21 Nitze in Mintzer and Leonard 1994, p. 193.
22 George H. W. Bush Library 1990a, p. 18.
23 Rich 2018.
24 DNC 1991.
25 Roberts 1989, p. 993.
26 Waldman and Hulac 2018.
27 Kerr 1992, p. 1138; for a review, see DeSmog, "John H. Sununu."
28 Gelb 1991.
29 Gore 1991.
30 Nielsen 2015.
31 Luntz 2002, p. 137; see Burkeman 2003.
32 Luntz 2002, p. 132; Luntz has since reversed his support of climate denial (see Montlake 2019).
33 United States Department of State 2002, p. 3; also see Dickinson 2007.
34 Seelye 2002.
35 Mackenzie 2002.
36 McCright and Dunlap 2010, p. 101; the article reviews the climate denial of the Bush administration.
37 McCright and Dunlap 2010, p. 101.
38 Criticism of the Bush administration policy on Kyoto and climate change is presented by Harrison 2007, pp. 104, 105, 111.
39 Kutney 2014, pp. 152–154; also see Newsweek 2007; Baker 2017.
40 Dickinson 2007.
41 Dickinson 2007; see the article for more details on the involvement of Dick Cheney on the politics of climate change.
42 Dickinson 2007.
43 Cheney 2018.
44 For a biography of James Inhofe, see DeSmog, "James Inhofe"; a collection of his speeches and press releases from 2003 to 2006 are provided in United States Senate 2007a, pp. 997–1053.
45 United States Senate 2007a, pp. 4–5.
46 Inhofe 2003, pp. 19933–19943.
47 Inhofe published a book called *The Greatest Hoax*, in 2012.
48 United States Senate 2003, p. 1.
49 United States Senate 2003, pp. 15–16; also see pp. 24–26. Schlesinger was a popular reference for Senator Inhofe; see, for example, Inhofe 2003, p. 19934; Inhofe 2004, p. S11292.
50 United States Senate 2003, p. 24.
51 United States Senate 2003, p. 29.
52 United States Senate 2003, pp. 178–194.
53 Mann 2012, p. 121.
54 Inhofe 2019b.
55 Inhofe 2019a.
56 Inhofe 2004, p. S11297.
57 Inhofe 2004, p. S11297.
58 See United States Senate 2007a, p. 1010; for another press release, see p. 1014.
59 United States Senate 2005, pp. 1–2; also see Mann 2012, p. 144.
60 The senior member of the minority party.
61 United States Senate 2007c, p. 7.
62 United States Senate 2013, p. 44.
63 Whitehouse 2015, pp. 771–772.
64 Inhofe 2015a; also see his YouTube video (Inhofe 2015b) and tweets (Inhofe 2015c–e).
65 Barrett 2015.

66 Inhofe 2015f; he used the Lindzen quote often; GovInfo lists 21 records; see, for example, United States Senate 2007d, p. 6.
67 Inhofe 2015g.
68 Inhofe 2022a.
69 Inhofe 2022b.
70 For example see United States Senate 2017a; Inhofe 2017.
71 Inhofe 2012, p. S5853.
72 Little 2011.
73 Eilperin and Dennis 2017.
74 D'Angelo 2022.
75 Brulle 2014.
76 Plait 2010; for a detailed report on the antics of Joe Barton against Dr. Mann, see Mann 2012, pp. 149–175.
77 United States House of Representatives 2007c, p. 8.
78 At another meeting of the Energy and Commerce Committee, Joe Barton observed that: "when it is hot, we get shade" (Barton 2009); for related comments by Barton, see United States House of Representatives 2010, p. 8.
79 Newton-Small and Steinmetz 2010; other quotes from Joe Barton are presented.
80 United States House of Representatives 2013a, pp. 66–67; also see Goldenberg 2013.
81 Plait 2010.
82 Newton-Small and Steinmetz 2010.
83 United States House of Representatives 2007b, p. 239.
84 United States House of Representatives 2007e, p. 8.
85 Mervis 2011; also, in a radio interview in 2012, James Inhofe expressed similar religious arguments: "God's still up there. The arrogance of people to think that we, human beings, would be able to change what He is doing in the climate is to me outrageous" (Tashman 2012).
86 United States House of Representatives 1995, p. 2; contrarians at the session included Patrick Michaels and William Nierenberg.
87 United States House of Representatives 1995, p. 3.
88 United States House of Representatives 2007f, pp. 4–5.
89 United States House of Representatives 2007g, pp. 78–79; this is one of the first mentions of "climate deniers" in a congressional hearing.
90 United States House of Representatives 2007g, p. 79, also see p. 82.
91 United States House of Representatives 2007g, pp. 90–93.
92 Rohrabacher 2011, p. H8310.
93 United States House of Representatives 2013b, p. 31.
94 For a review of Smith's climate change denial see DeSmog, "Lamar S. Smith."
95 Smith 2009, pp. 29845–29846.
96 Krauss 2016; also see Goldman 2015; Smith 2015; in one session of Congress, he issued 26 subpoenas (see Kelly 2018).
97 Mann 2015.
98 United States House of Representatives 2016b, p. 10.
99 United States House of Representatives 2016b, p. 124.
100 United States House of Representatives 2016a, p. 21.
101 United States House of Representatives 2017a, pp. 4–5; also see Nuccitelli 2017a.
102 Kolbert 2017.
103 United States House of Representatives 2018, p. 4.
104 Kelly 2018; for more propaganda from Mr. Smith see Hulac 2018; Schipani 2018.
105 See Chopra 2009.
106 Trump had been extremely popular, though controversial, on Twitter with 88 million followers (#6 in the global ranking; Obama was #1 at 126 million) at the end of 2020; some have suggested that he had been Twitter's saviour (Bloomberg 2017; Solon 2017).

107 For a discussion, see Coppins 2020.
108 Trump 2017b.
109 Milman 2020; also see Hern 2017; see Lavelle 2019 for the influence of bots on social media.
110 Trump 2011.
111 See, for example, Trump 2014; in the over 100 tweets related to climate change, the word "hoax" or "hoaxsters" appeared nine times, and "cold" weather 29 times.
112 Trump 2012.
113 Trump 2016.
114 See Carroll 2016.
115 See Anon. 2020 for a list of his tweets on climate change and/or global warming; number of tweets as of March 2020: climate change and global warming = 28, only climate change = 12, only global warming = 84.
116 Trump 2017a.
117 Trump 2018, 2019a–f; also see Trump 2020a.
118 Trump 2019g; also see Greve 2019; for other attacks on Ms. Thunberg see Trump 2019h,i; Gelin 2019. The week before Trump's tweet, Ms. Thunberg had appeared at a meeting of the Senate Climate Crisis Task force, chaired by Senator Ed Markey (Gambino 2019; United States House of Representatives 2019c).
119 Thunberg 2021.
120 Twitter 2021.
121 Musk 2022a.
122 Musk 2022b.
123 For a discussion on Trump's nominees, see Freedman 2016; Milman 2016; Sidahmed 2016.
124 Pence 2000; also see Schreyer 2016; Mayer 2017; Lavelle 2020.
125 Cillizza 2019.
126 Leber 2020.
127 Pompeo 2013; more recently, Pompeo tweeted: "If you stand for Climate Change First, you stand for America Last" (Pompeo 2021).
128 Gillman 2014; for more on Rick Perry see Murphy 2016.
129 Johnson 2014.
130 Holden 2019.
131 D'Angelo 2020.
132 United States Senate 2015a, p. 75; Jim Inhofe presided over the session.
133 Pruitt and Strange 2016; also see Grandoni, Dennis, and Mooney 2018; for actions in the EPA see Calma 2021.
134 Kaufman 2018.
135 Price 2010.
136 United States Senate 2017b.
137 Marinucci 2015.
138 Bridenstine 2013; also see Nuccitelli 2017c; for denier appointments at NOAA see Klein 2020.
139 Neate 2016.
140 Perdue 2017.
141 Waldman 2019.
142 Waldman 2019.
143 Waldman 2020.
144 Weaver and Lyons 2019.
145 Trump 2020b.
146 Papenfuss 2022.
147 Taylor 2018; also see Braun 2020 for more connection between Donald Trump and the Heartland Institute; for more on the Heartland Institute, see United States House of Representatives 2019b, p. 17.

148 Cassels 2021.
149 Milman 2023.
150 Manchin 2014.
151 United States Senate 2019a, pp. 7–9.
152 Manchin 2021.
153 Milman 2022.
154 Gabbatt, Pengelly, and Stein 2022.
155 United States Senate 1989, p. 1.
156 United States House of Representatives 2007a, pp. 2–3; for more on the Cooney affair, see Dickinson 2007; United States House of Representatives 2007d.
157 United States House of Representatives 2007a, p. 10.
158 United States House of Representatives 2007a, pp. 96–97.
159 United States Senate 2007b, p. 6.
160 United States Senate 2007b, p. 53.
161 United States Senate 2007b, p. 54.
162 United States House of Representatives 2007d, pp. 1–2.
163 United States House of Representatives 2007d, p. 304.
164 United States House of Representatives 2007f, p. 3; a witness at this hearing was Jeff Kueter, president of the right-wing think tank George C. Marshall Institute.
165 United States Senate 2007a, p. 91.
166 Obama 2011; for a comment on "science deniers" see Obama 2012b.
167 Obama 2014b.
168 Obama 2014c.
169 Obama 2015.
170 Obama 2016b; also present were the actor Leonardo DiCaprio (1974–) and the noted climate scientist Katharine Hayhoe (1972–).
171 See, for example, Obama 2013, 2014a.
172 Obama 2016a.
173 Obama 2016c.
174 Stahl 2018.
175 Karp 2020.
176 Wise 2018.
177 Ocasio-Cortez 2020.
178 McGrath 2020.
179 United States House of Representatives 2019a, pp. 147–148.
180 D'Angelo 2019; for examples of climate denial by Mr. Trump see Cheung 2020; Jackson 2020; Roberts 2020; Kahn 2021.
181 Chomsky 2020.
182 Krugman 2018.
183 Earthjustice 2020.
184 Corbyn 2019.
185 United States House of Representatives 2008b, pp. 4–5.
186 United States House of Representatives 2008a, p. 1.
187 United States Senate 2015b.
188 United States Senate 2019b, pp. 117–118; also see Markey 2021, p. S173.
189 See, for example, Markey 2000, pp. H8776; 2018; 2022, p. S1766.
190 Waxman 2013.
191 United States House of Representatives 2013c, pp. 5, 125; also see United States House of Representatives 2013a, p. 9.
192 He did this a few times in 2012 and 2013; see, for example, Obama 2012a.
193 United States Senate 2008, p. 40.
194 Schumer 2019.
195 IPCC 2001, p. 2.

196 Mann 2012, ch. 4.
197 United States House of Representatives 2006, p. 10.
198 United States House of Representatives 2006, p. 10.
199 United States House of Representatives 2006, p. 20.
200 For comments on this hearing see Nuccitelli 2017b.
201 Mervis 2017.
202 United States House of Representatives 2017b, pp. 4–5.
203 United States House of Representatives 2017b, pp. 8–9.
204 United States House of Representatives 2017b, p. 98; also see Leber 2017.
205 United States House of Representatives 2017b, p. 106.
206 United States House of Representatives 2017b, pp. 111–112.
207 United States House of Representatives 2017b, p. 109.
208 DNC 2022.
209 Cama 2019; see United States House of Representatives 2019b; for an overview of political climate deniers see Skeptical Science 2020; DeSmog undated, "Climate Disinformation Database."
210 Chomsky 2016.
211 Drennen and Hardin 2021; also see Drennen 2021.

References

Anon. (2020). *Trump Twitter Archive*; https://www.thetrumparchive.com/

Baker, Peter (2017). "16 Years Later, Bush's Climate Pact Exit Holds Lessons for Trump." *New York Times*, June 4; https://www.nytimes.com/2017/06/04/us/politics/trump-paris-accord-bush-kyoto.html

Barrett, Ted (2015). "Inhofe brings Snowball on Senate Floor as Evidence Globe Is Not Warming." *CNN Politics*, February 27; https://www.cnn.com/2015/02/26/politics/james-inhofe-snowball-climate-change/index.html

Barton, Joe (2009). *Rep. Barton on Global Warming: Find Some Shade*. YouTube, March 25; https://www.youtube.com/watch?time_continue=2&v=Z2bM5_Pe-rw&feature=emb_logo

Biello, Peter (2019). "The Sununu Family and Climate Change Over the Years." *New Hampshire Public Radio*, April 4; https://www.nhpr.org/post/sununu-family-and-climate-change-over-years#stream/0

Bloomberg (2017). "What is Trump Worth to Twitter? One Analyst Estimates $2 Billion." *Fortune*, August 17; http://fortune.com/2017/08/17/trump-worth-to-twitter/

Braun, Stuart (2020). "The Spin Machine Upending the Climate Consensus." *DW*, March 9; https://www.dw.com/en/trump-climate-change-denial-emissions-environment-germany-fake-heartland-seibt/a-52688933

Bridenstine, Jim (2013). *Trump NASA Nominee Rep Jim Bridenstine (R-OK) Demands Obama Apologize on Global Warming*. C-Span, June 11; https://www.youtube.com/watch?v=GUcsAFnwC7k

Brulle, Robert J. (2014). "Institutionalizing Delay: Foundation Funding and the Creation of U.S. Climate Change Counter-Movement Organizations." *Climatic Change* 122, p. 681; https://link.springer.com/article/10.1007/s10584-013-1018-7#page-1

Burkeman, Oliver (2003). "Memo Exposes Bush's New Green Strategy." *Guardian*, March 4; https://www.theguardian.com/environment/2003/mar/04/usnews.climatechange

Bush, George H. W. (1990). *Remarks to the Intergovernmental Panel on Climate Change*. George H.W. Bush Presidential Library & Museum, February 5; https://bush41library.tamu.edu/archives/public-papers/1514

Calma, Justine (2021). *How Scientists Scrambled to Stop Donald Trump's EPA from Wiping Out Climate Data*. The Verge, March 8; https://www.theverge.com/22313763/scientists-climate-change-data-rescue-donald-trump

Cama, Timothy (2019). "House Energy Panel to Dedicate First Hearing to Climate Change." *The Hill*, January 3; https://thehill.com/policy/energy-environment/423771-house-energy-panel-to-dedicate-first-hearing-to-climate-change

Carroll, Lauren (2016). "At New York Debate, Donald Trump Denies Saying Climate Change Is a Chinese Hoax." *Politifact*, September 26; https://www.politifact.com/truth-o-meter/statements/2016/sep/26/donald-trump/donald-trump-denies-saying-global-warming-chinese-/

Cassels, Laura (2021). "Avoiding Words 'Climate Change,' DeSantis Says Global-Warming Concerns Involve 'Left-Wing Stuff.'" *Florida Phoenix*, December 7; https://floridaphoenix.com/2021/12/07/avoiding-words-climate-change-desantis-says-global-warming-concerns-involve-left-wing-stuff/

Cheney, Dick (2018). *Cheney on Climate*. C-Span, November 21; https://www.c-span.org/video/?c4762059/cheney-climate

Cheung, Helier (2020). "What Does Trump Actually Believe on Climate Change?" *BBC News*, January 23; https://www.bbc.com/news/world-us-canada-51213003

Chomsky, Noam (2016). *Global Warming and the Future of Humanity*. "C. J. Polychroniou interviews Noam Chomsky and Graciela Chichilnisky." Chomsky.Info, September 17; https://chomsky.info/global-warming-and-the-future-of-humanity/

Chomsky, Noam (2020). *Noam Chomsky on Trump vs Biden, Threat of Nuclear War, & Julian Assange | Going Underground*. RT, Going Underground, YouTube, September 7; https://www.youtube.com/watch?v=4skx0XKAMqU (note: currently blocked as RT is a Russian state-funded news organization that has been blocked by Google).

Chopra, Deepak (2009). "Letter to President Obama from Business Leaders." *HuffPost*, December 7; https://www.huffpost.com/entry/letter-to-president-obama_b_383129

Cillizza, Chris (2019). "Mike Pence's Unbelievable Answer on Whether Climate Change is a Threat." *CNN Politics*, June 24; https://www.cnn.com/2019/06/24/politics/mike-pence-climate-crisis/index.html

Coppins, McKay (2020). "The Billion-Dollar Disinformation Campaign to Reelect the President." *The Atlantic*, February 10; https://www.theatlantic.com/magazine/archive/2020/03/the-2020-disinformation-war/605530/

Corbyn, Zoë (2019). "Naomi Oreskes: 'Discrediting Science is a Political Strategy.'" *Guardian*, November 3; https://www.theguardian.com/science/2019/nov/03/naomi-oreskes-interview-why-trust-science-climate-donald-trump-vaccine?CMP=share_btn_tw

D'Angelo, Chris (2019). "Trump's Attack on Climate Science Echoes Big Oil's 1998 Climate Campaign." *HuffPost*, June 27; https://www.huffingtonpost.ca/entry/trump-climate-panel-american-petroleum-institute-1998-campaign_n_5d129d6fe4b04f059e4b3d18

D'Angelo, Chris (2020). "Interior Dept. Consulted 'Junk Science' Troll on 'Transparency' Rule. Emails Show." *HuffPost*, June 18; https://www.huffingtonpost.ca/entry/steve-milloy-junk-science-interior-department_n_5eebbe63c5b6fb7ffc86626d?ri18n=true

D'Angelo, Chris (2022). "Farewell to the Senate's Biggest Climate Denier." *HuffPost*, December 31; https://www.huffpost.com/entry/jim-inhofe-retire-climate-denial_n_63a4c721e4b0d6f0b9ec07ec

DeSmog (undated). "Climate Disinformation Database"; https://www.desmogblog.com/global-warming-denier-database

DeSmog (undated). "James Inhofe"; https://www.desmogblog.com/james-inhofe

DeSmog (undated). "John H. Sununu"; https://www.desmogblog.com/john-h-sununu

DeSmog (undated). "Lamar S. Smith"; https://www.desmogblog.com/lamar-smith

Dickinson, Tim (2007). "The Secret Campaign of President Bush's Administration to Deny Global Warming." *Rolling Stone*, June 20; https://web.archive.org/web/20070626050118/http://www.rollingstone.com/politics/story/15148655/the_secret_campaign_of_president_george_bushs_administration_to_deny_global

DNC (1991). *Putting the Heat on the Bush Environmental Record.* Net '92 Issue of the Week, July 22, 1991; Council on Environmental Quality and Kathleen "Katie" McGinty, "Administration Global Warming Policy," Clinton Digital Library; https://clinton.presidentiallibraries.us/items/show/67816

DNC (2022). *Fact Check: The GOP Is STILL the Party of the Climate Deniers*. DNC, June 3; https://democrats.org/news/fact-check-the-gop-is-still-the-party-of-climate-deniers/

Drennen, Ari (2021). *A Look at the Rhetorical Evolutions of Congress' Former Climate Deniers*. Center for American Progress, May 7; https://www.americanprogress.org/issues/green/news/2021/05/07/499277/look-rhetorical-evolutions-congress-former-climate-deniers/

Drennen, Ari, and Hardin, Sally (2021). *Climate Deniers in the 117th Congress*. Center for American Progress, March 30; https://www.americanprogress.org/issues/green/news/2021/03/30/497685/climate-deniers-117th-congress/

Earthjustice (2020). *Environmental and Public Health Advocates Agree: Trump Is the Worst President for Our Environment in History*. February 4; https://earthjustice.org/news/press/2020/trump-worst-president-for-environment-in-history

Eilperin, Juliet, and Dennis, Brady (2017). "How James Inhofe Is Upending the Nation's Energy and Environmental Policies." *Washington Post*, March 14; https://www.washingtonpost.com/national/health-science/how-james-inhofe-is-upending-the-nations-energy-and-environmental-policies/2017/03/14/2bebdbfa-081c-11e7-a15f-a58d4a988474_story.html

Freedman, Andrew (2016). "A Guide to Trump's Alarming Cabinet Full of Climate Deniers." *Yahoo News*, December 18; https://www.yahoo.com/news/guide-trumps-alarming-cabinet-full-182739974.html

Gabbatt, Adam, Pengelly, Martin, and Stein, Chris (2022). "Biden Pledges Executive Action after Joe Manchin Scuppers Climate Agenda." *Guardian*, July 15; https://www.theguardian.com/us-news/2022/jul/15/joe-manchin-climate-funding-tax-hikes

Gambino, Lauren (2019). "Greta Thunberg to Congress: 'You're Not Trying Hard Enough. Sorry.'" *Guardian*, September 17; https://www.theguardian.com/environment/2019/sep/17/greta-thunberg-to-congress-youre-not-trying-hard-enough-sorry

Gelb, Leslie H. (1991). "Foreign Affairs; Sununu vs. Scientists." *New York Times*, Section 4, p. 17, February 10; https://www.nytimes.com/1991/02/10/opinion/foreign-affairs-sununu-vs-scientists.html

Gelin, Martin (2019). "The Misogyny of Climate Deniers." *The New Republic*, August 28; https://newrepublic.com/article/154879/misogyny-climate-deniers

George H. W. Bush Library (1990a). "Memorandum of Conversation, Meeting with Prime Minister Margaret Thatcher of Great Britain." *George H.W. Bush Presidential Library and Museum*, April 13; https://bush41library.tamu.edu/files/memcons-telcons/1990-04-13--Thatcher.pdf

George H. W. Bush Library (1990b). *Memorandum of Conversation, Bilateral Meeting with German Chancellor Helmut Kohl.* George H.W. Bush Presidential Library and Museum, July 9; https://bush41library.tamu.edu/files/memcons-telcons/1990-07-09--Kohl.pdf

Gillman, Todd J. (2014). "In DC, Gov. Rick Perry Talks 2016, Ted Cruz, Climate Change, Border Security." *The Dallas Morning News*, June 19; https://www.dallasnews.com/news/politics/2014/06/19/in-dc-gov-rick-perry-talks-2016-ted-cruz-climate-change-border-security

Goldenberg, Suzanne (2013). "U.S. Congressman Cites Biblical Flood to Dispute Human Link to Climate Change." *Guardian*, April 11; https://www.theguardian.com/environment/blog/2013/apr/11/republican-biblical-flood-climate-change

Goldman, Gretchen (2015). *What's the Deal with Rep. Lamar Smith's Subpoena to NOAA Over Climate Science? An FAQ Resource*. Union of Concerned Scientists, November 16; https://blog.ucsusa.org/gretchen-goldman/whats-the-deal-with-rep-lamar-smiths-subpoena-to-noaa-over-climate-science-an-faq-resource-958

Gore, Al (1991). *Untitled, Letter to the Editor of the 'New York Times.'* Council on Environmental Quality, Kathleen "Katie" McGinty, "Global Warming Skeptics." Clinton Digital Library, February 20; https://clinton.presidentiallibraries.us/items/show/76169

Grandoni, Dino, Dennis, Brady, and Mooney, Chris (2018). "Scott Pruitt Asks Whether Global Warming Is 'Necessarily Is a Bad Thing.'" *Washington Post*, February 7; https://www.denverpost.com/2018/02/07/scott-pruitt-views-on-climate-change/

Greve, Joan E. (2019). "Greta Thunberg Turns Tables on Trump and Quotes his Mockery in New Twitter Bio." *Guardian*, September 24; https://www.theguardian.com/environment/2019/sep/24/greta-thunberg-trump-twitter-bio

Harrison, Kathryn (2007). "The Road Not Taken: Climate Change Policy in Canada and the United States." *Global Environmental Politics* 7 (4), p. 99, November; https://muse.jhu.edu/article/224162/pdf

Hecht, Alan D., and Tirpak, Dennis (1995). "Framework Agreement on Climate Change: A Scientific and Policy History." *Climatic Change* 29, p. 371; https://link.springer.com/article/10.1007/BF01092424

Hern, Alex (2017). "Facebook and Twitter Are Being Used to Manipulate Public Opinion." *Guardian*, June 19; https://www.theguardian.com/technology/2017/jun/19/social-media-proganda-manipulating-public-opinion-bots-accounts-facebook-twitter

Hoffman, David (1990). "Bush Struggles for 'Balance' Between 2 Sharp Edges." *Washington Post*, April 20; https://www.washingtonpost.com/archive/politics/1990/04/20/bush-struggles-for-balance-between-2-sharp-edges/e30ad657-953c-4170-9bc2-8d0c8b05de42/?utm_term=.b37dc0925662

Holden, Emily (2019). "Trump's Interior Secretary: I Haven't 'Lost Sleep' Over Record CO_2 Levels." *Guardian*, May 15; https://www.theguardian.com/environment/2019/may/15/interior-secretary-david-bernhardt-c02

Howe, Joshua P. (2014). *Behind the Curve: Science and the Politics of Global Warming*. Seattle: University of Washington Press.

Hulac, Benjamin (2018). "A Ranch, an Oil Field, and a Career Colored by Skepticism." *E&E News*, December 21; https://www.eenews.net/stories/1060110337

Inhofe, James (2003). *Science of Climate Change*. Congressional Record – Senate, p. 19933, July 28; https://www.govinfo.gov/content/pkg/CRECB-2003-pt15/pdf/CRECB-2003-pt15-Pg19933.pdf

Inhofe, James (2004). *The Science of Climate Change*. Congressional Record – Senate, p. S11291, October 11; https://www.govinfo.gov/content/pkg/CREC-2004-10-11/pdf/CREC-2004-10-11-pt1-PgS11291-4.pdf

Inhofe, James (2012). *Global Warming*. Congressional Record – Senate, p. S5853 August 1; https://www.govinfo.gov/content/pkg/CREC-2012-08-01/pdf/CREC-2012-08-01-pt1-PgS5825.pdf

Inhofe, James (2015a). *Inhofe Speaks on Why Climate Change Is a Hoax, as Defined by Whitehouse Amendment #29*. YouTube, January 21; https://www.youtube.com/watch?v=pT9ZeEFDVyM&t=4s

Inhofe, James (2015b). Tweet, Twitter, January 21; https://twitter.com/JimInhofe/status/558047155072798725

Inhofe, James (2015c). Tweet, Twitter, January 21; https://twitter.com/JimInhofe/status/558011974500380672

Inhofe, James (2015d). Tweet, Twitter, January 21; https://twitter.com/JimInhofe/status/558033392944840704

Inhofe, James (2015e). Tweet, Twitter, January 21; https://twitter.com/JimInhofe/status/558036206433599489

Inhofe, James (2015f). *Examining State Perspectives of the EPA's Proposed Carbon Dioxide Emissions Rules for Existing Power Plants*. U.S. Senate Committee on Environment and Public Works, Hearings, March 11; https://www.epw.senate.gov/public/index.cfm/hearings?Id=61944108-BC2D-D8B0-D90A-09D5B909D464&Statement_id=62C7D2B7-F789-4361-8F47-530E74F936B6

Inhofe, James (2015g). *Inhofe Statement on Pope Francis' Encyclical on Climate Change*. U.S. Senate Committee on Environment and Public Works, Newsroom, June 18; https://www.epw.senate.gov/public/index.cfm/2015/6/inhofe-statement-on-pope-francis-encyclical-on-climate-change

Inhofe, James (2017). *Inhofe Leads Senators in Letter to Trump Calling for Withdrawal from Paris Climate Agreement*. James M. Inhofe, May 25; https://www.inhofe.senate.gov/newsroom/press-releases/inhofe-leads-senators-in-letter-to-trump-calling-for-withdrawal-from-paris-climate-agreement

Inhofe, James (2019a). *Inhofe Announces Climate Hypocrite Awards*. James M. Inhofe, U.S. Senator for Oklahoma, September 23; https://www.inhofe.senate.gov/climate-week

Inhofe, James (2019b). Tweet, Twitter, September 23; https://twitter.com/JimInhofe/status/1176262240594472960

Inhofe, James (2022a). Tweet, Twitter, July 20; https://twitter.com/JimInhofe/status/1549861710789427200

Inhofe, James (2022b). *Inhofe Gives Farewell Address on Senate Floor*. James M. Inhofe, U.S. Senator for Oklahoma, November 16; https://www.inhofe.senate.gov/newsroom/press-releases/inhofe-gives-farewell-address-on-senate-floor

IPCC (2001). *Climate Change 2001: The Scientific Basis. Summary for Policymakers. Contribution of Working Group I to the Third Assessment Report of the Intergovernmental Panel on Climate Change* [Houghton, J. T., Y. Ding, D. J. Griggs, M. Noguer, P. J. van der Linden, X. Dai, K. Maskell, and C. A. Johnson (eds.)]. Cambridge University Press, Cambridge and New York; https://www.ipcc.ch/site/assets/uploads/2018/07/WG1_TAR_SPM.pdf

Jackson, Derrick Z. (2020). *Election Day Is Nearing, But There's No End in Sight of Trump's Attacks on Science*. Grist, September 19; https://grist.org/politics/election-day-is-nearing-but-theres-no-end-in-sight-to-trumps-attacks-on-science/

Jimmy Carter Library (1977a). *Collection: Office of Staff Secretary; Series: Presidential Files; Folder: 7/18/77 [1]; Container 31*. Jimmy Carter Presidential Library and Museum; https://www.jimmycarterlibrary.gov/digital_library/sso/148878/31/SSO_148878_031_07.pdf

Jimmy Carter Library (1977b). *Folder Citation: Collection: Office of Staff Secretary; Series: Presidential Files; Folder: 8/1/77 [1]; Container 34*. Jimmy Carter Presidential Library and Museum; https://www.jimmycarterlibrary.gov/digital_library/sso/148878/34/SSO_148878_034_07.pdf

Johnson, Charles S. (2014). "Lewis, Zinke Debate Federal Budget, Health Care, Global Warming." *Billings Gazette*, October 4; https://billingsgazette.com/news/local/government-and-politics/lewis-zinke-debate-federal-budget-health-care-global-warming/article_d062bc3c-c8e9-5909-854f-5e78efc52868.html

Kann, Drew (2021). "'The Lost Years': Climate Damage That Occurred on Trump's Watch Will Endure Long after He Is Gone." *CNN*, January 18; https://www.cnn.com/2021/01/18/politics/trump-climate-legacy-bidens-challenge/index.html

Karp, Paul (2020). "Former Australian PM Malcolm Turnbull says Trump Is the World's 'Leading Climate Denier.'" *Guardian*, January 22; https://www.theguardian.com/science/2020/jan/22/former-australian-pm-malcolm-turnbull-says-trump-is-the-worlds-leading-climate-denier?platform=hootsuite

Kaufman, Alexander C. (2018). "Trump's Climate-Denying Coal Lobbyist Nominee Inches Closer to EPA's No. 2 Job." *HuffPost*, February 6; https://www.huffingtonpost.ca/entry/andrew-wheeler-epa_n_5a78dc2ce4b0164659c73e97

Kelly, Julie (2018). "Friend of Science." *National Review*, January 22; https://www.nationalreview.com/magazine/2018/01/22/lamar-smith-climate-change-texas-rep-friend-science/

Kerr, Richard A. (1992). "Greenhouse Science Survives Skeptics." *Science* 256 (5060), p. 1138, May 22; https://science.sciencemag.org/content/256/5060/1138.2

Klein, Charlotte (2020). "Scientists Who Don't Believe in Climate Change Are Now Leading a Top Environmental Agency." *Vanity Fair*, October 28; https://www.vanityfair.com/news/2020/10/trump-climate-change-denial-noaa

Kolbert, Elizabeth (2017). "The Congressman Who's Trying to Make the Environment Worse Again." *New Yorker*, February 8; https://www.newyorker.com/news/daily-comment/the-congressman-whos-trying-to-make-the-environment-worse-again

Krauss, Lawrence M. (2016). "The House Science's Committee Anti-Science Rampage." *New Yorker*, September 14; https://www.newyorker.com/news/news-desk/the-house-science-committees-anti-science-rampage

Krugman, Paul (2018). "The Depravity of Climate-Change Denial." *New York Times*, November 26; https://www.nytimes.com/2018/11/26/opinion/climate-change-denial-republican.html

Kutney, Gerald W. (2014). *Carbon Politics and the Failure of the Kyoto Protocol*. London: Routledge.

Lavelle, Marianne (2019). "'Trollbots' Swarm Twitter with Attacks on Climate Science Ahead of UN Summit." *Inside Climate News*, September16; https://insideclimatenews.org/news/16092019/trollbot-twitter-climate-change-attacks-disinformation-campaign-mann-mckenna-greta-targeted

Lavelle, Marianne (2020). "A Climate Change Skeptic, Mike Pence Brought to the Vice Presidency Deep Ties to the Koch Brothers." *Inside Climate News*, August 31; https://insideclimatenews.org/news/31082020/candidate-profile-mike-pence-climate-change-election-2020/

Leber, Rebecca (2017). *A Scientist Just Spent 2 Hours Debating the Biggest Global Warming Deniers in Congress*. Mother Jones, March 29; https://www.motherjones.com/environment/2017/03/michael-mann-lamar-smith-house-science-committee/

Leber, Rebecca (2020). *Pence Said Trump Listens to Scientists. That's a Joke, Right?* Mother Jones, October 7; https://www.motherjones.com/environment/2020/10/pence-said-trump-listens-to-scientists-thats-a-joke-right/

Little, Amanda (2010). *James Inhofe, Senate's Top Skeptic, Explains His Climate-hoax Theory*. Grist, February 28; https://grist.org/article/2010-02-25-james-inhofe-senate-top-skeptic-explains-climate-hoax-theory/

Luntz, Frank (2002). *The Environment: A Cleaner, Safer, Healthier America.* The Luntz Research Companies; http://www.sindark.com/NonBlog/Articles/LuntzResearch_environment.pdf

Mackenzie, Debora (2002). "Too Hot for Head of Climate Panel." *New Scientist* 174 (2399), April 20, p. 16; https://www.newscientist.com/article/mg17423392-400-too-hot-for-head-of-climate-panel/

Manchin, Joe (2014). Tweet, Twitter, June 25; https://twitter.com/Sen_JoeManchin/status/481885321685401600

Manchin, Joe (2021). Tweet, Twitter, January 21; https://twitter.com/Sen_JoeManchin/status/1352317399878479872

Mann, Michael E. (2012). *The Hockey Stick and the Climate Wars*. New York: Columbia University Press.

Mann, Michael E. (2015). "The Assault on Climate Science." *New York Times*, December 8; https://www.nytimes.com/2015/12/08/opinion/the-assault-on-climate-science.html

Marinucci, Carla (2015). "Ben Carson, the 'Anti-Trump,' Brings Conservative Message to S.F." *San Francisco Chronicle*, September 8; https://www.sfchronicle.com/bayarea/article/Ben-Carson-comes-to-SF-and-he-has-unfamiliar-6492055.php

Markey, Ed (2000). *Protecting Our Environment*. Congressional Record – House, p. H8774, October 4; https://www.govinfo.gov/content/pkg/CREC-2000-10-04/pdf/CREC-2000-10-04-pt1-PgH8774-2.pdf

Markey, Ed (2018). Tweet, Twitter, January 4; https://twitter.com/senmarkey/status/949037961668186112

Markey, Ed (2021). *Climate Change*. Congressional Record – Senate, p. S171, January 27; https://www.govinfo.gov/content/pkg/CREC-2021-01-27/pdf/CREC-2021-01-27-pt1-PgS171-4.pdf

Markey, Ed (2022). *Ukraine*. Congressional Record – Senate, p. S1760, March 24; https://www.govinfo.gov/content/pkg/CREC-2022-03-24/pdf/CREC-2022-03-24-pt1-PgS1760-3.pdf

Mayer, Jane (2017). "The Danger of President Pence." *New Yorker*, October 16; https://www.newyorker.com/magazine/2017/10/23/the-danger-of-president-pence

McCright, Aaron M., and Dunlap, Riley E. (2010). "Anti-Reflexivity." *Theory, Culture & Society* 27 (2–3), p. 100; https://journals.sagepub.com/doi/abs/10.1177/0263276409356001

McGrath, Matt (2020). "US Election 2020: What the Results Will Mean for Climate Change." *BBC News*, October 20; https://www.bbc.com/news/science-environment-54395534

Mervis, Jeffrey (2011). "Ralph Hall Speaks Out on Climate Change." *Science*, December 14; https://www.sciencemag.org/news/2011/12/ralph-hall-speaks-out-climate-change

Mervis, Jeffrey (2017). "Lamar Smith, Unbound, Lays Out Political Strategy at Climate Doubters' Conference." *Science*, March 24; https://www.sciencemag.org/news/2017/03/lamar-smith-unbound-lays-out-political-strategy-climate-doubters-conference

Milman, Oliver (2016). "Trump's Transition: Sceptics Guide Every Agency Dealing with Climate Change." *Guardian*, December 12; https://www.theguardian.com/us-news/2016/dec/12/donald-trump-environment-climate-change-skeptics

Milman, Oliver (2020). "Revealed: Quarter of All Tweets about Climate Crisis Produced by Bots." *Guardian*, February 21; https://www.theguardian.com/technology/2020/feb/21/climate-tweets-twitter-bots-analysis?CMP=Share_AndroidApp_Tweet

Milman, Oliver (2022). "A Modern-Day Villain': Joe Manchin Condemned for Killing US Climate Action." *Guardian*, July 15; https://www.theguardian.com/us-news/2022/jul/15/joe-manchin-condemned-us-climate-action

Milman, Oliver (2023). "DeSantis Accused of 'Catastrophic' Climate Approach after Campaign Launch." *Guardian*, May 28; https://www.theguardian.com/us-news/2023/may/28/ron-desantis-climate-crisis-campaign

Mintzer, Irving M., and Leonard, J. Amber (ed.) (1994). *Negotiating Climate Change*. Cambridge: Cambridge University Press.

Montlake, Simon (2019). "What Does Climate Change Have to Do with Socialism?" *Christian Science Monitor*, August 5; https://www.csmonitor.com/Environment/2019/0805/What-does-climate-change-have-to-do-with-socialism

Mooney, Chris (2005). "Stop Him Before He Writes Again." *The American Prospect*, August 23; https://prospect.org/article/stop-him-he-writes-again

Murphy, Tim (2016). *Rick Perry's War on Science*. Mother Jones, December 13; https://www.motherjones.com/environment/2016/12/rick-perry-energy-secretary-climate-censorship/

Musk, Elon (2022a). Tweet, Twitter, November 18; https://twitter.com/elonmusk/status/1593767953706921985

Musk, Elon (2022b). Tweet, Twitter, November 19; https://twitter.com/elonmusk/status/1594131768298315777

Neate, Rupert (2016). "ExxonMobil CEO: Ending Oil Production 'Not Acceptable for Humanity.'" *Guardian*, May 25; https://www.theguardian.com/business/2016/may/25/exxonmobil-ceo-oil-climate-change-oil-production

Newsweek Staff (2007). "Global Warming Deniers Well Funded." *Newsweek*, August 12; https://www.newsweek.com/global-warming-deniers-well-funded-99775

Newton-Small, Jay, and Steinmetz, Katy (2010). "Eight More Deep Thoughts from Rep. Joe Barton." *Time*, June 18; http://content.time.com/time/politics/article/0,8599,1997963,00.html

Nielsen, Aly (2015). "Sununu Hits Back at Climate Charge Alarmism on HuffPost Live." *HuffPost*, June 16; https://www.newsbusters.org/blogs/aly-nielsen/2015/06/16/sununu-hits-back-climate-change-alarmism-huffpost-live

Nuccitelli, Dana (2017a). "Whistleblower: 'I Knew People Would Misuse This.' They Did – To Attack Climate Change." *Guardian*, February 9; https://www.theguardian.com/environment/climate-consensus-97-per-cent/2017/feb/09/whistleblower-i-knew-people-would-misuse-this-they-did-to-attack-climate-science

Nuccitelli, Dana (2017b). "Inconceivable! The Latest Theatrical House 'Science' Committee Hearing." *Guardian*, April 4; https://www.theguardian.com/environment/climate-consensus-97-per-cent/2017/apr/04/inconceivable-the-latest-theatrical-house-science-committee-hearing

Nuccitelli, Dana (2017c). "We Have Every Reason to Fear Trump's Pick to Head Nasa." *Guardian*, November 6; https://www.theguardian.com/environment/climate-consensus-97-per-cent/2017/nov/06/we-have-every-reason-to-fear-trumps-pick-to-head-nasa

Obama, Barack (2011). *Remarks by the President at a DNC Event*. White House, President Barack Obama, April 21; https://obamawhitehouse.archives.gov/the-press-office/2011/04/21/remarks-president-dnc-event

Obama, Barack (2012a). *Remarks by the President on Energy*. White House, President Barack Obama, March 21; https://obamawhitehouse.archives.gov/the-press-office/2012/03/21/remarks-president-energy

Obama, Barack (2012b). *Remarks by the President at a Campaign Event*. White House, President Barack Obama, March 30; https://obamawhitehouse.archives.gov/the-press-office/2012/03/30/remarks-president-campaign-event-0

Obama, Barack (2013). Tweet, Twitter, October 27; https://twitter.com/BarackObama/status/394598788393742336

Obama, Barack (2014a). Tweet, Twitter, April 25; https://twitter.com/BarackObama/status/459767415929188352

Obama, Barack (2014b). *President Obama Speaks on American Energy*. The White House, President Barack Obama, May 9; https://obamawhitehouse.archives.gov/photos-and-video/video/2014/05/09/president-obama-speaks-american-energy#transcript

Obama, Barack (2014c). *Remarks by the President at the University of California–Irvine Commencement Ceremony*. The White House, President Barack Obama, June 14; https://obamawhitehouse.archives.gov/the-press-office/2014/06/14/remarks-president-university-california-irvine-commencement-ceremony

Obama, Barack (2015). *Remarks by the President at the GLACIER Conference–Anchorage, AK*. The White House, President Barack Obama, August 31; https://obamawhitehouse.archives.gov/the-press-office/2015/09/01/remarks-president-glacier-conference-anchorage-ak

Obama, Barack (2016a). *Remarks by the President at Hillary for America Campaign Event*. The White House, President Barack Obama, July 5; https://obamawhitehouse.archives.gov/the-press-office/2016/07/05/remarks-president-hillary-america-campaign-event

Obama, Barack (2016b). *Remarks by the President in South by South Lawn Panel Discussion on Climate Change*. The White House, President Barack Obama, October 3; https://obamawhitehouse.archives.gov/the-press-office/2016/10/03/remarks-president-south-south-lawn-panel-discussion-climate-change

Obama, Barack (2016c). *Remarks by the President at North Carolina Democratic Party Rally*. The White House, President Barack Obama, October 11; https://obamawhitehouse.archives.gov/the-press-office/2016/10/11/remarks-president-north-carolina-democratic-party-rally

Ocasio-Cortez, Alexandria (2020). Tweet, Twitter, September 29; https://twitter.com/AOC/status/1311133187251359744

Papenfuss, Mary (2022). "Trump Still Can't Wrap Head Around Climate Change: 'You Have A Thing Called Weather.'" *HuffPost*, March 21; https://www.huffpost.com/entry/trump-weather-climate-change-denial-fox-business_n_6238f3c3e4b0f1e82c4d58b0

Pence, Mike (2000). *Global Warming Disaster.* Mike Pence Congress; https://web.archive.org/web/20010415121513/http:/mikepence.com/warm.html

Perdue, Sonny (2017). "Agriculture Secretary Perdue talks TPP, Nafta and Climate Change." *The Wall Street Journal*, October 15; https://www.wsj.com/articles/agriculture-secretary-perdue-talks-tpp-nafta-and-climate-change-1508119261

Plait, Phil (2010). "Exclusive: Michael Mann Responds to Rep. Barton." *Discover*, October 14; https://www.discovermagazine.com/the-sciences/exclusive-michael-mann-responds-to-rep-barton

Pompeo, Mike (2013). *News Review with Representative Mike Pompeo*. C-Span, June 25; https://www.c-span.org/video/?c4456999/mike-pompeo-washington-journal

Pompeo, Mike (2021). Tweet, Twitter, May 29; https://twitter.com/mikepompeo/status/1398765659039977472

Price, Tom (2010). *Republicans Continue to Fight National Energy Tax*. Vote Smart, March 2; https://votesmart.org/public-statement/490191/republicans-continue-to-fight-national-energy-tax#.XIO8yVNKiu5

Pruitt, Scott, and Strange, Luther (2016). "The Climate-Change Gang." *National Review*, May 17; https://www.nationalreview.com/2016/05/climate-change-attorneys-general/

Rich, Nathaniel (2018). "Losing Earth: The Decade We Almost Stopped Climate Change." *New York Times*, August 1; https://www.nytimes.com/interactive/2018/08/01/magazine/climate-change-losing-earth.html

Roberts, David (2020). "A Second Trump Term Would Mean Severe and Irreversible Changes in the Climate." *Vox*, November 3; https://www.vox.com/energy-and-environment/2020/8/27/21374894/trump-election-second-term-climate-change-energy-russia-china

Roberts, Leslie (1989). "Global Warming: Blaming the Sun." *Science* 246 (4933), p. 992, November 24; https://science.sciencemag.org/content/246/4933/992

Rohrabacher, Dana (2011). *The Specter of Global Governance*. Congressional Record - House, p. H8309, December 8; https://www.govinfo.gov/content/pkg/CREC-2011-12-08/pdf/CREC-2011-12-08-pt1-PgH8309.pdf

Schipani, Vanessa (2018). *Smith's Error-Filled Climate Op-Ed*. FactCheck.org, March 23; https://www.factcheck.org/2018/03/smiths-error-filled-climate-op-ed/

Schlesinger, James (2003). "Climate Change: The Science Isn't Settled." *Washington Post*, July 7; https://www.washingtonpost.com/archive/opinions/2003/07/07/climate-change-the-science-isnt-settled/62c458de-a6d4-4046-a873-43f070b7c6c9/?utm_term=.23e763055306

Schlesinger, James (2005). "The Theology of Global Warming." *Wall Street Journal*, August 8, p. A10; http://www.mitosyfraudes.org/Polit/Theology.html

Schreyer, Natalie (2016). *On Climate Change, Pence and Trump Are a Perfect Match*. Mother Jones, July 15; https://www.motherjones.com/environment/2016/07/mike-pence-climate-change/

Schumer, Chuck (2019). *Climate Change*. Congressional Record – Senate, p. S227, June 5; https://www.govinfo.gov/content/pkg/CREC-2019-06-05/pdf/CREC-2019-06-05-pt1-PgS3227.pdf

Seelye, Katharine Q. (2002). "President Distances Himself from Global Warming Report." *New York Times*, June 5; https://www.nytimes.com/2002/06/05/us/president-distances-himself-from-global-warming-report.html

Sidahmed, Mazin (2016). "Climate Change Denial in the Trump Cabinet: Where Do His Nominees Stand?" *Guardian*, December 15; https://www.theguardian.com/environment/2016/dec/15/trump-cabinet-climate-change-deniers

Skeptical Science (2020). *Climate Myths from Politicians*; https://www.skepticalscience.com/skepticquotes.php

Smith, Lamar (2009). *Networks Ignore Climategate Scandal*. Congressional Record - House, p. 29845, December 8; https://www.govinfo.gov/content/pkg/CRECB-2009-pt22/pdf/CRECB-2009-pt22-Pg29845-11.pdf

Smith, Lamar (2015). "Lamar Smith: NOAA's Climate Change Science Fiction." *Washington Times*, November 26; https://www.washingtontimes.com/news/2015/nov/26/lamar-smith-noaas-climate-change-science-fiction/

Solon, Olivia (2017). "Can Donald Trump Save Twitter?" *Guardian*, January 5; https://www.theguardian.com/technology/2017/jan/05/can-donald-trump-save-twitter

Stahl, Lesley (2018). "President Trump on Christine Blasey Ford, His Relationships with Vladimir Putin and Kim Jong Un and More." *CNN, 60 Minutes*, October 15; https://www.cbsnews.com/news/donald-trump-full-interview-60-minutes-transcript-lesley-stahl-2018-10-14/

Tashman, Brian (2012). *James Inhofe Says the Bible Refutes Climate Change*. Right Wing Watch, March 8; https://www.rightwingwatch.org/post/james-inhofe-says-the-bible-refutes-climate-change/

Taylor, James (2018). *Trump, Educated by Heartland, Makes Bold Pitch for Climate Realism*. Heartland Institute, News and Opinion, October 19; https://www.heartland.org/news-opinion/news/trump-educated-by-heartland-makes-bold-pitch-for-climate-realism#.W83lXvIJiLs.twitter

Thunberg, Greta (2021). Tweet, Twitter, January 20; https://twitter.com/GretaThunberg/status/1351890941087522820

Trump, Donald (2011). Tweet, Twitter, November 1; https://twitter.com/realdonaldtrump/status/131456348822061056

Trump, Donald (2012). Tweet, Twitter, November 6; https://twitter.com/realdonaldtrump/status/265895292191248385?lang=en

Trump, Donald (2014). Tweet, Twitter, January 25; https://twitter.com/realdonaldtrump/status/427226424987385856?lang=en

Trump, Donald (2016). "Donald Trump Slams Iran Deal." *Fox News*, January 18; https://www.foxnews.com/video/4709867547001?intcmp=hphz01

Trump, Donald (2017a). Tweet, Twitter, December 28; https://twitter.com/realDonaldTrump/status/946531657229701120

Trump, Donald (2017b). Tweet, Twitter, December 30; https://twitter.com/realDonaldTrump/status/947235015343202304

Trump, Donald (2018). Tweet, Twitter, November 21; https://twitter.com/realDonaldTrump/status/1065400254151954432

Trump, Donald (2019a). Tweet, Twitter, January 20; https://twitter.com/realDonaldTrump/status/1086971499725160448

Trump, Donald (2019b). Tweet, Twitter, January 28; https://twitter.com/realDonaldTrump/status/1090074254010404864

Trump, Donald (2019c). Tweet, Twitter, February 9; https://twitter.com/realDonaldTrump/status/1094375749279248385

Trump, Donald (2019d). Tweet, Twitter, February 10; https://twitter.com/realDonaldTrump/status/1094718856197799936

Trump, Donald (2019e). Tweet, Twitter, March 12; https://twitter.com/realdonaldtrump/status/1105445788585467904?lang=en

Trump, Donald (2019f). Tweet, Twitter, September 4; https://twitter.com/realDonaldTrump/status/1169356703126773762

Trump, Donald (2019g). Tweet, Twitter, September 23; https://twitter.com/realDonaldTrump/status/1176339522113679360

Trump, Donald (2019h). Tweet, Twitter, October 3; https://twitter.com/realDonaldTrump/status/1179710734030249984

Trump, Donald (2019i). Tweet, Twitter, December 12; https://twitter.com/realDonaldTrump/status/1205100602025545730

Trump, Donald (2020a). Tweet, Twitter, August 18; https://twitter.com/realDonaldTrump/status/1295792202526973952

Trump, Donald (2020b). *President Trump: "I Don't Think Science Knows, Actually."* C-Span, September 14; https://www.youtube.com/watch?app=desktop&v=tRQwqWN5k_M

Twitter (2021). *Permanent Suspension of @realDonaldTrump*. Twitter Inc., January 8; https://blog.twitter.com/en_us/topics/company/2020/suspension.html

United States Department of State (2002). *U.S. Climate Action Report – 2002.* EPA, May; https://unfccc.int/resource/docs/natc/usnc3.pdf

United States House of Representatives (1995). *Scientific Integrity and Public Trust: The Science Behind Federal Policies and Mandates: Case Study2-Climate Models and Projections of Potential Impacts of Global Climate Change*. Hearing before the Subcommittee on Energy and Environment of the Committee on Science, U.S. House of Representatives, 104th Congress, First Session, November 16; https://babel.hathitrust.org/cgi/pt?id=uc1.31210012872972;view=1up;seq=1

United States House of Representatives (2006). *Questions Surrounding the 'Hockey Stick' Temperature Studies: Implications for Climate Change Assessment*. Hearings before the Subcommittee on Oversight and Investigations of the Committee on Energy and Commerce, House of Representatives, 109th Congress, Second Session, July 18 and 27; https://www.govinfo.gov/content/pkg/CHRG-109hhrg31362/pdf/CHRG-109hhrg31362.pdf

United States House of Representatives (2007a). *Allegations of Political Interference with the Work of Government Climate Change Scientists*. Hearing before the Committee on Oversight and Government Reform, House of Representatives, 110th Congress, First Session, January 30; https://babel.hathitrust.org/cgi/pt?id=pst.000061491786&view=1up&seq=1

United States House of Representatives (2007b). *The State of Climate Change Science 2007*. Hearings before the Committee on Science and Technology, 110th Congress, First Session, February 8, April 17, and May 15; https://babel.hathitrust.org/cgi/pt?id=pst.000063517842;view=1up;seq=1

United States House of Representatives (2007c). *Climate Change: Are Greenhouse Gas Emissions from Human Activities Contributing to the Warming of the Planet*? Hearing before the Subcommittee on Energy and Air Quality of the Committee on Energy and Commerce, House of Representatives, 110th Congress, First Session, March 7; https://babel.hathitrust.org/cgi/pt?id=pst.000063522273&view=1up&seq=1

United States House of Representatives (2007d). *Allegations of Political Interference with Government Climate Change Science*. Hearing before the Committee on Oversight and Government Reform, House of Representatives, 110th Congress, First Session, March 19; https://babel.hathitrust.org/cgi/pt?id=pst.000063504637&view=1up&seq=3

United States House of Representatives (2007e). *Climate Change: Perspectives of Utility CEOs*. Hearing before the Subcommittee on Energy and Air Quality of the Committee on Energy and Commerce, House of Representatives, 110th Congress, First Session, March 20; https://babel.hathitrust.org/cgi/pt?id=pst.000065512944;view=1up;seq=1

United States House of Representatives (2007f). *Shaping the Message, Distorting the Science: Media Strategies to Influence Science Policy*. Hearing before the Subcommittee on Investigations and Oversight, Committee on Science and Technology, House of Representatives, 110th Congress, First Session, March 28; https://babel.hathitrust.org/cgi/pt?id=pst.000061515956&view=1up&seq=3

United States House of Representatives (2007g). *Disappearing Polar Bears and Permafrost: Is a Global Warming Tipping Point Embedded in the Ice?* Hearing before the Subcommittee on Investigations and Oversight, Committee on Science and Technology, House of Representatives, 110th Congress, First Session, October 17; https://babel.hathitrust.org/cgi/pt?id=pst.000063523263&view=1up&seq=1

United States House of Representatives (2008a). *Massachusetts V. U.S. EPA, PART II: Implications of the Supreme Court Decision*. Hearing before the Select Committee on Energy Independence and Global Warming, House of Representatives, 110th Congress, Second Session, March 13; https://babel.hathitrust.org/cgi/pt?id=uc1.c106280134&view=1up&seq=3

United States House of Representatives (2008b). *Final Staff Report for the 110th Congress together with Additional Views Submitted by Mr. Markey, Chairman, Select Committee on Energy Independence and Global Warming*. House of Representatives, 110th Congress, Second Session, November 19; https://babel.hathitrust.org/cgi/pt?id=mdp.39015089033834&view=1up&seq=1

United States House of Representatives (2010). *Clean Energy Policies that Reduce Our Dependence on Oil*. Hearing before the Subcommittee on Energy and the Environment of the Committee on Energy and Commerce, House of Representatives, 111th Congress, Second Session, April 28; https://babel.hathitrust.org/cgi/pt?id=umn.31951d036344201&view=1up&seq=5

United States House of Representatives (2013a). *H.R. 3, the Northern Route Approval Act*. Hearing before the Subcommittee on Energy and Power of the Committee of Energy and

Commerce, House of Representatives, 113th Congress, First Session, April 10; https://babel.hathitrust.org/cgi/pt?id=uc1.b5626296&view=1up&seq=1&q1=%22Great%20flood%22

United States House of Representatives (2013b). *Department of Energy Science and Technology Priorities*. Hearing before the Committee on Science, Space, and Technology, House of Representatives, 113th Congress, First Session, June 18; https://www.govinfo.gov/content/pkg/CHRG-113hhrg81725/pdf/CHRG-113hhrg81725.pdf

United States House of Representatives (2013c). *The Obama Administration's Climate Change Policies and Activities*. Hearing before the Subcommittee on Energy and Power of the Committee on Energy and Commerce, House of Representatives, 113th Congress, First Session, September 18; https://www.govinfo.gov/content/pkg/CHRG-113hhrg87109/pdf/CHRG-113hhrg87109.pdf

United States House of Representatives (2016a). *Ensuring Sound Science at EPA*. Hearing before the Committee on Science, Space, and Technology, House of Representatives, 114th Congress, Second Session, June 22; https://www.govinfo.gov/content/pkg/CHRG-114hhrg20914/pdf/CHRG-114hhrg20914.pdf

United States House of Representatives (2016b). *Affirming Congress' Constitutional Oversight Responsibilities: Subpoena Authority and Recourse for Failure to Comply with Lawfully Issued Subpoenas*. Hearing before the Committee on Science, Space, and Technology, House of Representatives, 114th Congress, Second Session, September 14; https://www.govinfo.gov/content/pkg/CHRG-114hhrg22561/pdf/CHRG-114hhrg22561.pdf

United States House of Representatives (2017a). *Make EPA Great Again*. Hearing before the Committee on Science, Space, and Technology, House of Representatives, 115th Congress, First Session, February 7; https://www.govinfo.gov/content/pkg/CHRG-115hhrg24628/pdf/CHRG-115hhrg24628.pdf

United States House of Representatives (2017b). *Climate Science: Assumptions, Policy Implications, and the Scientific Method*. Hearing before the Committee on Science, Space, and Technology, House of Representatives, 115th Congress, First Session, March 29; https://www.govinfo.gov/content/pkg/CHRG-115hhrg25098/pdf/CHRG-115hhrg25098.pdf

United States House of Representatives (2018). *Using Technology to Address Climate Change*. Hearing before the Committee on Science, Space, and Technology, House of Representatives, 115th Congress, Second Session, May 16; https://www.govinfo.gov/content/pkg/CHRG-115hhrg30322/pdf/CHRG-115hhrg30322.pdf

United States House of Representatives (2019a). *The Costs of Climate Change: Risks to the U.S. Economy and the Federal Budget*. Hearing before the Committee on the Budget, House of Representatives, 116th Congress, First Session, June 11; https://www.govinfo.gov/content/pkg/CHRG-116hhrg37609/pdf/CHRG-116hhrg37609.pdf

United States House of Representatives (2019b). *Recovery, Resiliency and Readiness: Contending with Natural Disasters in the Wake of Climate Change*. Hearing before the Subcommittee on Environment of the Committee on Oversight and Reform, House of Representatives, 116th Congress, First Session, June 25; https://www.govinfo.gov/content/pkg/CHRG-116hhrg37262/pdf/CHRG-116hhrg37262.pdf

United States House of Representatives (2019c). *Voices Leading the Next Generation on the Global Climate Crisis*. Hearing before the Subcommittee on Europe, Eurasia, Energy, and the Environment of the Committee on Foreign Affairs, House of Representatives, 116th Congress, First Session, September 18; https://www.govinfo.gov/content/pkg/CHRG-116hhrg37705/pdf/CHRG-116hhrg37705.pdf

United States Senate (1989). *Climate Surprises*. Hearing before the Subcommittee of Science, Technology, and Space of the Committee on Commerce, Science, and Transportation, United States Senate, 101st Congress, First Session, May 8; https://babel.hathitrust.org/cgi/pt?id=uc1.b5176796&view=1up&seq=1

United States Senate (2000). *The Science Behind Global Warming*. Hearing before the Committee on Commerce, Science, and Transportation, United States Senate, 106th Congress, Second Session, May 17; https://babel.hathitrust.org/cgi/pt?id=uc1.b5183343&view=1up&seq=1

United States Senate (2003). *Climate History and the Science Underlying Fate, Transport, Health Effects of Mercury Emissions*. Hearing before the Committee on Environment and Public Works, United States Senate, 108th Congress, First Session, July 29; https://babel.hathitrust.org/cgi/pt?id=uc1.b5139310;view=1up;seq=1

United States Senate (2005). *The Role of Science in Environmental Policy Making*. Hearing before the Committee on Environment and Public Works, United States Senate, 109th Congress, First Session, September 28; https://babel.hathitrust.org/cgi/pt?id=pst.000063509977;view=1up;seq=1

United States Senate (2007a). *Senators' Perspectives on Global Warming*. Hearing before the Committee on Environment and Public Works, United States Senate, 110th Congress, First Session, January 30; https://babel.hathitrust.org/cgi/pt?id=uc1.31822034529552;view=1up;seq=1

United States Senate (2007b). *Climate Change Research and Scientific Integrity*. Hearing before the Committee on Commerce, Science, and Transportation United States Senate, 110th Congress, First Session, February 7; https://www.govinfo.gov/content/pkg/CHRG-110shrg35039/pdf/CHRG-110shrg35039.pdf

United States Senate (2007c). *Vice President Al Gore's Perspective on Global Warming*. Hearing before the Committee on Environment and Public Works, United States Senate, 110th Congress, First Session, March 21; https://babel.hathitrust.org/cgi/pt?id=uc1.31822037831047;view=1up;seq=1

United States Senate (2007d). *The Implications of the Supreme Court's Decision Regarding EPA's Authorities with Respect to Greenhouse Gases Under the Clean Air Act*. Hearing before the Committee on Environment and Public Works, United States Senate, 110th Congress, First Session, April 24; https://babel.hathitrust.org/cgi/pt?id=uc1.31822038354130;view=1up;seq=1

United States Senate (2008). *Effects of Climate Change on Marine and Coastal Ecosystems in Washington State*. Field Hearing before the Subcommittee on Oceans, Atmosphere, Fisheries, and Coast Guard of the Committee on Commerce, Science, and Transportation, United States Senate, 110th Congress, Second Session, May 27; https://www.govinfo.gov/content/pkg/CHRG-110shrg75204/pdf/CHRG-110shrg75204.pdf

United States Senate (2013). *Climate Change: It's Happening Now*. Hearing before the Committee on Environment and Public Works, United States Senate, 114th Congress, First Session, July 18; https://www.govinfo.gov/content/pkg/CHRG-113shrg95976/pdf/CHRG-113shrg95976.pdf

United States Senate (2015a). *Oversight Hearing: The President's Fiscal Year 2016 Budget Request for the U.S. Environmental Protection Agency*. Hearing before the Committee on Environment and Public Works, United States Senate, 114th Congress, First Session, March 4; https://www.govinfo.gov/content/pkg/CHRG-114shrg94980/pdf/CHRG-114shrg94980.pdf

United States Senate (2015b). *Data or Dogma? Promoting Open Inquiry in the Debate over the Magnitude of Human Impact on Earth's Climate*. Hearing before the Subcommittee on Space, Science, and Competitiveness of the Committee on Commerce, Science, and Transportation, United States Senate, 114th Congress, First Session, December 8; https://www.govinfo.gov/content/pkg/CHRG-114shrg21644/html/CHRG-114shrg21644.htm

United States Senate (2017a). *Letter to President Trump.* May 25; https://www.inhofe.senate.gov/imo/media/doc/Paris%20letter.pdf

United States Senate (2017b). *Nomination of Kirstjen M. Nielsen.* Hearing before the Committee on Homeland Security and Governmental Affairs, United States Senate, 115th Congress, First Session, November 8; https://www.govinfo.gov/content/pkg/CHRG-115shrg30099/pdf/CHRG-115shrg30099.pdf

United States Senate (2019a). *The Electricity Sector in a Changing Climate.* Hearing before the Committee on Energy and Natural Resources, United States Senate, 116th Congress, First Session, March 5; https://www.govinfo.gov/content/pkg/CHRG-116shrg35558/pdf/CHRG-116shrg35558.pdf

United States Senate (2019b). *States' Role in Protecting Air Quality: Principles of Cooperative Federalism.* Hearing before the Subcommittee on Clean Air and Nuclear Safety of the Committee on Environment and Public Works, United States Senate, 116th Congress, First Session March 5; https://www.govinfo.gov/content/pkg/CHRG-116shrg35947/pdf/CHRG-116shrg35947.pdf

Waldman, Scott (2019). "Scientist Who Rejects Warming Is Named to EPA Advisory Board." *Scientific American, E&E News,* February 1; https://www.scientificamerican.com/article/scientist-who-rejects-warming-is-named-to-epa-advisory-board/

Waldman, Scott (2020). "Trump Team Delaying Work on Major Climate Report." *E&E News,* October 5; https://www.eenews.net/stories/1063715433

Waldman, Scott, and Hulac, Benjamin (2018). "This is When the GOP Turned Away from Climate Policy." *E&E News,* December 5; https://www.eenews.net/stories/1060108785/

Waxman, Harry A. (2013). "Energy and Water Development and Related Agencies Appropriations Act, 2014." *Congressional Record – Extensions of Remarks,* p. E1048, July 9; https://www.govinfo.gov/content/pkg/CREC-2013-07-11/pdf/CREC-2013-07-11-pt1-PgE1048.pdf

Weaver, Matthew, and Lyons, Kate (2019). "Donald Trump Tells Prince Charles US has 'Clean Climate.'" *Guardian,* June 5; https://www.theguardian.com/us-news/2019/jun/05/donald-trump-tells-prince-charles-us-is-clean-on-climate-change

Whitehouse, Sheldon (2015). "Sense of the Senate Regarding Climate Change." *Congressional Record – Senate,* p. 771, January 20; https://www.govinfo.gov/content/pkg/CRECB-2015-pt1/pdf/CRECB-2015-pt1-Pg760-2.pdf

Wise, Justin (2018). "Al Gore Rips Trump as the Face of 'Climate Denial.'" *The Hill,* November 29; https://thehill.com/policy/energy-environment/418841-al-gore-tears-into-trump-he-is-the-face-of-climate-denial

Woodwell, George M., MacDonald, Gordon J., Revelle, Roger, and Keeling, C. David (1979). *The Carbon Dioxide Problem: Implications for Policy in the Management of Energy and Other Resources.* Council on Environmental Quality, July; http://graphics8.nytimes.com/packages/pdf/science/woodwellreport.pdf

5 The Denial Machine

The climate brawl has been examined in Twitter and in the political arena with the contrarians and the denial cabal. In this last chapter, the focus is on the corporate climate deniers, the rise of their political influence and the denial machine. Unlike with the contrarians and the denial cabal, Democrats and scientific witnesses were more willing to challenge members of the energy-industrial complex, which are discussed in detail here to highlight the climate denial counter movement, known as climate brawl. The energy-industrial complex was the source of the propaganda for climate denial that had targeted conservative ideology and first whipped the right into a frenzy against policies to restrict fossil fuels over three decades ago and which continues today. Until their propaganda is muted, the necessary political negotiations will not happen on the climate crisis.

The Rise of Climate Denialism within the Energy-Industrial Complex

Pollution associated with petroleum has long been a focus of the American Petroleum Institute (API) since its formation in 1919. After World War II, atmospheric pollutants, such as smog, became an issue for the API. In response, in 1951, the API formed the Smoke and Fumes Committee which was created to protect their interests and ensure[1]:

> proper decisions with regard to the control of atmospheric pollutants could not be made without accurate scientific information … based on reliable research rather than on theory or hypothesis should be available to government organizations to avert restrictive and uneconomic rulings of the type that had proved unnecessary in the past.

The Smoke and Fumes Committee warned of "Panicky People," for[2]:

> The worst thing that can happen, in many instances, is the passage of a law or laws for the control of a given air pollution situation. For in many such cases their passage results from a panicky feeling … And passing the wrong law can also often be the most expensive and irritating way to approach the problem.

DOI: 10.4324/9781003455417-6

Soon, the API and their members were looking at a different type of atmospheric pollution also caused by the burning of fossil fuels that could not only affect their profits but put their entire industry at risk. The scientific prowess of the petroleum industry was top notch, and their astute researchers had identified the threat to their industry from the modern science of climate change which started to be published in the 1950s.[3] In 1957, scientists from Humble Oil had studied[4]: "the fate of the enormous quantity of carbon dioxide which has been introduced into the atmosphere since the beginning of the industrial revolution;" these researchers had discussed their work in private communications with Roger Revelle.[5]

Meanwhile, the famous scientist Edward Teller (1908–2003)[6] had warned about the danger of carbon dioxide being spewed into the atmosphere from the burning of fossil fuels (that this would cause a greenhouse effect and warm the planet) at a conference, which was picked up by the media.[7] His warnings of a future catastrophe had caught the attention of the vigilant petroleum industry. On October 8, 1959, a rebuttal from Marcus Aurelius Matthews of Shell appeared in the popular science magazine *New Scientist*, where the Shell researcher argued that global warming would not be a problem for the foreseeable future and wrote that there was[8]: "no reason to prophesy any drastic upset in the Earth's carbon cycles due to the amount of fuel burned so far." Also, the Shell article concluded[9]: "There seem few grounds for believing that our furnaces and motor car engines will have any large effect on the carbon dioxide balance. The only alarming feature is that fossil fuels are presumably being used up."

As the petroleum industry was keen to hear directly from the nuclear physicist himself, Dr. Teller had been invited to speak about energy in the future at the special gala of the 100th anniversary of the American petroleum industry, on November 4, 1959. He presented the same message from the earlier conference, except now for the executives of Big Oil[10]: "It has been calculated that a temperature rise corresponding to a 10 per cent increase in carbon dioxide will be sufficient to melt the icecap and submerge New York. All the coastal cities would be covered." Well, that must have muted their celebrations, and more depressing news for the petroleum industry was on the way.

A few years later, the petroleum industry got another bitter dose of greenhouse gas medicine; in 1965, at another meeting of the American Petroleum Institute, the president Frank Ikard (1913–91) brought the following message about a study sent to President Johnson[11] to its members[12]:

> One of the most important predictions of the report is that carbon dioxide is being added to the atmosphere by the burning of coal, oil, and natural gas at such a rate that by the year 2000 the heat balance will be so modified as possibly to cause marked changes in climate.

A 1969 study was prepared for the American Petroleum Institute called "Sources, Abundance, and Fate of Gaseous Atmospheric Pollutants," which raised more concerns[13]:

> On the basis of our present knowledge, significant temperature changes could be expected to occur by the year 2000 as a result of increased CO_2. in the atmosphere. These could bring about long term climatic changes … the prospect for the future must be of serious concern.

The 1969 prediction that atmospheric levels of carbon dioxide would reach 370 ppm with a corresponding temperature rise of 0.5°C in 2000 has since been observed to be correct.

By the time of President Reagan, even scientists who worked for Big Oil were informing senior management of the alarming consequences of climate change. In a study by the Exxon Research and Engineering Company on November 12, 1982, an internal report of the greenhouse effect had concluded that carbon levels in the atmosphere would double by the end of the 21st century (using Exxon energy forecasts), which would increase the temperature by 1.3°C–3.1°C, and the analysis included this warning[14]: "The 'greenhouse effect' is not likely to cause substantial climatic changes until the average global temperature rises at least 1°C above today's levels." A graph in the Exxon study projected a rise in global temperatures of 1°C by 2020, when carbon dioxide in the atmosphere would reach 420 ppm (which has turned out to be very close to the mark). Big Oil knew global warming was around the corner and understood only too well the threats to their business and the global community. Unfortunately, the decisions made by the energy-industrial complex ensured that global warming would arrive on schedule, as predicted in the Exxon report.[15]

At the end of the 1980s, an internal study from Shell predicted more grim consequences from global warming[16]:

> There would be more violent weather – more storms, more droughts, more deluges … The changes would, however, most impact on humans … The potential refugee problem in GLOBAL MERCANTILISM could be unprecedented … Conflicts would abound. Civilisation could prove a fragile thing.

In 1980, the American Petroleum Institute made its first serious foray against the science of climate change in a book called *Two Energy Futures: A National Choice for the 80s*.[17] The book downplayed the threat of climate change and promoted the use of coal for synthetic fuels in response to the energy crisis that had hit America in the late 1970s. The API was just getting started in what would become a core activity for them – producing propaganda. This new modus operandi may have been encouraged by a business philosophy that had been introduced in the previous decade.

An article in the *New York Times* appeared on September 13, 1970, "A Friedman Doctrine – The Social Responsibility of Business is to Increase Its Profits," which outlined this new direction. Milton Friedman (1912–2006), an economist, wrote that corporations which had a[18]: "social conscience … were preaching pure and unadulterated socialism … social responsibilities … are the responsibilities of individuals, not of businesses." He specifically mentioned that pollution was the responsibility of government legislation and not for businesses to voluntarily control. The American economist, in short, promoted profits over people. Either by coincidence or design, the propaganda campaign against the science of climate change implemented by the energy-industrial complex came right out of the Friedman doctrine.[19] The details on how corporate science denialism created the tragedy of the climate crisis are discussed below.

The Denial Machine

At the end of Reagan's term, the energy-industrial complex was being hammered by a perfect storm on the climate front: Hansen's high-profile Senate presentation, the formation of the IPCC, and the upcoming Earth Summit for Rio. Big Oil had to react as a consensus on the science of climate change was being forged that greenhouse gas emissions would have to be reduced, creating a grave outlook for their industry. Rather than planning for an organized transition away from fossil fuels to non-carbon emitting technologies, the energy-industrial complex made the decision to fight dirty against the science: deny the harm, deny the science, deny the need for policies, deny, deny, deny. The energy-industrial complex chose profits over people; Milton Friedman would have approved.

In order to protect their profits, a cunning and devious strategy was rolled out by the energy-industrial complex to undermine the science of climate change. The multi-million-dollar ploy involved a slick PR campaign, the funding of politicians, contrarians and right-wing think tanks, media manipulation, and lobbying. Later, the Union of Concerned Scientists chastised Big Oil[20]: "It is not acceptable for fossil fuel companies to deny the science, nor is it acceptable for them to publicly accept the science while funding climate contrarian scientists or front groups that distort or deny the science."

The well-oiled operations behind one of the greatest propaganda campaigns in history became known as the denial machine in 2006, when *The Fifth Estate* of CBC broadcasted an episode on their program called the "Denial Machine:"[21]

> The *Denial Machine* investigates the roots of the campaign to negate the science and the threat of global warming … The documentary shows how fossil fuel corporations have kept the global warming debate alive long after most scientists believed that global warming was real and had potentially catastrophic consequences.

What does the denial machine manufacture? Simply put, it produces tons of doubt: doubt about the science, doubt about scientists, and doubt about the need to reduce fossil fuels.[22] The objective of the campaign was not to win the climate change debate (which could not be done) but to delay any policies that would restrict fossil fuels for as long as possible; this was a war of attrition that would involve many groups[23]: "This counter-movement involves a large number of organizations, including conservative think tanks, advocacy groups, trade associations and conservative foundations, with strong links to sympathetic media outlets and conservative politicians."

Big Oil poured money into right-wing think tanks,[24] financing these special-interest groups which lobbied directly or through trade associations, such as the American Petroleum Institute (API).[25] The two most aggressive think tanks against the climate sciences were the George C. Marshall Institute[26] and the Heartland Institute[27]; ironically, both were founded in 1984, a year made famous by the dystopian novel by George Orwell (1903–50), *Nineteen Eighty-Four*, about an authoritarian state propped up by a sophisticated denial machine, known as the "Ministry of Truth."

Politicians had been the initial target of the propaganda campaign by the energy-industrial complex, but, after Kyoto, a "Global Climate Science Communications Action Plan" had been proposed to the API that included the public as well[28]:

Victory Will Be Achieved When

- Average citizens "understand" (recognize) uncertainties in climate science; recognition of uncertainties become part of the "conventional wisdom"
- Media "understands" (recognize) uncertainties in climate science
- Media coverage reflects balance on climate science and recognition of the validity of the viewpoints that challenge the current "conventional wisdom"
- Industry senior leadership "understands" uncertainties in climate sciences, making them strong ambassadors to those who shape climate policy
- Those promoting the Kyoto treaty on the basis of extant science appear to be out of touch with reality.

Furthermore, a major aspect of the strategy was to recruit "true climate scientists" (which was basically "code" for contrarians)[29]:

> Identify, recruit and train a team of five independent scientists to participate in media outreach. These will be individuals who do not have a long history of visibility and/or participation in the climate change debate. Rather, this team will consist of new faces who will add their voices to those recognized scientists who already are vocal.
>
> Develop a global climate science information kit for media including peer-reviewed papers that undercut the "conventional wisdom" on climate science …

The API officially denied implementing the proposed plan.

Global Climate Coalition

A powerful industry juggernaut, the Global Climate Coalition (GCC)[30] whose members were from the energy-industrial complex, as well as related corporations and trade associations, was a lobbying group extraordinaire with immense political clout. The GCC had been set up in 1989 mainly to sabotage the Kyoto negotiations,[31] and they funded a massive PR campaign to cast doubt on the science.[32]

The GCC was represented by a witness at a Senate hearing in 1997. The session was opened by John Chafee (1922–99) (Republican from Rhode Island) with the following remarks[33]: "This morning, we will receive testimony on one of the most important and challenging environmental, economic and political matters of our time." Contrarians at the session included John Christy, Richard Lindzen, and William O'Keefe – the chairman of the Global Climate Coalition – who stated[34]:

> Climate policy is simply not a dichotomy of action versus no action. We agree that action is justified but reject the course being pursued in international negotiations. It is an unjustified rush to judgment.
>
> The major difference between the GCC and our understanding of the Clinton administration is over approach, not need. We believe that a wise policy on climate change is akin to driving in a thick fog. The prudent course of action is to proceed at a speed consistent with how well the car's headlights illuminate the road ahead.
>
> The Administration approach appears akin to driving full speed on the Autobahn on a clear day with no reason for caution. That approach is flawed and risks a fatal crash.

The GCC message was that the Clinton administration and the Kyoto negotiations were moving too quickly to restrict greenhouse gas emissions.

The GCC pursued this slow-down strategy even though their own experts agreed with the climate sciences. A draft report from the GCC – "Predicting Future Climate Change: A Primer" issued in 1995 – contained a side-by-side comparison of "contrarian theory" versus "counter-arguments" (i.e., mainstream science).[35] The internal document stated that the contrarians did "not offer convincing arguments" which just confirmed that the battle with science could never be won directly, hence the necessity for a propaganda campaign to create doubt.

At the start of the millennium, the GCC issued "The Global Climate Coalition's 21st Century Climate Action Agenda," which defined its raison d'être and included the following platitude that would later become common among political climate deniers (emphasis added)[36]: "The GCC will continue working with policy makers in support of a balanced international approach to address concerns about the climate that **does not harm the U.S. economy or hurt American workers and families**." The GCC strategy was presented to a Senate hearing, "Solutions to Climate Change," in September 2000.[37]

The above document also had been forwarded to the United States Department of State in 2001 as background material for a meet-and-greet between the GCC and the Under Secretary of Global Affairs Paula Dobriansky (1955–) at the offices of the American Petroleum Institute. Prior to the meet-and-greet, the Department of State had been informed by the GCC that[38]:

> GCC members are completely supportive of the Administration's position on climate change and the rejection of the Kyoto Protocol ... GCC participants, some of whom are scientific experts, will state that they are 100% behind the remarks articulated by the President [George W. Bush] on climate change policy.

The GCC closed operations the next year as its mission, by then, had been largely completed, and the Kyoto Protocol was dead as far as the Americans were concerned.

Dark Money

The energy-industrial complex continued their spending spree but, by the end of George W. Bush's second term, had switched to "dark money." An analysis of the tax returns of conservative think tanks and trade associations was carried out for 2003–2010,[39] and found that funding from the ExxonMobil Foundation had ceased, and from the Koch Foundations[40] had sharply declined, towards the end of that period. Nevertheless, Senator Ed Markey (1946–) (Democrat from Massachusetts) tweeted[41]: "In 2009/10, the Koch brothers, #BigOil, and others spent a fortune to kill climate change legislation." So, what was going on? A protected tax class, known as "donor directed" foundations, meant that the identities of all donors were no longer available.[42] Untraceable philanthropic funding was becoming more common place, leading to "immense influence on U.S. politics," especially on the issue of climate change.[43]

The flow of dark money into Washington was harshly denounced by Senator Sheldon Whitehouse (1955–) (Democrat from Rhode Island)[44]: "The fossil fuel industry's dark money has polluted our politics as badly as its carbon emissions have polluted our atmosphere and oceans." He associated the failure of Congress to act on climate change with the "secret empire of dark money." More criticism came from the report by the Senate Democrats' Special Committee on the Climate Crisis[45]:

> These executives used weak American laws and regulations governing election spending, lobbying, and giving to advocacy groups to mount a massive covert operation. Their goal was to spread disinformation about climate change and obstruct climate action. The covert campaign spanned at least three decades …
>
> Fossil fuel executives realized that they only needed to keep one party in line, especially given Senate procedural rules that make it difficult to pass legislation without at least some bipartisan support. They made the strategic decision to target Republican officials. As the party traditionally more aligned with business interests – and more dependent on business interests for political funding – they were also an easier target …
>
> Under the pressure of reality, the climate denial and obstruction strategy is finally crumbling. But it is not yet dead.

The report summarized the shadowy history of Big Oil distorting the truth about the science of climate change.

The petroleum industry belongs to trade associations that do the dirty work of undermining climate policymaking[46]; in October 2020, the American Petroleum

Institute (API), for example, rolled out more propaganda to influence the electorate, and its president explained[47]: "The wrong policy choices could cost hundreds of thousands of good-paying jobs and millions in revenue for states, all at a time when Americans are working toward an economic recovery." Stoking the fears of job losses and economic calamity remain a common strategy of the denial machine. In 2021, Senator Whitehouse was again pointing out the "climate misinformation" from the API[48]: "The truth is that they are driving our planetary crisis toward a breaking point ... don't be fooled by this happy talk."

The Energy-Industrial Complex Engineered the Climate Crisis

World leaders met first at Rio in 1992 and agreed that climate change was serious enough to be addressed by all nations, then again in Kyoto in 1997, and in Paris in 2015. Two presidents of the United States (one Republican and one Democrat) had signed international agreements that America would do its part to reign in greenhouse gas emissions, but the American government's inaction has allowed climate change to run amok, morphing into a climate crisis. What went wrong?

By the 1980s, the energy-industrial complex knew that the science was irrefutable, and they could not win the fight over climate change, but they could drag out the discussions for as long as possible, which they have done for the last few decades, especially by promoting climate denial. The laissez-faire, pro-business policies of Ronald Reagan and the Friedman doctrine granted the energy-industrial complex license to wage a no-holds-barred war against the climate sciences, as petro-dollars were poured into the denial machine. Ultimately, the energy-industrial complex gained undue influence, if not control, over the energy and environmental agenda of Republican administrations.

In a 2021 interview, Noam Chomsky exposed what the Republican Party had become[49]:

> The only organization in human history that is dedicated with passion to ensuring human, that survival, survival of organized human society, will be impossible; that is exactly their program, since 2009 when they shifted to a denialist position under the impact of the Koch Brothers juggernaut.

He reiterated their wicked intent the following year[50]:

> [The Republican Party] is a political organization that is committed, dedicated to destroying organized human life as quickly as possible. They are a denialist party, either denying that global warming is taking place or saying we shouldn't do anything about it.

By promoting climate denial in American politics, powerful vested interests have been successful in blocking political action for so long that climate change has become the climate crisis.

The Denial Reign of Terror Against Science

The evidence is clear on the role of the energy-industrial complex[51]:

> Corporations, especially the fossil fuel industry, have spent huge sums attempting to obfuscate the reality of climate change … To some extent, then, we are the victims of a well-funded and sophisticated misinformation campaign that attempts to keep us in the dark about climate change.

A prime target of the denial machine was the IPCC, where every spelling mistake in their reports was used to create doubt about the science among the general public and politicians. The slightest hint of any irregularity was taken as proof of the corruption of leading climate scientists. A series of conspiracy theories against climate scientists (and, especially, the IPCC) emerged during the George W. Bush administration and reached a peak in the early term of President Obama.

Conspiracy theories have been a standard ploy of climate denialism, but the most underhanded of them all was the purported "climategate." In late 2009, a malicious attack had taken place against authors associated with the AR4 of the IPCC, when emails had been illegally hacked from the Climate Research Unit of the University of East Anglia. In an attempt to discredit leading climate scientists, snippets of these cherry-picked personal communications were then broadcasted across the internet. Several official investigations in the UK and the U.S. found no misconduct by the scientists.[52] "Climategate" happened just before the important COP meeting in Copenhagen which was an interesting coincidence. The culprits responsible for the data breach have never been identified, but the energy-industrial complex certainly gained the most from the illegal hacking activity.

The "climategate" conspiracy was followed by a campaign of intimidation against leading climate scientists by the political elite.[53] Had such harassment been carried out by one rogue politician, it could have been dismissed, but the enthusiastic persecution of scientists by several high-ranking Republican officials[54] suggested a level of command and control for the benefit of Big Oil and other members of the energy-industrial complex. There was one scientist, the prize-winning Michael E. Mann (1965–), who has felt the full wrath of the denial cabal.[55] Dr. Mann may well be the most hated scientist in conservative America; climate scientist Ben Santer explained[56]: "You go after the things that are important, that are iconic, that are visual, visceral, powerful, and easily interpretable … And if you can't attack the underlying science, you go after the scientist."

The high-level political bullying launched a literal reign of terror against scientists. According to NASA climate scientist Gavin Schmidt[57]: "You can't have a spelling mistake in a paper without it being evidence on the floor of the Senate that the system is corrupt," and another scientist Paul Ehrlich confessed[58]: "Everyone is scared shitless, but they don't know what to do." The science-hate speech from the denial cabal members of the House, Senate, and even the White House, turned public opinion against science.[59] Social scientists raised the alarm over the politically-driven assault on scientists[60]:

> Conservatives attempted to intimidate these high-profile scientists, and sully their image, with the aim of disparaging mainstream climate science more generally by association. While their tactics produced no evidence of scientific wrongdoing, their public accusations nevertheless were enough to cast doubt on climate science within Congress.

Members of the National Academy of Sciences were even compelled to issue an extraordinary public appeal about the abuse against the scientists[61]:

> We also call for an end to McCarthy-like threats of criminal prosecution against our colleagues based on innuendo and guilt by association, the harassment of scientists by politicians seeking distractions to avoid taking action, and the outright lies being spread about them.

In *Carbon Politics and the Failure of the Kyoto Protocol*, I expressed my feelings on the troubling series of events[62]:

> Does it seem right that American citizens, many internationally recognized for their contributions to science, are bullied by elected officials? At this political level, the skeptics movement is a travesty of justice, not only against the scientific community but the rights of citizens in general, that is unprecedented in a democracy.

Years later, the attacks by the denial cabal on scientists have persisted. Naomi Oreskes warned about this on-going malicious campaign against science in America[63]:

> There is a crisis in US science … The crisis is the attempt to discredit scientific findings that threaten powerful corporate interests … One tactic exploits the idea of scientific uncertainty to imply there is no scientific consensus. Another, seen in the latest efforts, insinuates that relevant research might be flawed.

The Bulletin of the Atomic Scientists also protested the horrendous inquisition of climate scientists by the denial cabal[64]:

> In the United States, there is active political antagonism toward science and a growing sense of government-sanctioned disdain for expert opinion, creating fear and doubt regarding well-established science about climate change and other urgent challenges. Countries have long attempted to employ propaganda in service of their political agendas. Now, however, the internet provides widespread, inexpensive access to worldwide audiences, facilitating the broadcast of false and manipulative messages to large populations and enabling millions of individuals to indulge in their prejudices, biases, and ideological differences.

Why would political elites participate in the character assassination of leading scientists and seek to destroy the credibility of science? History has documented similar government-led persecutions of science and scientists, but only in totalitarian states, such as Stalinist Russia and Nazi Germany. How could this have happened in the most powerful democracy in the world?

The Oiligarchs and the Deep State of Denial

Propaganda is the voice of tyranny, not democracy. How far has the siege on American democracy by the propaganda of climate denial progressed? During the administration of George W. Bush, climate denialism infiltrated the Republican Party, and the political power of the energy-industrial complex in Washington reached new heights which has been called an "oiligarchy."[65] While the oiligarchs of the energy-industrial complex had been developing unprecedented power inside the George W. Bush administration, they appeared to be even controlling aspects of governance under President Trump. In his first State of the Union Address, President Trump announced[66]: "We have ended the war on American energy, and we have ended the war on beautiful, clean coal." Trump's vision for America was clear: Science would not be allowed to interfere with the profits of the fossil fuel industry. There is no doubt that the powerful energy-industrial complex with its climate denial tentacles had penetrated deep within the political establishment of the United States of America.

The rise of the oiligarchs did not go unnoticed and resulted in several climate brawls. An op-ed by the acclaimed economist Paul Krugman (1953–) criticized the Bush administration and Big Oil for their failure to act on climate change[67]: "But the fact is that whatever small chance there was of action to limit global warming became even smaller because Exxon Mobil chose to protect its profits by trashing good science." Similar protests were later made by Al Gore (1948–) about the funding of right-wing think tanks by the energy-industrial complex and their control of Congress.[68]

Congress was an active site for such climate brawls. Henry Waxman (1939–) (Democrat from California) exposed the nature of the propaganda campaigns of the energy-industrial complex[69]:

> For decades the tobacco industry mounted a disinformation campaign to create doubt about the dangers of smoking. Major energy industries are now trying the same approach about the consequences of global warming. But no one should be deceived: global warming is real and it is an enormous threat to our Nation … We are literally mortgaging our children's future so that we can continue to consume unlimited amounts of fossil fuels.

At another hearing, Mr. Waxman accused the administration of being in cahoots with the oil industry[70]:

> President Bush and Vice President Cheney came into office determined to radically change the Nation's energy policy, and that is what they did. They

crafted their policy with oil companies like Exxon and Mobil and refused to meet with consumer or environmental groups. Their plan bestowed countless favors on oil, coal, and other polluting industries and it abandoned the President's pledge to reduce greenhouse gas emissions. In fact, under the plan they developed, we have wasted precious years and exacerbated global warming.

Bush's vice president Dick Cheney had welcomed the American Petroleum Institute and the Global Climate Coalition to help guide national energy and climate policy.[71] In the Staff Report of the Select Committee on Energy Independence and Global Warming, accusations were made directly against the vice president[72]: "Oil industry lobbyists argued against regulatory action with the support of the Office of Vice President Cheney." Later, Michael E. Mann denounced Cheney's connections to the fossil fuel industry.[73]

Speaking further about the energy-industrial complex, Dr. Mann condemned them for charting a devastating course[74]:

> They have used their immense resources to create fake scandals and to fund a global disinformation campaign aimed at vilifying the scientists, discrediting the science, and misleading the public and policymakers. Arguably, it is the most villainous act in the history of human civilisation.

And, in a hearing, Senator Bernie Sanders (1941–) (independent from Vermont) voiced his bewilderment over their influence[75]:

> This truly is an Alice in Wonderland hearing. Within this little room we are clearly living in two separate planets. Two separate worlds. And whether the differences are influenced by the fact that the Koch brothers at ExxonMobil and the petroleum industry and fossil fuel industry is pouring hundreds of millions of dollars into think tanks trying to confuse the American people or are we just dealing with politics here, I really do not know.
>
> But to deny the fact that the overwhelming, overwhelming majority, some 98 percent of scientists who have published peer reviewed articles, believe not only that global warming is real, but that it is manmade.

In a later Senate hearing, Senator Sanders again challenged the fossil fuel industry[76]:

> In my view, we have a fundamental choice to make. We can listen to the fossil fuel industry and climate deniers and not worry about the impact of climate change. Or we can listen to the scientists who tell us that we have got to act boldly and aggressively to prevent a climate catastrophe.
>
> In my view, we have spent far too long and wasted too much time discussing whether or not climate change is real. This debate was not driven by science but by a decades-long campaign of lies, distortion, and deceit funded by the fossil fuel industry …

> In my view, we have got to make it clear to the fossil fuel industry that their short-term profits are not more important than the future of our planet.

Besides climate policymaking, American democracy itself was under attack. Chuck Schumer (1950–) (Democrat from New York) defended the Constitution when he tweeted[77]: "We're launching a campaign to highlight how courts—with help from Sen. McConnell & Senate GOP—have been captured by dark money. There's a twisted web of dark money invested to plant ideological activist judges & rewrite the Constitution." In fact, climate denialism had infiltrated all three branches of government – the legislative (Congress) and the executive (White House) as discussed previously, and even the judicial (including the Supreme Court). Democrats prepared a damning report on the judicial branch called "Captured Courts, the GOP's Big Money Assault On the Constitution, Our Independent Judiciary and the Rule of Law:"[78]

> Under the Trump Administration, the Mitch McConnell-led Senate has produced few significant legislative accomplishments. Instead, it has prioritized packing the judiciary with far-right extremists, who then enjoy life tenure as federal judges. Working hand-in-hand with the administration and anonymously-funded outside groups, the Senate has confirmed **200 new life-tenured federal judges to aggressively remake the federal courts and rewrite the Constitution**. Most of these judges were chosen not for their qualifications or experience—which are often lacking—but for their demonstrated allegiance to Republican Party political goals. These judges have already begun rolling back the clock on civil rights, consumer protections, and the rights of ordinary Americans, reliably putting a thumb on the scale in favor of corporate and Republican political interests. From the Supreme Court on down, the special interests responsible for these judges' selection and confirmation are effectively capturing the judicial branch, packing our courts with politicians in robes … [many pages]
>
> The Chamber also actively lobbies Congress to confirm judges and justices who it believes are likely to rule in line with its pro-business, anti-climate, anti-worker agenda.

Subsequently, Republicans rammed through the appointment of climate deniers to the Supreme Court.[79] The stacking of the Supreme Court with conservatives had immediate benefits to the energy-industrial complex, as the EPA was stripped of its power to regulate greenhouse gas emissions from power plants.[80]

The Brawling Senator

Climate brawls from Senator Sheldon Whitehouse have been discussed previously, but the climate brawls for which he is especially known are those with the energy-industrial complex; he was appalled by these special interest groups as the following shows[81]:

> Let me just briefly welcome our witnesses to this chamber, one in which reality is so often suspended, one in which science is so often twisted and mocked and one in which the power of special interests to manipulate American democracy is often so nakedly revealed. My belief is that the propaganda machine behind the climate denial effort will go down in history as one of our great American scandals.

At the same hearing, Barbara Boxer (1940–) (Democrat from California) had connected the energy-industrial complex to the contrarians[82]:

> And I can tell you that most scientists who say no to climate change have ties to big oil and coal polluters, including the scientist [John Christy] who was mentioned here today by Senator Wicker. We checked it out. He is from a think tank that is funded by the Koch brothers.

Senator Boxer added[83]: "If you are a climate denier, you are out of touch. So I wish this committee would find the common ground with the American people."

In 2015, Sheldon Whitehouse was back with another hard-hitting brawl (one of my favorites), this time from the Senate floor[84]:

> But the Senate is jammed by persistent, meretricious climate denial. The denial comes in many guises, but, like a compass, all the denial points in the same direction: whatever helps the fossil fuel industry keep polluting. That is the true north of climate denial– whatever helps the fossil fuel industry. Look at the fossil fuel money pouring into the Republican Party and tell me this is a coincidence.
>
> We have Senators who deny that anything is happening, who say it is a hoax. We have Senators who deny that we can solve this. We have Senators who deny their faith in the American economy to win if we innovate. We have Senators who simply shrug and say: I am not a scientist. A bunch of Senators say: Don't even worry about it; climate change has stopped …
>
> The enterprise that performs that evil task of feeding mischief into this debate is perhaps the biggest and the most complex racket in American history. It is phony. They cherry-pick a handful of statistically insignificant data points and tell us the whole problem went away on its own. Then the real scientists take a look at it and say that is bunk. But in the meantime, the polluter enterprise notched a public relations victory. It bought some time to keep polluting for free, and sadly it got some of our colleagues to be party to it.

In 2019, he continued his crusade against the power of the oiligarchy in Washington[85]:

> After *Citizens United*, we see immensely powerful, climate denying, dark-money front groups, all likely funded by fossil-fuel interests. And we see no

> Republican senator willing to cross them. The spending, and the more silent threat of spending, is a blockade. It reeks.
>
> There's a case study in how dark and unlimited money played in the 2016 Senate elections …
>
> If the administration's 2018 regulatory actions read like a fossil fuel industry wish list, it's because they are.

The persistent assault by the energy-industrial complex on climate science and American democracy never abated, nor did the climate brawls from Senator Whitehouse[86]:

> This is the cliff fossil fuel companies are driving the world toward, while executives in the rest of corporate America wring their hands. Even the companies that have committed to cutting their emissions have failed to prompt any serious action in one place that matters: Washington. In the halls of Congress, I see firsthand where big companies are putting their lobbying might, and it's not behind climate legislation. To get us to where the science tells us we need to be, current corporate efforts are not going to cut it. In fact, taken as a whole, corporate America is doing more political harm than good in Congress.
>
> That's because the fossil fuel industry still funds a vast anti-climate lobbying and political spending machinery …
>
> The bigger problem, however, is that America's top corporate trade associations follow the fossil fuel industry's lead. Such trade groups wield massive power in Washington because of the money they spend on lobbying, elections, and other influence activities, and because of their carefully cultivated relationships with key policy makers.

In 2021, Senator Whitehouse was pushing for a congressional investigation of "climate corruption" by the Trump administration, for their ties to the fossil fuel industry.[87] The tenacious senator from Rhode Island marked his 280th climate speech on the Senate floor in early 2022 with another blistering attack against the fossil fuel industry, which included[88]:

> The fossil fuel industry is still at it; they've just changed it up a bit.
>
> They can't debate the science any more, and they can't argue against the urgency, but they can still write checks. They can fund phony front groups and fill Republican campaign coffers. And though they can't sell climate denial, they can buy climate delay …
>
> The industry doesn't just lie, and pay politicians — fossil fuel companies also use trade associations and dark-money front groups to whip up opposition to climate legislation …

Then, on April 23, 2023, in another speech in the Senate, he added[89]: "One near constant in these speeches has been the oily, often covert hand of the fossil fuel industry lurking behind the opposition to climate action through its campaign of climate denial, delay, and obstruction."

Sheldon Whitehouse is an exceptional champion of the climate brawl with the energy-industrial complex.

The Brawling Congressional Committee

An unprecedented series of climate hearings on the oiligarchs were carried out by the Committee on Oversight and Reform, under the leadership of Carolyn B. Maloney (1946–) (Democrat from New York). In a 2019 hearing called, "Examining the Oil Industry's Efforts to Suppress the Truth about Climate Change," Jamie Raskin (1962–) (Democrat of Maryland) opened the session with an informative and masterful climate brawl on the threat of the energy-industrial complex to democracy[90]:

> Oil companies like Exxon knew the scientific reality 40 years ago but waged a war of deception that cost us precious time in the fight to save our planet …
>
> But this was not the path that Exxon chose. Instead, it sold off its renewable energy companies, it doubled down on fossil fuels, and along with other big oil companies like, Shell and Mobil, it launched an extensive and sinister campaign of climate denial, undermining the work and the warnings of its own scientists …
>
> This revealing course of conduct simply gives the game away. They used their knowledge of climate change to protect their future profits, while preventing the American people from acting together to protect our collective future. They used their knowledge of climate change for purposes of corporate planning, but publicly denied the reality of climate change for purposes of national planning …
>
> The decades-long denial campaign has twisted and perverted our democracy. By funding climate denial and lobbying against governmental action, big oil has not only achieved a loud and distorting voice in the climate change debate, it has also deprived voters and policymakers of the materials and the ability necessary to make informed decisions about this fundamental challenge to the future of human existence …
>
> The people have been denied the power that knowledge gives, which means that we've effectively been governed by Big Oil with respect to climate change …

A witness at the Raskin hearing was Naomi Oreskes, co-author of the *Merchants of Doubt*, who offered her own passionate plea concerning the "cost to our democracy" in her written statement[91]:

> The costs of climate change cannot be measured only in houses damaged, people evacuated, dollars spent, or even lives lost. There has been another cost: the cost to our democracy. The long history of fossil fuel funded denial has resulted in a profound, and deliberate, distortion of our democracy.

On April 22, 2021, another memorable session took place called "The Role of Fossil Fuel Subsidies in Preventing Action on the Climate Crisis;" an extraordinary feature of this virtual Congressional hearing was the presence of the famous activist Greta Thunberg (2003–) as a witness. Her opening remarks went straight to the point[92]:

> It is the year 2021. The fact that we are still having this discussion and, even more, that we are still subsidizing fossil fuels, directly or indirectly, using taxpayer money, is a disgrace. It is a clear proof that we have not understood the climate emergency at all …
>
> How long do you honestly believe that people in power like you will get away with it? How long do you think you can continue to ignore the climate crisis, the global aspect of equity and historic emissions without being held accountable?
>
> You get away with it now, but sooner or later people are going to realize what you have been doing all this time. That's inevitable. You still have time to do the right thing and to save your legacies, but that window of time is not going to last for long.

More hearings by the Committee of Oversight and Reform followed called "Fueling the Climate Crisis;" Carolyn Maloney, the chair, continued the allegations against the oiligarchs, and these executives were at the session this time[93]:

> For far too long, Big Oil has escaped accountability for its central role in bringing our planet to the brink of a climate catastrophe. That ends today.
>
> Big Oil has known the truth about climate change for decades. In the 1970's and 80's, Exxon's own scientists privately told top executives that burning fossil fuel was changing the global climate. Exxon and other Big Oil companies had the opportunity to tell the truth and lead the way to find alternative energy sources, but instead, Big Oil doubled down on fossil fuels.

Afterwards, Alexandria Ocasio-Cortez (1989–) (Democrat from New York) tweeted[94]:

> Yesterday, we learned that Exxon's CEO had calls with members of Congress during the reconciliation and infrastructure process. The fossil fuel industry can't keep writing our laws. Some of us have to actually live the future that fossil fuel companies are setting on fire.

And on February 8, 2022, at a follow-up session by the committee, Ms. Maloney gave this summary[95]:

> For decades, the fossil fuel industry waged a multimillion-dollar disinformation campaign to cast doubt on the science and prevent action to reduce emissions, all to protect its bottom line …
>
> At our October hearing, fossil-fuel executives admitted for the first time to Congress that climate change is real, it is happening now, and burning fossil

> fuels is the primary cause, four decades after they first learned the truth. Because of their lies, humanity lost four decades. We are now on the brink of a climate catastrophe.

A witness at the session was Michael E. Mann who laid out the problem[96]:

> They [ExxonMobil] even used the word "catastrophic" to describe the potential impacts of that warming, but rather than come forward with what their own scientists had concluded, they engaged in a campaign of denial and delay, which continues on today. We are now paying the cost of that delay in the form of extreme weather events exacerbated by climate change.

Another session of "Fueling the Climate Crisis" was held on September 15, 2022, when Ms. Maloney again attacked the oiligarchs[97]:

> Finally, we invited board members from Exxon, Chevron, Shell, and BP to testify today. We wanted them to answer for the record profits their companies are raking in, while fleecing consumers at the pump and refusing to take meaningful action on climate change… But Big Oil needs to do its part. They must end their greenwashing and finally take climate change seriously, before more Americans are harmed.

Attached to her opening remarks was a 22-page memorandum outlining "the fossil fuel industry's decades-long campaign to mislead the American people about the industry's role in climate change," which was followed by more documentation of Big Oil's greenwashing on December 8th, when Chairwoman Maloney stated[98]:

> Even though Big Oil CEOs admitted to my Committee that their products are causing a climate emergency, today's documents reveal that the industry has no real plans to clean up its act and is barreling ahead with plans to pump more dirty fuels for decades to come … Today's new evidence makes clear that these companies know their climate pledges are inadequate, but are prioritizing Big Oil's record profits over the human costs of climate change. It's time for the fossil fuel industry to stop lying to the American people and finally take serious steps to reduce emissions and address the global climate crisis they helped create.

These sessions of the Committee of Oversight and Reform placed oil executives on the hot seat for their complicity in the climate crisis and were among the most impressive climate brawls in Congress.

The Deep State of Denial

Propaganda has enabled the energy-industrial complex to subvert democratic processes for policy development on climate change. Democrats had even warned about the threat to American democracy in general that this posed,[99] and there are signs that the level of political interference from the energy-industrial complex has characteristics of a deep state as suggested by the many examples of its undue influence discussed above. What is a deep state? The definition from the online Merriam-Webster dictionary is[100]: "an alleged secret network of especially nonelected government officials and sometimes private entities (as in the financial services and defense industries) operating extralegally to influence and enact government policy." The deep state of climate denial in America has been created by "private entities" – the energy-industrial complex – and politicians, even presidents, have become puppets of an oiligarchy[101]; so, the will of the people has been supplanted by the will of the corporate elite. American democracy is slowly being crushed under the weight of propaganda, a death by a thousand lies.

The Climate Brawl Handbook, Politics Edition

Governments are fortunate to have global expert assessments to guide them in their policymaking options for the climate crisis. The reports of the IPCC (and the National Climate Assessment) represent the consensus view of the peer-reviewed science of climate change and are essential resources on how to deal with the climate crisis. However, there is a huge obstacle to moving forward – the politics. The NASA climate scientist Peter Kalmus described what is happening[102]:

> We have to look look very carefully, figure out the causes and say, oh, yeah, okay: it's the fossil fuel industry … But for some reason, there isn't a kind of public collective will to end the fossil fuel industry, which is the reason the politicians aren't doing it.

And a study by psychologists found that[103]: "Neither scientific knowledge nor political rhetoric has thus far resulted in global commitments that are sufficient to address the climate crisis." Consequently, something else must be done.

The necessary global commitments will not happen without the climate policy leadership of the United States. America and the rest of the world are gradually moving away from fossil fuels, but climate denialism has still been successful at slowing this social and economic transition to a snail's pace which has been a disaster in the making, and this is the climate delaying strategy. Unfortunately, we no longer have the luxury of time. To accelerate the progress, political action is required, but political will to act is still missing, especially in the United States, primarily because the propaganda[104] is so effective. We desperately need to get the ball rolling, but we don't even have a ball yet. As discussed in this study, climate brawls have a major role to play in muting the propaganda of climate denialism, helping to clear the path ahead for political action on climate change.

In the early part of *Climate Denial in American Politics: #ClimateBrawl*, I stated my personal quest in this work was to find the "solution to the politics." As Congress is still arguing over the science of climate change, which puts America decades behind most of the rest of the world, the first order of business has to be putting an end to these futile hearings and foolish debates over the science of climate change, which means a couple of bipartisan ground rules must exist:

- Support for the Paris Climate Agreement
- Acceptance of the reports of the IPCC.

In other words, climate denialism, created by decades of propaganda by the energy-industrial complex, has to be overcome before these two conditions can be met, and climate policymaking can commence. Senator Ed Markey has tweeted about the dilemma[105]: "We don't have time for climate deniers," and[106]: "You cannot negotiate a climate bill with climate deniers." Sadly, this will likely remain a challenge, as the former U.S. chief negotiator at the Paris Agreement Todd Stern (1951–) has observed[107]: "Congress seems to be the last bastion of climate denialism left in America."

The difficulty in overcoming climate denial within the political ranks of the United States is a formidable task but still the first hurdle that must be cleared. The methods for challenging climate denial were discussed in the first chapter, in the section "The Climate Brawl Handbook, The Twitter Edition." The twitterverse, of course, is not Congress, but climate denial tends to be the same wherever it rears its ugly head; I can attest to the similarities from my own experiences on Twitter and my research on the Congressional Record. Such tools need to be applied more rigorously and frequently at congressional hearings against contrarians and the denial cabal. Climate brawls are needed to break apart the insidious deep state of denial in Washington until the basic science of climate change is accepted by all parties, and climate negotiations to reach a bipartisan agreement on the climate crisis can finally begin.

What is a climate brawl? It is not about the peer-reviewed science of climate change, since a political debate over an accepted science supported by decades of research would be a waste of time. Primarily, a climate brawl is a challenge to the hard-core climate deniers – the political elite, contrarians, and major Twitter influencers – who are leading the spread of climate denial propaganda. The denial cabal and their contrarian allies must be forced to defend their false claims, which they cannot. The objective is to discredit climate deniers and their sources which, in turn, mutes the influence of the propaganda. The hard-core climate deniers are unlikely to change; so, they must be marginalized, and in the particular case of political climate deniers, they become "marginalized" by being voted out of office. Climate denial will always lose in a climate brawl.

A congressional hearing is a high-profile opportunity to make a case that is often lost by climate scientists who are reluctant to criticize their contrarian peers publicly (with a few notable exceptions as discussed). While they may have been furious or dumbfounded over the testimony of the contrarians, climate experts have seldom displayed any grit at congressional hearings. However, the climate crisis is far too important for a passive approach. Any cordial protocols were broken over 30 years ago when contrarians first showed up in Congress – that is when the climate brawls should have begun. The very presence of contrarians as witnesses at a hearing creates an adversarial environment, and such climate hearings should resemble an intense rhetorical brawl, not an afternoon social tea. And, if politicians give denial-type comments and questions, scientists must also respond to them with a stern rebuke rather than with polite silence; there are no better examples of these science brawls than between Dr. Mann with Lamar Smith (discussed above) and Dr. Anthony Fauci (1940–), in a COVID brawl, with Senator Rand Paul (1963–) (Republican from Kentucky).[108] Scientists who are unwilling to engage in this manner are not suitable witnesses for climate hearings.

Democrats must also do more, much more. Even though the science, literally all of the science, supports the concept of legislation on climate change (as provided, for example, in the IPCC reports) endorsed by the Democrats, they lost this huge advantage by allowing a handful of contrarians to steal the show. Despite growing hostility towards climate scientists by committee members, Democrats have often left their scientific guests hanging high and dry, saying nothing in the face of inappropriate comments by their colleagues from across the floor, perhaps in the name of congressional etiquette. Lately, there has been some improvement but there is still too much apathy on the part of most Democratic politicians in congressional hearings.

Once climate denial has been discredited and climate deniers have been marginalized in American politics, then Congress can begin the real job of passing bipartisan legislation and identifying the best options to halt the climate crisis. Climate brawls are not the whole "solution to the politics" but a vital piece to getting us there.

Notes

1 Jones 1958, p. 268; the Smoke and Fumes Committee was based on the Petroleum Industry Committee on Smog formed in late 1946 (Jenkins 1954, p. 146).
2 Jenkins 1954, p. 145.
3 For discussions on when the petroleum industry knew about climate change, see Glaser 1982; Banerjee, Song, and Hasemyer 2015; Hall 2015; Mulvey and Shulman 2015, pp. 9–12; Cushman 2018; Franta 2018b; Cook et al. 2019; United States House of Representatives 2019b, pp. 10–12; Watts et al. 2019; Yoder 2020. For the coal industry, see Young 2019, and the natural gas industry see Drugmand 2020.
4 Brannon et al. 1957, p. 463.
5 Brannon et al. 1957, p. 649.
6 Teller, himself, later became a contrarian (see Oreskes and Conway 2010, pp. 54–56).
7 See Anon. 1957, p. 30; also see Matthews 1959.
8 Matthews 1959, p. 645.
9 Matthews 1959, p. 646.
10 Nevins 1960 ("Energy Patterns of the Future" by Edward Teller, no page numbers); also see Franta 2018a.
11 For the report, see White House 1965, pp. 126–127.
12 Ikard 1965, p. 13.
13 Robinson and Robbins 1969, pp. 24–25.
14 Glaser 1982, p. 2; the Exxon report was discussed in Congress, see questioning by Alexandria Ocasio-Cortez United States House of Representatives 2019b, p. 37.

15 For a detailed analysis on the Exxon climate research, see Supran, Rahmstorf, and Oreskes 2023.
16 Shell 1989, p. 36.
17 American Petroleum Institute 1980; for a critique of the API report see Franta 2021a.
18 Friedman 1970.
19 See, for example, Posner 2019; Ward 2020; Dolšak, Griffin, and Prakash 2020.
20 Mulvey and Shuman 2015, p. 29.
21 Government Accountability Project 2006b; the term "denial machine" had been mentioned a few months before (Government Accountability Project 2006a), and a few months later, the "global warming denial machine" was briefly mentioned at a Senate hearing (United States Senate 2007, p. 39). An article on the denial machine appeared in *Newsweek* a year later (Newsweek Staff 2007); while it presented an excellent exposé, it failed to acknowledge the CBC broadcast.
22 For examples of propaganda campaigns from the denial machine, see Gelbspan 1998; Oreskes and Conway 2010; Mann and Toles 2016; Mann 2017; Oreskes 2018; Freese 2020; Pomerantsev 2020.
23 Brulle 2014.
24 For a study of the messaging of conservative think tanks see Coan et al. 2021.
25 D'Angelo 2019.
26 For a review of this organization, see DeSmog, "George C. Marshall Institute;" Oreskes and Conway 2010, pp. 54–59, 78; Franken 2011, p. S8590. The Institute closed in 2015.
27 For a review of this organization, see DeSmog, "Heartland Institute;" Oreskes and Conway 2010, p. 247; Franken 2011, p. S8590; Montlake 2019.
28 Walker 1998, p. 3.
29 Walker 1998, p. 4. A similar tactic was to hire contrarian economists (Franta 2021b).
30 DeSmog, "Global Climate Coalition;" also see Rich 2018, *Epilogue*.
31 See Kutney 2014, ch. 2–3; Hope and Savage 2019.
32 McMullen 2022.
33 United States Senate 1997, p. 1.
34 United States Senate 1997, pp. 230–231. For other congressional hearings with the GCC see, for example, United States House of Representatives 1992, pp. 174–181; 1993, pp 11–14; 1995, pp. 73–74; United States Senate 1991, pp. 65–68.
35 Global Climate Coalition 1995, pp. 13–16; also see Mulvey and Shuman 2015, pp. 25–27.
36 United States Department of State 2001. By this time, the GCC had shed its direct corporate members and just contained trade associations.
37 United States Senate 2000, p. 71.
38 United States Department of State 2001.
39 Brulle 2014; also see Coan et al. 2021.
40 For an investigative report on "dark money" and the Koch brothers see Mayer 2016; also see DeSmog, "Koch Network Database"; Mann and Toles 2016, pp. 108–113; Hertel-Fernandez, Tervo, and Skocpol 2018; Leonard 2019.
41 Markey 2019.
42 Brulle 2014; for a recent analysis of non-profit climate denial funders see Kotch 2022.
43 Farrell 2019.
44 Showstack 2019.
45 Senate Democrats' Special Committee on the Climate Crisis 2020, pp. 199–201.
46 See Frumhoff and Oreskes 2015; Farand 2018; Crowe 2019; McCarthy 2019; Statista 2019; Thacker 2019; Brigham 2020; also see the "Oil/Gas Sector Climate Lobbying Update" InfluenceMap (https://influencemap.org/index.html).
47 American Petroleum Institute 2020.
48 Whitehouse 2021a.
49 Chomsky 2021.

50 Chomsky 2022.
51 Gorman and Gorman 2019.
52 For a review, see Kutney 2014, pp. 98–107; for more on climategate see Mashey 2010.
53 For examples, see Kutney 2014, pp. 102–114.
54 As discussed in the previous chapter.
55 See Kutney 2014, ch. 2.
56 Banerjee 2013; the article provides an excellent review of the early attacks against Dr. Mann.
57 Kintisch 2009.
58 Anon. 2010.
59 Funk and Kennedy 2016.
60 McCright and Dunlap 2010, p. 119.
61 Gleick et al. 2010, p. 689.
62 Kutney 2014, p. 116.
63 Oreskes 2018.
64 Bulletin of the Atomic Scientists 2020.
65 For a definition, see Urban Dictionary (2008): "A government literally controlled by oil companies … The US is an oiligarchy thanks to Bush."
66 Trump 2018; also see Trump 2019.
67 Krugman 2006; he also labelled ExxonMobil "a worse environmental villain than other big oil companies" and "an enemy of the planet" for their campaigns against the science of climate change.
68 Gore 2007, 2011.
69 United States House of Representatives 2006a, p. 9.
70 United States House of Representatives 2006b, p. 10.
71 Dickinson 2007.
72 See United States House of Representatives 2008, p. 107.
73 Mann 2019; also see Mann 2012, p. 59.
74 Page 2019.
75 United States Senate 2013, p. 49.
76 United States Senate 2021, p. 3.
77 Schumer 2020.
78 Stabenow, Schumer, and Whitehouse 2020, pp. 3, 27; also see Whitehouse and Mueller 2022.
79 See Guardian Staff and Agency 2020; Millhiser 2021; Whitehouse 2021b; Taft 2022.
80 Supreme Court of the United States 2022.
81 United States Senate 2014, pp. 8–9.
82 United States Senate 2014, p. 94.
83 United States Senate 2014, p. 115.
84 Whitehouse 2015.
85 Whitehouse 2019a; also see Whitehouse 2016, 2019b.
86 Whitehouse 2020.
87 Waldman 2021.
88 Whitehouse 2022.
89 Whitehouse 2023.
90 United States House of Representatives 2019b, pp. 2–4; also see Mulvey and Shulman 2015.
91 Oreskes 2019.
92 United States House of Representatives 2021a, p. 12; for an earlier testimony of Ms. Thunberg see United States House of Representatives 2019a, pp. 10–11, where she presented a copy of the IPCC SR15 report (Milman and Smith 2019; also see Gambino 2019).
93 United States House of Representatives 2021b.

94 Ocasio-Cortez 2021.
95 United States House of Representatives 2022a, p. 1.
96 United States House of Representatives 2022a, p. 8.
97 United States House of Representatives 2022b, p. 2.
98 Maloney 2022.
99 For a review of the influence of propaganda on recent American politics see Coppins 2020; for other studies on propaganda by the oil industry, see Hiar 2021 (based on United States House of Representatives 2018); Dembicki 2022.
100 Merriam-Webster 2020.
101 See, for example, Pierson 2017; Robinson 2019.
102 Mortillaro 2023.
103 Hornsey and Lewandowsky 2022, p. 1461; also see Lewandowsky 2020, pp. 8–13.
104 Some reports argue that focus should not be on climate denial; see, for example, the opinion piece by Bretter and Schulz 2023.
105 Markey 2020.
106 Markey 2021.
107 See Milman 2021; also see Gross 2021.
108 See, for example, Foley and Leonard 2022.

References

American Petroleum Institute (1980). *Two Energy Futures: A National Choice for the 80s*. Washington, DC: American Petroleum Institute; https://www.documentcloud.org/documents/21094733-1980-api-two-energy-futures-a-national-choice-for-the-80s

American Petroleum Institute (2020). *API Expands Advertising Efforts in Key Battleground States*. API, October 6; https://www.api.org/news-policy-and-issues/news/2020/10/02/api-expands-advertising-efforts-in-key-battleground-states

Anon. (1957). "Scientist Sees World Overheated Soon." *The Steubenville Herald-Star*, Steubenville, OH, December 7, p. 20; https://newspaperarchive.com/steubenville-herald-star-dec-07-1957-p-20/

Anon. (2010). "Climate of Fear." *Nature* 464 (March 10), p. 141; https://www.nature.com/articles/464141a

Banerjee, Neela (2013). "The Most Hated Climate Scientist in the US fights Back." *Yale Alumni Magazine*, March/April; https://yalealumnimagazine.org/articles/3648-the-most-hated-climate-scientist-in-the-us-fights-back

Banerjee, Neela, Song, Lisa, and Hasemyer, David (2015). "Exxon's Own Research Confirmed Fossil Fuels' Role in Global Warming Decades Ago." *Inside Climate News*, September 16; https://insideclimatenews.org/news/15092015/Exxons-own-research-confirmed-fossil-fuels-role-in-global-warming

Brannon Jr., H. R., Daughtry, A. C., Perry, D., Whitaker, W. W., and Williams, M. (1957). "Radiocarbon Evidence on the Dilution of Atmospheric and Oceanic Carbon by Carbon from Fossil Fuels." *Transactions American Geophysical Union* 38 (5), p. 643; https://agupubs.onlinelibrary.wiley.com/doi/abs/10.1029/TR038i005p00643

Bretter, Christian, and Schulz, Felix (2023). "Why Focusing on 'Climate Change Denial' is Counterproductive." *Proceedings of the National Academy of Sciences* 120 (10), March 1; https://www.pnas.org/doi/10.1073/pnas.2217716120

Brigham, Katie (2020). "The Money Behind Climate Change Denial." *CNBC*, December 20; https://www.cnbc.com/video/2020/12/20/why-climate-change-denial-still-exists-in-the-us.html

Brulle, Robert J. (2014). "Institutionalizing Delay: Foundation Funding and the Creation of U.S. Climate Change Counter-Movement Organizations." *Climatic Change* 122, p. 681; https://link.springer.com/article/10.1007/s10584-013-1018-7#page-1

Bulletin of the Atomic Scientists (2020). *Closer Than Ever: It Is 100 Seconds to Midnight, 2020 Doomsday Clock Statement*. January 23; https://thebulletin.org/doomsday-clock/2020-doomsday-clock-statement/

Chomsky, Noam (2021). "Legendary Activist Noam Chomsky on Biden's Presidency and the Modern GOP." The Mehdi Hasan Show, *MSNBC*, April 18; https://www.msnbc.com/mehdi-on-msnbc/watch/legendary-activist-noam-chomsky-on-biden-s-presidency-and-the-modern-gop-110429765669

Chomsky, Noam (2022). *Noam Chomsky: On the Pandemic, Ukraine Crisis & Climate Change*. Europe Matter, January 10; https://www.youtube.com/watch?v=QC9MJssZWt4

Coan, Travis G., Boussalis, Constantine, Cook, John, and Nanko, Mirjam O. (2021). "Computer-Assisted Classification of Contrarian Claims about Climate Change." *Scientific Reports* 11, 22320; https://www.nature.com/articles/s41598-021-01714-4#citeas

Cook, John, Supran, Geoffrey, Lewandowsky, Stephan, Oreskes, Naomi, and Maibach, Ed (2019). *America Misled: How the Fossil Fuel Industry Deliberately Misled Americans about Climate Change*. Fairfax, VA: George Mason University Center for Climate Change Communication; https://www.climatechangecommunication.org/america-misled/

Coppins, McKay (2020). "The Billion-Dollar Misinformation Campaign to Reelect the President." *The Atlantic*, February 10; https://www.theatlantic.com/magazine/archive/2020/03/the-2020-disinformation-war/605530/

Crowe, Kelly (2019). "How 'Organized Climate Change Denial' Shapes Public Opinion on Global Warming." *CBC News*, September 27; https://www.cbc.ca/news/technology/climate-change-denial-fossil-fuel-think-tank-sceptic-misinformation-1.5297236

Cushman Jr., John H. (2018). "Shell Knew Fossil Fuels Created Climate Change Risks Back in 1980s, Internal Documents Show." *Inside Climate News*, April 5; https://insideclimatenews.org/news/05042018/shell-knew-scientists-climate-change-risks-fossil-fuels-global-warming-company-documents-netherlands-lawsuits

D'Angelo, Chris (2019). "Big Business Spent $1.4 Billion on PR, Advertising over the Last Decade." *HuffPost*, March 14; https://www.huffingtonpost.ca/entry/industry-trade-groups-public-relations-spending_n_5c89aa69e4b0fbd76620a0a3

Dembicki, Geoff (2022). *The Petroleum Papers*. Vancouver: Greystone Books.

DeSmog (undated). "George C. Marshall Institute"; https://www.desmogblog.com/george-c-marshall-institute

DeSmog (undated). "Global Climate Coalition"; https://www.desmogblog.com/global-climate-coalition

DeSmog (undated). "Heartland Institute"; https://www.desmogblog.com/heartland-institute

DeSmog (undated). "Koch Network Database"; https://www.desmogblog.com/koch-network-database

Dickinson, Tim (2007). "The Secret Campaign of President Bush's Administration to Deny Global Warming." *Rolling Stone*, June 20; https://web.archive.org/web/20070626050118/http://www.rollingstone.com/politics/story/15148655/the_secret_campaign_of_president_george_bushs_administration_to_deny_global

Dolšak, Nives, Griffin, Jennifer J., and Prakash, Aseem (2020). *Milton Friedman Versus Jeff Bezos on Climate Leadership*. The Regulatory Review, December 28; https://www.theregreview.org/2020/12/28/dolsak-griffin-prakash-milton-friedman-versus-jeff-bezos-climate-leadership/

Drugmand, Dana (2020). *Unplugged: How the Gas Industry Is Fighting Efforts to Electrify Buildings*. DeSmog, July 22; https://www.desmogblog.com/2020/07/22/unplugged-how-gas-industry-fighting-efforts-electrify-buildings

Farand, Chloe (2018). *Report: 90 Percent of World's Largest 200 Industrial Firms Are Using Trade Associations to Oppose Climate Policy*. DeSmog, September 12; https://www.desmogblog.com/2018/09/12/report-90-percent-world-s-largest-200-industrial-companies-are-using-trade-associations-oppose-climate-policy

Farrell, Justin (2019). "The Growth of Climate Change Misinformation in US Philanthropy: Evidence from Natural Language Processing." *Environmental Research Letters* 14 (3), March 15; https://iopscience.iop.org/article/10.1088/1748-9326/aaf939/meta

Foley, Katherine E., and Leonard, Ben (2022). "Covid Doesn't Stop Anthony Fauci from Taking on Rand Paul – Again." *Politico*, June 16; https://www.politico.com/news/2022/06/16/covid-fauci-rand-paul-00040210

Franken, Al (2011). *Climate Change*. Congressional Record – Senate, p. S8589, December 14; https://www.govinfo.gov/content/pkg/CREC-2011-12-14/pdf/CREC-2011-12-14-pt1-PgS8589.pdf

Franta, Benjamin (2018a). "On Its 100th Birthday in 1959, Edward Teller Warned the Oil Industry about Global Warming." *Guardian*, January 1; https://www.theguardian.com/environment/climate-consensus-97-per-cent/2018/jan/01/on-its-hundredth-birthday-in-1959-edward-teller-warned-the-oil-industry-about-global-warming

Franta, Benjamin (2018b). "Shell and Exxon's Secret 1980s Climate Change Warnings." *Guardian*, September 19; https://www.theguardian.com/environment/climate-consensus-97-per-cent/2018/sep/19/shell-and-exxons-secret-1980s-climate-change-warnings

Franta, Benjamin (2021a). "Early Oil Disinformation on Global Warming." *Environmental Politics* 30 (4), p. 663, January 5; https://www.tandfonline.com/doi/full/10.1080/09644016.2020.1863703#

Franta, Benjamin (2021b). "Weaponizing Economics: Big Oil, Economic Consultants, and Climate Policy Delay." *Environmental Politics* 31 (4), p. 555, August 25; https://www.tandfonline.com/doi/full/10.1080/09644016.2021.1947636

Freese, Barbara (2020). *Industrial-Strength Denial*. Oakland: University of California Press.

Friedman, Milton (1970). "A Friedman Doctrine – The Social Responsibility of Business Is to Increase Its Profits." *New York Times*, September 13; https://www.nytimes.com/1970/09/13/archives/a-friedman-doctrine-the-social-responsibility-of-business-is-to.html

Frumhoff, Peter C., and Oreskes, Naomi (2015). "Fossil Fuel Firms Are Still Bankrolling Climate Denial Lobby Groups." *Guardian*, March 25; https://www.theguardian.com/environment/2015/mar/25/fossil-fuel-firms-are-still-bankrolling-climate-denial-lobby-groups

Funk, Cary, and Kennedy, Brian (2016). *The Politics of Climate*. Pew Research Center, October 4; http://www.pewresearch.org/science/2016/10/04/the-politics-of-climate/

Gambino, Lauren (2019). "Greta Thunberg to Congress: 'You're Not Trying Hard Enough. Sorry.'" *Guardian*, September 17; https://www.theguardian.com/environment/2019/sep/17/greta-thunberg-to-congress-youre-not-trying-hard-enough-sorry

Gelbspan, Ross (1998). *The Heat Is On: The Climate Crisis, the Cover-Up, the Prescription*. Cambridge: Perseus Books; https://archive.org/details/heatisonclim00gelb/page/183/mode/1up?q=denial

Glaser, M. B. (1982). *CO_2 "Greenhouse" Effect*. November 12, see ClimateFiles, *1982 Memo to Exxon Management about CO2 Greenhouse Effect*; http://www.climatefiles.com/exxonmobil/1982-memo-to-exxon-management-about-co2-greenhouse-effect/

Gleick, P. H., et al. (2010). "Climate Change and the Integrity of Science." *Science* 328 (5979), p. 689, May 7; http://science.sciencemag.org/content/328/5979/689

Global Climate Coalition (1995). *Approval Draft, Predicting Future Climate Change: A Primer*; https://www.ucsusa.org/sites/default/files/attach/2015/07/Climate-Deception-Dossier-7_GCC-Climate-Primer.pdf

Gore, Al (2007). *The Assault on Reason*. New York: Penguin Books; https://books.google.ca/books?id=AGFDIwcyQ4AC&dq=Gore,+Al+(2007).+The+Assault+on+Reason&source=gbs_navlinks_s.

Gore, Al (2011). "Al Gore: Climate of Denial, Can Science and the Truth Withstand the Merchants of Poison?" *Rolling Stone*, June 22; https://www.rollingstone.com/politics/politics-news/al-gore-climate-of-denial-244124/

Gorman, Sara, and Gorman Jack M. (2019). "Climate Change Denial." *Psychology Today*, January 12; https://www.psychologytoday.com/us/blog/denying-the-grave/201901/climate-change-denial

Government Accountability Project (2006a). *Refuting a Global Warming Denier*. Government Accountability Project, June 11; https://whistleblower.org/politicization-of-climate-science/global-warming-denial-machine/refuting-a-global-warming-denier/

Government Accountability Project (2006b). *'The Denial Machine' Airs on CBC-TV*. Government Accountability Project, November 17; https://whistleblower.org/politicization-of-climate-science/global-warming-denial-machine/the-denial-machine-airs-on-cbc-tv/

Gross, Samantha (2021). *Republicans in Congress are Out of Step with the American Public on Climate*. Brookings, May 10; https://www.brookings.edu/blog/planetpolicy/2021/05/10/republicans-in-congress-are-out-of-step-with-the-american-public-on-climate/

Guardian Staff and Agency (2020). "Amy Coney Barrett Refuses to Tell Kamala Harris If She Thinks Climate Change Is Happening." *Guardian*, October 15; https://www.theguardian.com/us-news/2020/oct/15/amy-coney-barrett-refuses-to-tell-kamala-harris-if-she-thinks-climate-change-is-happening

Hall, Shannon (2015). "Exxon Knew about Climate Change Almost 40 Years Ago." *Scientific American*, October 26; https://www.scientificamerican.com/article/exxon-knew-about-climate-change-almost-40-years-ago/?redirect=1

Hertel-Fernandez, Alexander, Tervo, Caroline, and Skocpol, Theda (2018). "How the Koch Brothers Built the Most Powerful Rightwing Group You've Never Heard of." *Guardian*, September 26; https://www.theguardian.com/us-news/2018/sep/26/koch-brothers-americans-for-prosperity-rightwing-political-group

Hiar, Corbin (2021). "Exxon Sting Ensnares Think Tanks with Climate Credentials." *E&E News*, ClimateWire, July 29; https://www.eenews.net/articles/exxon-sting-ensnares-think-tanks-with-climate-credentials/

Hope, Mat, and Savage, Karen (2019). *Global Climate Coalition: Documents Reveal How Secretive Fossil Fuel Lobby Group Manipulated UN Climate Programs*. DeSmogUK, April 25; https://desmog.co.uk/2019/04/25/global-climate-coalition-documents-secretive-fossil-fuel-lobby-un-programs

Hornsey, Matthew J., and Lewandowsky, Stephan (2022). "A Toolkit for Understanding and Addressing Climate Scepticism." *Nature Human Behaviour* 6, p. 1454; https://www.nature.com/articles/s41562-022-01463-y

Ikard, Frank N. (1965). "Meeting the Challenges of 1966." *Proceedings* 45(1), p. 12. American Petroleum Institute; http://www.climatefiles.com/trade-group/american-petroleum-institute/1965-api-president-meeting-the-challenges-of-1966/

Jenkins, Vance N. (1954). "The Petroleum Industry Sponsors Air Pollution Research." *Air Repair* 3 (3), p. 144; https://www.tandfonline.com/doi/pdf/10.1080/00966665.1954.10467615

Jones, Charles A. (1958). "A Review of the Air Pollution Research Program of the Smoke and Fumes Committee of the American Petroleum Institute." *Journal of Air Pollution Control Association* 8 (3), p. 268; https://www.tandfonline.com/doi/pdf/10.1080/00966665.1958.10467854

Kintisch, Eli (2009). "Stolen E-mails Turn Up Heat on Climate Change Rhetoric." *Science* 326 (5958), p. 1329, December 4; http://science.sciencemag.org/content/326/5958/1329

Kotch, Alex (2022). *The Dirty Dozen: The Biggest Nonprofit Funders of Climate Denial*. The Center for Media and Democracy, Exposed by CMD, March 21; https://www.exposedbycmd.org/2022/03/21/the-dirty-dozen-the-biggest-nonprofit-funders-of-climate-denial/

Krugman, Paul (2006). "Enemy of the Planet." *New York Times*, April 17; https://www.nytimes.com/2006/04/17/opinion/enemy-of-the-planet.html

Kutney, Gerald W. (2014). *Carbon Politics and the Failure of the Kyoto Protocol*. London: Routledge.

Leonard, Christopher (2019). *Kochland: The Secret History of Koch Industries and Corporate Power in America*. New York: Simon & Schuster; https://books.google.ca/books?id=kwL8DwAAQBAJ&dq=Kochland:+The+Secret+History+of+Koch+Industries+and+Corporate+Power+in+America&source=gbs_navlinks_s.

Lewandowsky, Stephan (2020). "Climate Change, Disinformation, and How to Combat It." *Annual Review of Public Health, Forthcoming*, September 16; https://papers.ssrn.com/sol3/papers.cfm?abstract_id=3693773

Maloney, Carolyn B. (2022). *Oversight Committee Releases New Documents Showing Big Oil's Greenwashing Campaign and Failure to Reduce Emissions*. House of Representatives, Committee on Oversight and Reform, 117th Congress, December 8; https://oversight.house.gov/news/press-releases/oversight-committee-releases-new-documents-showing-big-oil-s-greenwashing (note that less than a month later, all references to this document had been purged from the Committee website, which was now under a Republican chair).

Mann, Michael E. (2012). *The Hockey Stick and the Climate Wars*. New York: Columbia University Press.

Mann, Michael E. (2017). "'Anatomy of a Smear' or 'How the Right Wing Denial Machine Distorts the Climate Change Discourse.'" *HuffPost*, July 16; https://www.huffingtonpost.com/michael-e-mann/anatomy-of-a-smear-or-how-the-right-wing-denial-machine_b_10997280.html

Mann, Michael E. (2019). *Vice: A Commentary on the Politics of Climate Denial*. Michaelmann.net, January 7; https://michaelmann.net/content/vice-commentary-politics-climate-denial

Mann, Michael, E., and Toles, Tom (2016). *The Madhouse Effect*. New York: Columbia University Press.

Markey, Ed (2019). Tweet, Twitter, February 8; https://twitter.com/SenMarkey/status/1093909303168299008

Markey, Ed (2020). Tweet, Twitter, October 14; https://twitter.com/EdMarkey/status/1316492210884489217

Markey, Ed (2021). Tweet, Twitter, May 23; https://twitter.com/EdMarkey/status/1396469958226370566

Mashey, John R. (2010). *Crescendo to Climategate Cacophony*. DeSmog; https://www.desmog.com/wp-content/uploads/files/crescendo%20climategate%20cacophony%20v1%200.pdf#page=93

Matthews, M. A. (1959). "The Earth's Carbon Cycle." *New Scientist* 6 (151), p. 644, October 8; https://books.google.ca/books?id=qYkP2bThsFEC&dq=%22The+Earth%27s+Carbon+Cycle%22,+teller&source=gbs_navlinks_s

Mayer, Jane (2016). *Dark Money*. New York: Doubleday; https://books.google.ca/books?id=X5nrCgAAQBAJ&dq=Mayer,+Jane+(2016).+Dark+Money&source=gbs_navlinks_s

McCarthy, Niall (2019). "Oil and Gas Giants Spend Millions Lobbying to Block Climate Change Policies [Infographic]." *Forbes*, March 25; https://www.forbes.com/sites/niallmccarthy/2019/03/25/oil-and-gas-giants-spend-millions-lobbying-to-block-climate-change-policies-infographic/#329d24ba7c4f

McCright, Aaron M., and Dunlap, Riley E. (2010). "Anti-reflexivity." *Theory, Culture & Society* 27 (2–3), p. 100; https://journals.sagepub.com/doi/abs/10.1177/0263276409356001

McMullen, Jane (2022). "The Audacious PR Plot that Seeded Doubt about Climate Change." *BBC News*, July 23; https://www.bbc.com/news/science-environment-62225696?fbclid=IwAR3cu2ilM5Sv1N6p7t96JOtgVnDvLarlKLu5eQV0EuN0C6si8zzh3y1YtJo

Merriam-Webster (2020). *Deep State*. Merriam-Webster.com Dictionary; https://www.merriam-webster.com/dictionary/deep%20state

Millhiser, Ian (2021). "The Supreme Court's Coming War with Joe Biden, Explained." *Vox*, March 27; https://www.vox.com/22276279/supreme-court-war-joe-biden-agency-regulation-administrative-neil-gorsuch-epa-nondelegation

Milman, Oliver (2021). "Dizzying Pace of Biden's Climate Action Sounds Death Knell for Era of Denialism." *Guardian*, January 30; https://www.theguardian.com/environment/2021/jan/30/joe-biden-climate-change-action

Milman, Oliver, and Smith, David (2019). "'Listen to the Scientists': Greta Thunberg Urges Congress to Take Action." *Guardian*, September 19; https://www.theguardian.com/us-news/2019/sep/18/greta-thunberg-testimony-congress-climate-change-action

Montlake, Simon (2019). "What Does Climate Change Have to Do with Socialism?" *Christian Science Monitor*, August 5; https://www.csmonitor.com/Environment/2019/0805/What-does-climate-change-have-to-do-with-socialism

Mortillaro, Nicole (2023). "We Are Heading Toward IPCC's 1.5 C Threshold of Warming, but All Is Not Lost." *CBC*, March 20; https://www.cbc.ca/news/science/ipcc-climate-target-1.6782625

Mulvey, Kathy, and Shulman, Seth (2015). *The Climate Deception Dossiers*. Union of Concerned Scientists, July; https://www.ucsusa.org/sites/default/files/attach/2015/07/The-Climate-Deception-Dossiers.pdf

Nevins, Allan (1960). *Energy and Man: A Symposium*. New York: Appleton-Century-Crofts; https://books.google.ca/books?id=gl2LDwAAQBAJ&source=gbs_navlinks_s

Newsweek Staff (2007). "Global Warming Deniers Well Funded." *Newsweek*, August 12; https://www.newsweek.com/global-warming-deniers-well-funded-99775

Ocasio-Cortez, Alexandria (2021). Tweet, Twitter, October 29; https://twitter.com/RepAOC/status/1454099976003506176

Oreskes, Naomi (2018). "Beware: Transparency Rule Is a Trojan Horse." *Nature* 557, p. 469, May 22; https://www.nature.com/articles/d41586-018-05207-9

Oreskes, Naomi, and Conway, Erik M. (2010). *Merchants of Doubt*. New York: Bloomsbury Press.

Page, Samantha (2019). "Climate Change: 'The Most Villainous Act in History …'" *Cosmos*, February 12; https://cosmosmagazine.com/earth/climate/the-most-villainous-act-in-the-history-of-human-civilisation-michael-e-mann-speaks-out/

Pierson, Paul (2017). "American Hybrid: Donald Trump and the Strange Merger of Populism and Plutocracy." *British Journal of Sociology* 68 (S1), p. S105, November 8; https://onlinelibrary.wiley.com/doi/full/10.1111/1468-4446.12323

Pomerantsev, Peter (presenter) (2020). "How They Made Us Doubt Everything." *BBC News*, July 27; https://www.bbc.co.uk/sounds/play/m000l7q0?fbclid=IwAR3-ec4PO2xX_ebutEpVchMOZB3TtSJh4zUYhwkCCRRyEnpKRIPweCBS-bk

Posner, Eric (2019). "Milton Friedman Was Wrong." *The Atlantic*, August 22; https://www.theatlantic.com/ideas/archive/2019/08/milton-friedman-shareholder-wrong/596545/

Rich, Nathaniel (2018). "Losing Earth: The Decade We Almost Stopped Climate Change." *New York Times*, August 1; https://www.nytimes.com/interactive/2018/08/01/magazine/climate-change-losing-earth.html

Robinson, E., and Robbins R. C. (1969). *Sources, Abundance, and Fate of Gaseous Atmospheric Pollutants Supplement*. Stanford Research Institute, June; http://chr.gov.ph/wp-content/uploads/2019/11/Exhibit-3I-Sources-Abundance-and-Fate-of-Gaseous-Atmospheric-Pollutants-Supplement.pdf

Robinson, Nathan (2019). "The Koch Brothers Tried to Build a Plutocracy in the Name of Freedom." *Guardian*, August 28; https://www.theguardian.com/commentisfree/2019/aug/28/the-koch-brothers-tried-to-build-a-plutocracy-in-the-name-of-freedom

Schumer, Chuck (2020). Tweet, Twitter, May 27; https://twitter.com/SenSchumer/status/1265742299176341506

Senate Democrats' Special Committee on the Climate Crisis (2020). *The Case for Climate Action, Building a Clean Economy for the American People*. Senate Democrats, August 25; https://www.schatz.senate.gov/imo/media/doc/SCCC_Climate_Crisis_Report.pdf

Shell (1989). *Scenarios 1989–2010, Challenge and Response*. Confidential Group Planning, PL89 S01, October; https://www.documentcloud.org/documents/23735737-1989-oct-confidential-shell-group-planning-scenarios-1989-2010-challenge-and-response-disc-climate-refugees-and-shift-to-non-fossil-fuels

Showstack, Randy (2019). "Senator Urges Ending Dark Money's Stifling of Climate Action." *Eos*, June 20; https://eos.org/articles/senator-urges-ending-dark-moneys-stifling-of-climate-action?fbclid=IwAR1ZTn21ARbblUbD-tQ6P4Jt8_GNIGTePIDViXoRDL_vKQ6esDa4Si_wC8I

Stabenow, Debbie, Schumer, Chuck, and Whitehouse, Sheldon (2020). *Captured Courts, the GOP's Big Money Assault on the Constitution, Our Independent Judiciary, and the Rule of Law*. Democratic Policy & Communications Committee, May; https://www.democrats.senate.gov/imo/media/doc/Courts%20Report%20-%20FINAL.pdf

Statista (2019). *Climate Lobbying Expenditures of Oil Supermajors in 2018, by Company*. April 25; https://www.statista.com/statistics/995930/climate-lobbying-spending-oil-supermajors-by-company/

Supran, G., Rahmstorf, S., and Oreskes, Naomi (2023). "Assessing ExxonMobil's Global Warming Projections." *Science* 379 (6628), January 13; https://www.science.org/doi/10.1126/science.abk0063

Supreme Court of the United States (2022). *West Virginia et al. v. Environmental Protection Agency et al. No. 20-1539*, June 30; https://www.supremecourt.gov/opinions/21pdf/20-1530_n758.pdf

Taft, Molly (2022). *How Oil and Gas Helped Create the Conservative Supreme Court*. Gizmodo, May 4; https://gizmodo.com/how-oil-and-gas-helped-create-the-conservative-supreme-1848874699

Thacker, Paul D. (2019). "Fossil Fuel Giants Claim to Support Climate Science, Yet Still Fund Denial." *HuffPost*, December 18; https://www.huffingtonpost.ca/entry/climate-denial-energy-in-depth_n_5df7eff6e4b0ae01a1e59371?ri18n=true

Trump, Donald (2018). *Address before a Joint Session of the Congress on the State of the Union*. govinfo, January 30; https://www.govinfo.gov/app/details/DCPD-201800064

Trump, Donald (2019). *President Donald J. Trump Is Unleashing American Energy Dominance*. The White House, Donald J. Trump, Archives, Fact Sheets, May 14; https://trumpwhitehouse.archives.gov/briefings-statements/president-donald-j-trump-unleashing-american-energy-dominance/

United States Department of State (2001). *Briefing Memorandum*. June 21; http://www.greenpeace.org/usa/wp-content/uploads/legacy/Global/usa/report/2009/10/global-climate-coalition-meeti.pdf

United States House of Representatives (1992). *Global Warming*. Hearing before the Subcommittee on Energy and Power of the Committee on Energy and Commerce, House of Representatives, 102nd Congress, Second Session, March 3; https://babel.hathitrust.org/cgi/pt?id=uc1.31210014950537&view=1up&seq=1

United States House of Representatives (1993). *Global Climate Change: Adequacy of the National Action Plan*. Hearing before the Subcommittee on Economic Policy, Trade and Environment of the Committee on Foreign Affairs, House of Representatives, 103rd Congress, First Session, March 1; https://babel.hathitrust.org/cgi/pt?id=pst.000021575259&view=1up&seq=1

United States House of Representatives (1995). *International Global Climate Change Negotiations*. Hearings before the Subcommittee on Energy and Power of the Committee on Commerce, House of Representatives, 104th Congress, First Session, March 21; https://babel.hathitrust.org/cgi/pt?id=pst.000024720847&view=1up&seq=1

United States House of Representatives (2006a). *Climate Change: Understanding the Degree of the Problem*. Hearing before the Committee on Government Reform, House of Representatives, 109th Congress, Second Session, July 20; https://babel.hathitrust.org/cgi/pt?id=pst.000058948934;view=1up;seq=1

United States House of Representatives (2006b). *Climate Change Technology Research: Do We Need a "Manhattan Project" for the Environment?* Hearing before the Committee on Government Reform, House of Representatives, 109th Congress, Second Session, September 21; https://babel.hathitrust.org/cgi/pt?id=pst.000058934647&view=page&seq=1

United States House of Representatives (2008). *Final Staff Report for the 110th Congress Together with Additional Views Submitted by Mr. Markey, Chairman, Select Committee on Energy Independence and Global Warming*. House of Representatives, 110th Congress, Second Session, November 19; https://babel.hathitrust.org/cgi/pt?id=mdp.39015089033834&view=1up&seq=1

United States House of Representatives (2018). *Geopolitics of U.S. Oil and Gas Competitiveness*. Hearing before the Subcommittee on Terrorism, Nonproliferation, and Trade of the Committee on Foreign Affairs, House of Representatives, 115th Congress, Second Session, May 22; https://www.govinfo.gov/content/pkg/CHRG-115hhrg30173/pdf/CHRG-115hhrg30173.pdf

United States House of Representatives (2019a). *Voices Leading the Next Generation on the Global Climate Crisis*. Hearing before the Subcommittee on Europe, Eurasia, Energy, and the Environment of the Committee on Foreign Affairs, House of Representatives, 116th Congress, First Session, September 18; https://www.govinfo.gov/content/pkg/CHRG-116hhrg37705/pdf/CHRG-116hhrg37705.pdf

United States House of Representatives (2019b). *Examining the Oil Industry's Efforts to Suppress the Truth about Climate Change*. Hearing before the Subcommittee on Civil Rights and Civil Liberties of the Committee on Oversight and Reform, House of Representatives, 116th Congress, First Session, October 23; https://www.govinfo.gov/content/pkg/CHRG-116hhrg38304/pdf/CHRG-116hhrg38304.pdf

United States House of Representatives (2021a). *The Role of Fossil Fuel Subsidies in Preventing Action on the Climate Crisis*. Hearing before the Subcommittee on Environment of the Committee on Oversight and Reform, House of Representatives, 117th Congress, First Session, April 22; https://www.govinfo.gov/content/pkg/CHRG-117hhrg44383/pdf/CHRG-117hhrg44383.pdf

United States House of Representatives (2021b). *Fueling the Climate Crisis: Exposing Big Oil's Disinformation Campaign to Prevent Climate Action.* Hearing before the Committee on Oversight and Reform, House of Representatives, 117th Congress, First Session, October 28; https://www.govinfo.gov/content/pkg/CHRG-117hhrg46026/pdf/CHRG-117hhrg46026.pdf

United States House of Representatives (2022a). *Fueling the Climate Crisis: Examining Big Oil's Climate Pledges*. Hearing before the Committee on Oversight and Reform, House of Representatives, 117th Congress, Second Session, February 8; https://www.govinfo.gov/content/pkg/CHRG-117hhrg46902/pdf/CHRG-117hhrg46902.pdf

United States House of Representatives (2022b). *Fueling the Climate Crisis: Examining Big Oil's Prices, Profits, and Pledges*. Hearing before the Committee on Oversight and Reform, House of Representatives, 117th Congress, Second Session, September 15; https://www.govinfo.gov/content/pkg/CHRG-117hhrg48612/pdf/CHRG-117hhrg48612.pdf

United States Senate (1991). *Policy Implications of Greenhouse Warming*. Hearing before the Committee on Commerce, Science, and Transportation, 102nd Congress, First Session, April 25; https://babel.hathitrust.org/cgi/pt?id=pst.000018818604&view=1up&seq=1

United States Senate (1997). *Global Climate Change*. Hearings before the Committee on Environment and Public Works, United States Senate, 105th Congress, First Session, July 10, 17; https://babel.hathitrust.org/cgi/pt?id=uc1.b5131476&view=1up&seq=1

United States Senate (2000). *Solutions to Climate Change*. Hearing before the Committee on Commerce, Science, and Transportation, United States Senate, 106th Congress, Second Session, September 21; https://www.govinfo.gov/content/pkg/CHRG-106shrg85521/pdf/CHRG-106shrg85521.pdf

United States Senate (2007). *Climate Change Research and Scientific Integrity*. Hearing before the Committee on Commerce, Science, and Transportation, United States Senate, 110th Congress, First Session, February 7; https://www.govinfo.gov/content/pkg/CHRG-110shrg35039/pdf/CHRG-110shrg35039.pdf

United States Senate (2013). *Climate Change: It's Happening Now*. Hearing before the Committee on Environment and Public Works, United States Senate 113th Congress, First Session, July 18; https://www.govinfo.gov/content/pkg/CHRG-113shrg95976/pdf/CHRG-113shrg95976.pdf

United States Senate (2014). *Review of the President's Climate Action Plan*. Hearing before the Committee on Environment and Public Works, United States Senate, 113th Congress, Second Session, January 16; https://www.govinfo.gov/content/pkg/CHRG-113shrg97581/pdf/CHRG-113shrg97581.pdf

United States Senate (2021). *The Cost of Inaction on Climate Change*. Hearing before the Committee on the Budget, United States Senate, 117th Congress, First Session, April 15; https://www.govinfo.gov/content/pkg/CHRG-117shrg45191/pdf/CHRG-117shrg45191.pdf

Urban Dictionary (2008). *Oiligarchy*. Urban Dictionary, November 12; https://www.urbandictionary.com/define.php?term=Oiligarchy

Waldman, Scott (2021). "Senator Seeks Criminal Climate Probe of Trump." *E&E News*, ClimateWire, February 1; https://www.eenews.net/stories/1063723979

Walker, Joseph (1998). *1998 American Petroleum Institute Global Climate Science Communications Team Action Plan.* Climatefiles, April 3; http://www.climatefiles.com/

trade-group/american-petroleum-institute/1998-global-climate-science-communications-team-action-plan/

Ward, Marguerite (2020). "As the Pandemic, Fires, and Inequity All Rage, Free Market Icon Milton Friedman's Declaration that the Sole Responsibility of Business 'Is to Increase Its Profits' Sounds Emptier Than Ever." *Insider*, September 13; https://www.businessinsider.com/milton-friedmans-theory-role-of-business-feels-emptier-than-ever-2020-9

Watts, Jonathan, Blight, Garry, McMullan, Lydia, and Gutiérrez, Pablo (2019). "Half a Century of Dither and Denial – A Climate Crisis Timeline." *Guardian*, October 9; https://www.theguardian.com/environment/ng-interactive/2019/oct/09/half-century-dither-denial-climate-crisis-timeline?CMP=share_btn_tw

White House (1965). *Restoring the Quality of Our Environment*. p. 112, November; https://babel.hathitrust.org/cgi/pt?id=uc1.b4116127&view=1up&seq=130

Whitehouse, Sheldon (2015). *Time to Wakeup: The Pause that Wasn't*. Sheldon Whitehouse, Senator for Rhode Island, September 29; https://www.whitehouse.senate.gov/news/speeches/time-to-wake-up-the-pause-that-wasnt

Whitehouse, Sheldon (2016). *The Future of Climate Action*. senwhitehouse.medium.com, November 14; https://medium.com/@senwhitehouse/the-future-of-climate-action-5e35f8c573c9

Whitehouse, Sheldon (2019a). *Time to Wake Up: 2018 Year in Review*. Sheldon Whitehouse U.S. Senator for Rhode Island, Home, News, Speeches; January 10; https://www.whitehouse.senate.gov/news/speeches/time-to-wake-up-2018-year-in-review

Whitehouse, Sheldon (2019b). *Time to Wake Up: Corruption Scheme*. senwhitehouse.medium.com, November 20; https://senwhitehouse.medium.com/time-to-wake-up-corruption-scheme-8cffe0d7effb

Whitehouse, Sheldon (2020). *Is Your Trade Group Blocking Climate Action?* Sheldon Whitehouse U.S. Senator for Rhode Island, Home, News, In the News; January 31; https://www.whitehouse.senate.gov/news/in-the-news/is-your-trade-group-blocking-climate-action

Whitehouse, Sheldon (2021a). Untitled. Video attached to NowThis tweet, May 14; https://twitter.com/nowthisnews/status/1393178781712535553

Whitehouse, Sheldon (2021b). "A Flood of Judicial Lobbying: Amicus Influence and Funding Transparency." *The Yale Law Journal Forum*, p. 141, October 24; https://www.yalelawjournal.org/pdf/F7.WhitehouseFinalDraftWeb_v4zwakx5.pdf

Whitehouse, Sheldon (2022). *With Climate Bill Stalled, Whitehouse Delivers 280th 'Time to Wake Up' Speech*. Sheldon Whitehouse, United States Senator for Rhode Island, News, Press Releases; February 2; https://www.whitehouse.senate.gov/news/release/with-climate-bill-stalled-whitehouse-delivers-280th-time-to-wake-up-speech

Whitehouse, Sheldon (2023). *Climate Change*. Congressional Record – Senate, p. S1375. April 26; https://www.govinfo.gov/content/pkg/CREC-2023-04-26/pdf/CREC-2023-04-26-pt1-PgS1370-5.pdf

Whitehouse, Sheldon, and Mueller, Jennifer (2022). *The Scheme, How the Right Wing Used Dark Money to Capture the Supreme Court*. New York: The New Press; https://www.google.ca/books/edition/The_Scheme/v-hZEAAAQBAJ?hl=en&gbpv=1&dq=How+the+Right+Wing+Used+Dark+Money+to+Capture+the+Supreme+Court,+whitehouse&printsec=frontcover

Yoder, Kate (2020). "How the Oil Industry Pumped Americans Full of Fake News." *Grist*; February 7; https://grist.org/climate/how-the-oil-industry-pumped-americans-full-of-fake-news/?platform=hootsuite

Young, Élan (2019). "Coal Knew, Too." *HuffPost*, November 22; https://www.huffingtonpost.ca/entry/coal-industry-climate-change_n_5dd6bbebe4b0e29d7280984f?ri18n=true#click=https://t.co/M2hijypVec

Index

Note: **Bold** page numbers refer to tables and page numbers followed by "n" denote endnotes

Printed in the United States
by Baker & Taylor Publisher Services